OPNET LAB MANUAL TO ACCOMPANY

BUSINESS DATA COMMUNICATIONS
FIFTH EDITION
BY WILLIAM STALLINGS

LEANN CHRISTIANSON AND KEVIN BROWN

PEARSON
Prentice Hall

Upper Saddle River, NJ 07458

Vice President and Editorial Director, ECS: *Marcia J. Horton*
Senior Acquisitions Editor: *Alan Apt*
Associate Editor: *Sarah E. Parker*
Editorial Assistant: *Patrick Lindner*
Vice President and Director of Production and Manufacturing, ESM: *David W. Riccardi*
Executive Managing Editor: *Vince O'Brien*
Managing Editor: *Camille Trentacoste*
Production Editor: *Mary C. Massey*
Director of Creative Services: *Paul Belfanti*
Creative Director: *Carole Anson*
Cover Designer: *Daniel Sandin*
Manufacturing Manager: *Trudy Pisciotti*
Manufacturing Buyer: *Ilene Kahn*
Executive Marketing Manager: *Pamela Hersperger*
Marketing Assistant: *Barrie Reinhold*

© 2005 Pearson Education, Inc.
Pearson Prentice Hall
Pearson Education, Inc.
Upper Saddle River, NJ 07458

All rights reserved. No part of this book may be reproduced in any form or by any means, without permission in writing from the publisher.

Pearson Prentice Hall® is a trademark of Pearson Education, Inc.

The author and publisher of this book have used their best efforts in preparing this book. These efforts include the development, research, and testing of the theories and programs to determine their effectiveness. The author and publisher make no warranty of any kind, expressed or implied, with regard to these programs or the documentation contained in this book. The author and publisher shall not be liable in any event for incidental or consequential damages in connection with, or arising out of, the furnishing, performance, or use of these programs.

Printed in the United States of America

10 9 8 7 6 5 4 3 2 1

ISBN: 0-13-148253-X

Pearson Education Ltd., *London*
Pearson Education Australia Pty. Ltd., *Sydney*
Pearson Education Singapore, Pte. Ltd.
Pearson Education North Asia Ltd., *Hong Kong*
Pearson Education Canada, Inc., *Toronto*
Pearson Educación de Mexico, S.A. de C.V.
Pearson Education—Japan, *Tokyo*
Pearson Education Malaysia, Pte. Ltd.
Pearson Education, Inc., *Upper Saddle River, New Jersey*

Lab 0 Introduction to OPNET IT Guru Academic Edition

Objectives
1. To download and install OPNET IT Guru Academic Edition.
2. To learn basic skills for creating OPNET projects, running simulations, and gathering results.

Description
OPNET IT Guru is an application that allows you to model a wide variety of networks and situations. The application can be used to test the performance of a modeled network configured with predefined parameters. After model construction, a simulation can be run to gather user-defined statistics. Results are presented as graphs for easy evaluation.

Downloading OPNET IT Guru
In order to download OPNET IT Guru Academic Edition, you must first register with OPNET Technologies. You can do this by visiting the website at

http://www.opnet.com/services/university/itguru_academic_edition.html

Click on the link labeled DownLoad Now

You will be asked to fill in a form in order to create your account. A user ID and password will be emailed to you along with instructions for downloading the software.

The systems requirements for this application are the following:

- Intel Pentium III, 4, or compatible (500 MHz or more)
- 256 MB RAM
- 400 MB disk space
- Display 1024 x 768 or higher resolution, 256 or more colors
- The English language version of the following Operating Systems are supported:
    - Microsoft Windows NT (Service Pack 3, 5, or 6a; Service Packs 4 and 6 are not supported)
    - Windows 2000 (Service Pack 1 and 2 are supported, but not required)
    - Windows XP (Service Pack 1 is required)
- PDF viewer for reading the help files and tutorial

When you receive your user account and password, download the application and install it on your computer.

Using OPNET IT Guru Academic Edition

1. **Click on** Start/Programs/ OPNET IT Guru Academic Edition x.x / OPNET IT Guru Academic Edition **(where x.x is the software version).**

2. Read the **Restricted Use Agreement** and click on the button **I have read this SOFTWARE AGREEMENT and understand and accept the terms and conditions described herein**.

3. You should then see the main menu, which looks like the following:

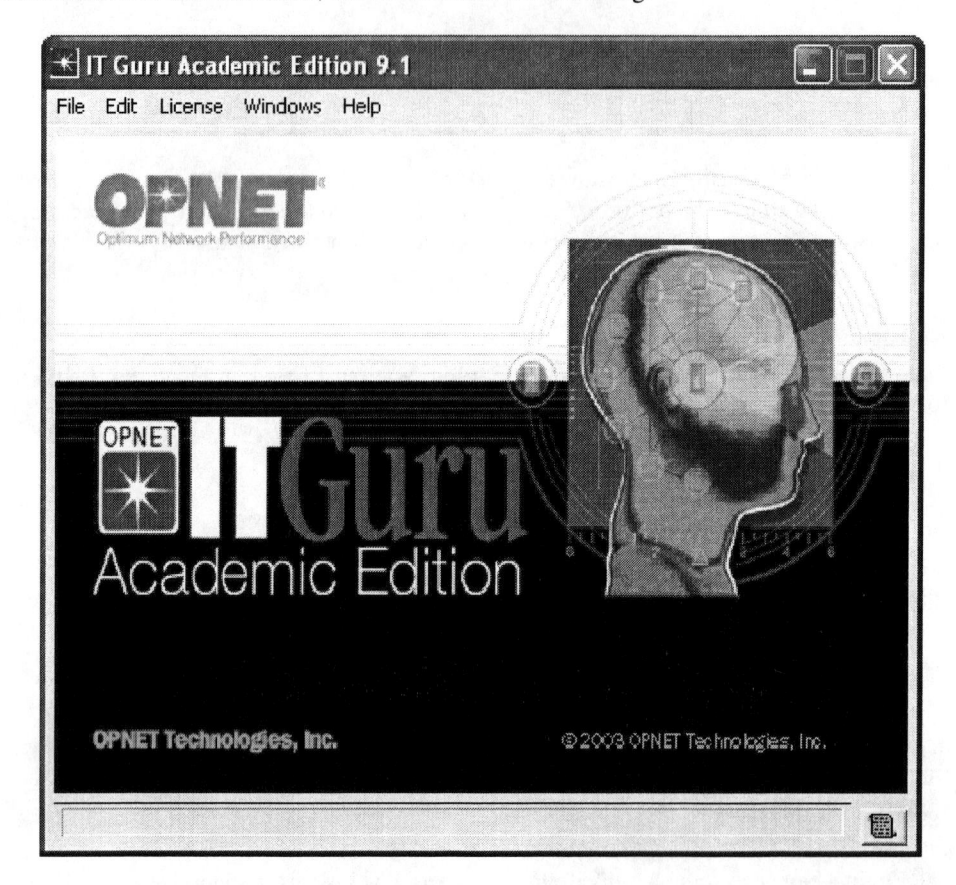

4. From the Main Menu, click on **Help** and select **Tutorial** from the pull down menu. Your PDF viewer will display the Tutorial pages. Under **Basic Lessons**, choose **Introduction.** In the Table of Contents, choose **About IT Guru.** Read through the **Introduction.** It will describe the Project Editor, Menu Bar, Tool Buttons, Workspace, Message Area, and Tool Tips.

5. Now, in the Table of Contents, choose **Tutorials Menu.** Under **Basic Lessons**, choose **Small Internetworks**. Follow the steps described in the Tutorial. You will build a small LAN and then expand it by connecting it to another LAN. The goal of the project is to compare the performance of the original LAN and the expanded LAN. You will build a project, run a simulation to gather statistics, and observe results.

Questions

1. Was the server able to handle the additional load of the second network?
2. How was delay affected when the second network was connected?
3. Try adding a third network similar to the second one you added. Does delay significantly increase or decrease? What is the new delay value? What is the new value for load at the server?
4. Based on what you observed, would it be advisable to expand the network to include two or three more LANs?

LAB 1   CENTRAL VERSUS DISTRIBUTED PROCESSING

Objectives
1. To build two campus networks, one with centralized services and one with distributed services.
2. To compare application response times for the centralized and distributed network architectures.

Motivation
Four departments at California University have formed advertising committees to promote their academic programs. The committees will use database, email, file transfer, and web applications to locate and correspond with potential students. At the same time, the university's Computer Services Department is considering modifying their network architecture. Currently, there is one centralized server for all applications. Computer Services would like to know if distributing applications across multiple servers would improve response time. Their plan is to create a network simulation of the centralized and distributed systems, and to capture response time statistics for the applications that the advertising committees will use. After evaluating performance statistics for each simulation, they will be able to choose the best network architecture for the university.

Description
Traditionally, system resources were centrally located on one computer. All hosts would communicate with this computer (typically a mainframe) for all application services. One benefit of a centralized system is that resources are easy to manage. A disadvantage, however, is that application response time may suffer when there are a large number of hosts. With the decline in price of computer equipment, distributed processing has become more popular. In a distributed system, resources are divided among multiple computers and locations. Distributed architectures have multiple benefits. Resources can be placed near those in need of the resources, applications can be optimized for a particular type of computer, and systems are easily scalable. A disadvantage of distributed systems, however, is that they are more difficult to manage.

In this lab, we will create two scenarios that model a campus network. The first will contain one centralized server that handles database, email, file transfer, and web services for four different departments. The second scenario will be a duplicate of the first; however, each application will be located on a separate server in a distributed fashion. Statistics regarding application response time and data sent and received will be gathered. We will also look at the server's CPU utilization. The results will be used to evaluate which architecture, centralized or distributed, provides the best application response time.

Create a New Project

1. Start the **OPNET IT Guru Academic Edition Application.** Create a **New Project** by selecting **File/New/Project** from the Main Menu. Click **OK.**

2. Give the project a name such as **your_initials_CentralvsDistrib.** Give the scenario a name such as **Scenario1_Central.** Click **OK.** You should see the **Startup Wizard Initial Topology** dialog box.

3. Verify that **Create Empty Scenario** is selected in the **Initial Topology Dialog Box.** Click **Next.** You will now see the **Network Scale** dialog box.

4. Choose **Campus** from the **Network Scale** list. Click **Next.**

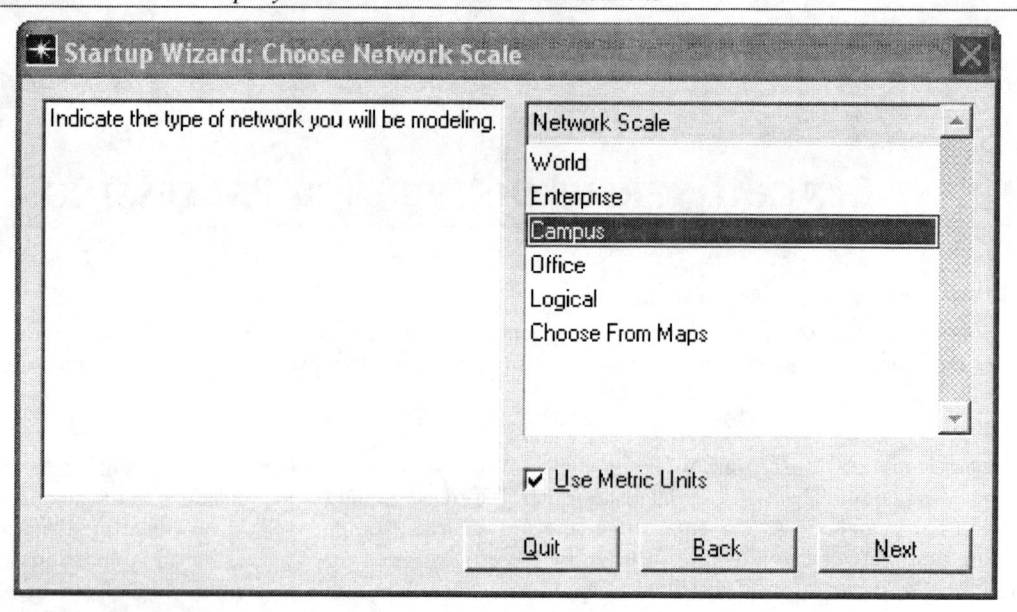

OPNET Lab Manual to accompany Business Data and Communications

[Startup Wizard: Choose Network Scale dialog box showing Network Scale options: World, Enterprise, Campus (selected), Office, Logical, Choose From Maps. Use Metric Units checkbox is checked.]

5. Choose **Miles** in the **size** dropdown box. Type **2** for the **X span** and **2** for the **Y span.** Click **Next**. You should see the **Select Technologies** dialog box.

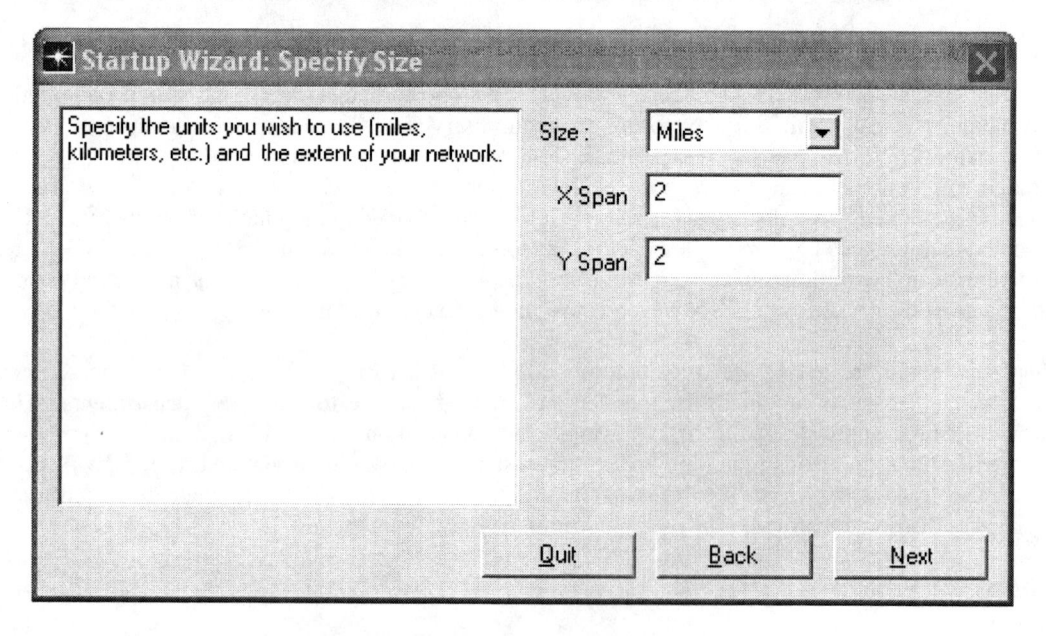

[Startup Wizard: Specify Size dialog box with Size: Miles, X Span: 2, Y Span: 2.]

6. Scroll down until you see **internet_toolbox** under the **Model Family** column. Under the **Include column**, click on the box next to **internet_toolbox**. This will change the "no" to "yes" indicating that **internet_toolbox** technologies will be included. Click **Next.**

Lab 1   Central versus Distributed Processing

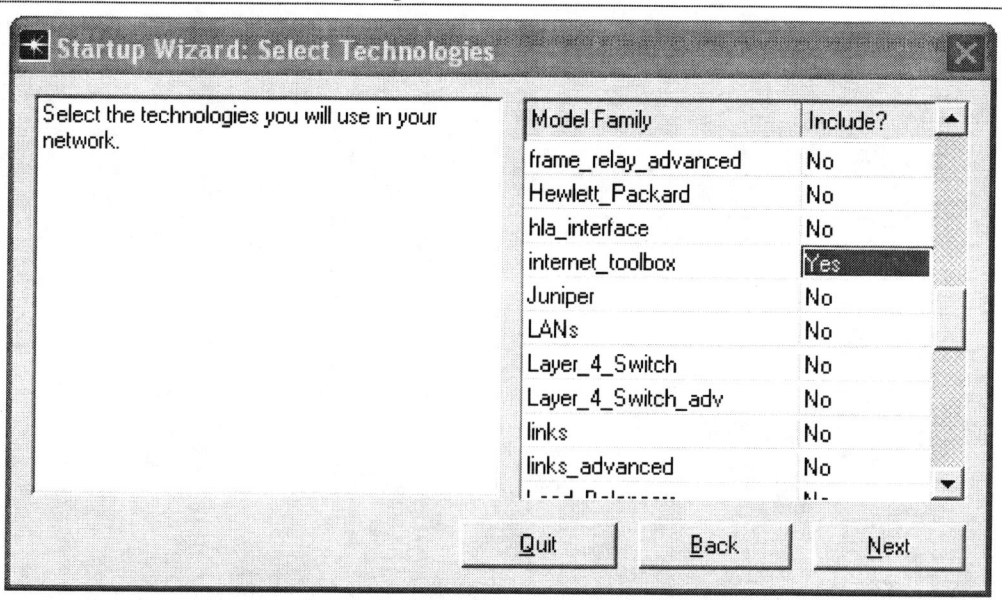

7. In the **Startup Wizard** Review dialog box, verify that the **internet_toolbox** is chosen and that the scale is Campus 2 miles x 2 miles. Click **OK**. You should see the **Object Palette** window.

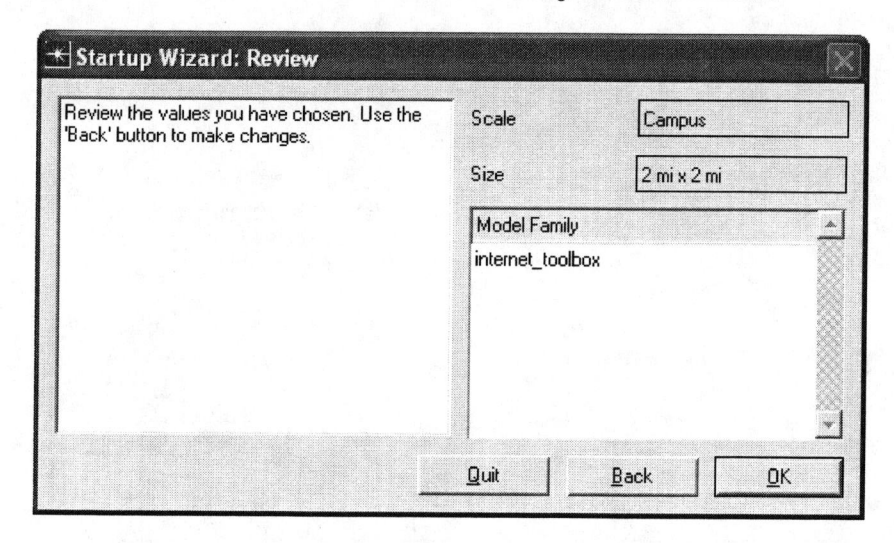

Building the Centralized Campus Network

We will begin by creating a campus network which links four departments to a centralized server that provides database, email, file transfer, and web services. The college networks will be 10BaseT LANs. These LANs will each connect to a 10BaseT switch. The switches and centralized server will be connected via a faster 100BaseT switch.

1. From the **Object Palette**, click on the **10BaseT_LAN** object to select it. Click to paste **four** copies of it in each corner of the project grid. Right click to release the object.

2. Click on the **10BaseT_LAN** in the upper left of the project grid to select it. Right click, and choose **Set Name** from the pull-down menu. Name the LAN **College of Arts**. Repeat the same step to name each of the other LANs in the following manner. Name the bottom left LAN **College of Science**, the upper right LAN **Admissions**, and the lower left LAN **College of Business**. Verify that your project looks like the following figure.

7

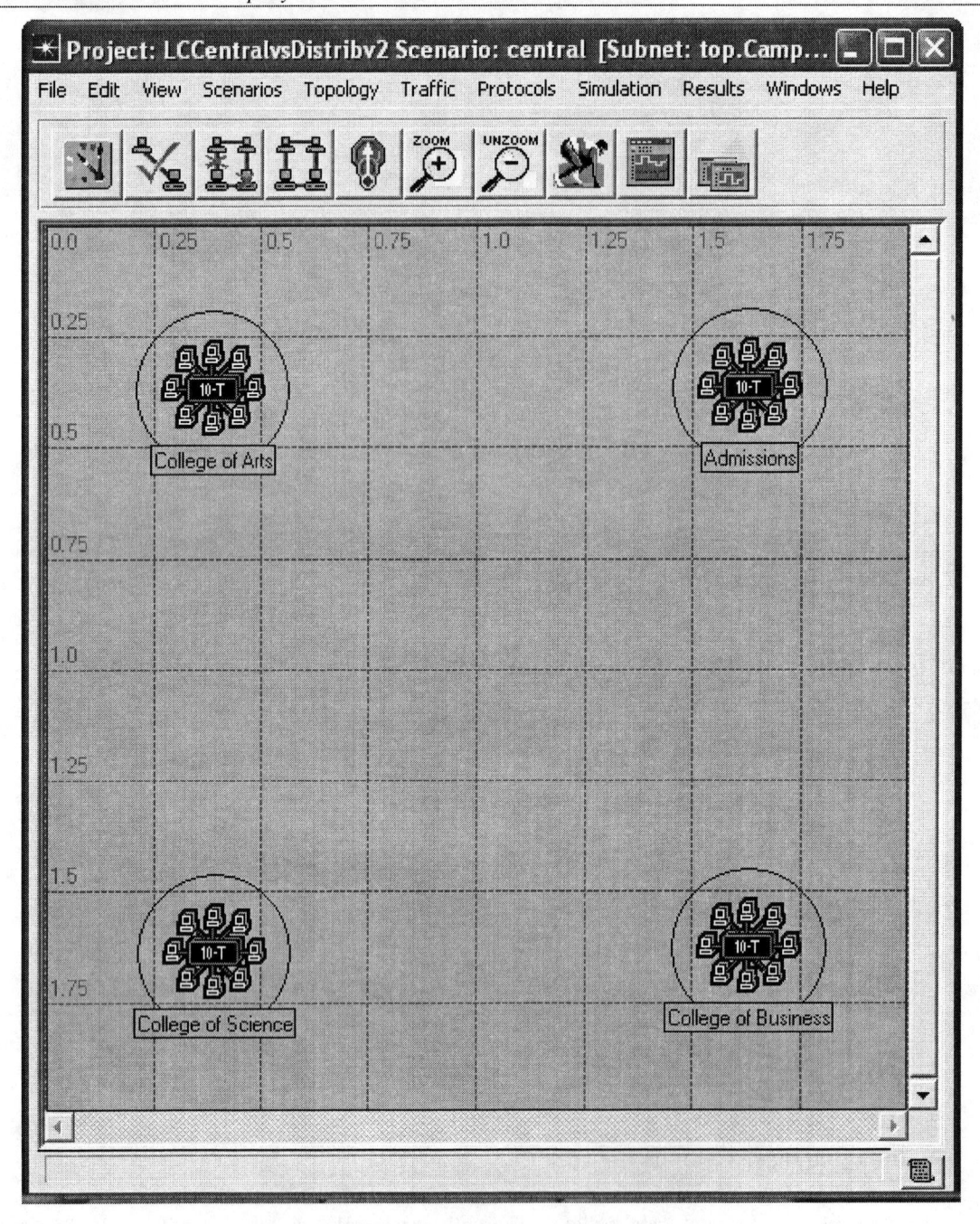

OPNET Lab Manual to accompany Business Data and Communications

3. From the **Object Palette**, click on the **ethernet15_switch** object to select it. Click to paste **four** copies of it in near each LAN on the project grid. Right click to release the object.

4. Click on the **ethernet15_switch** near the **College of Arts LAN** in the upper left-hand corner of the project grid to select it. Right click, and choose **Set Name** from the pull-down menu. Name the switch **ArtSwitch**. Repeat the same step to name each of the other switches. Name the bottom left switch **ScienceSwitch**, the upper right switch **AdmissionsSwitch**, and the lower right switch **BusinessSwitch**.

5. We will now connect the switches with 10BaseT links. 10BaseT links have a speed of 10 Mbps. From the **Object Palette**, click on the **10BaseT link** object. Click on a LAN and connect it to the appropriately named switch. Right click to release the link object.

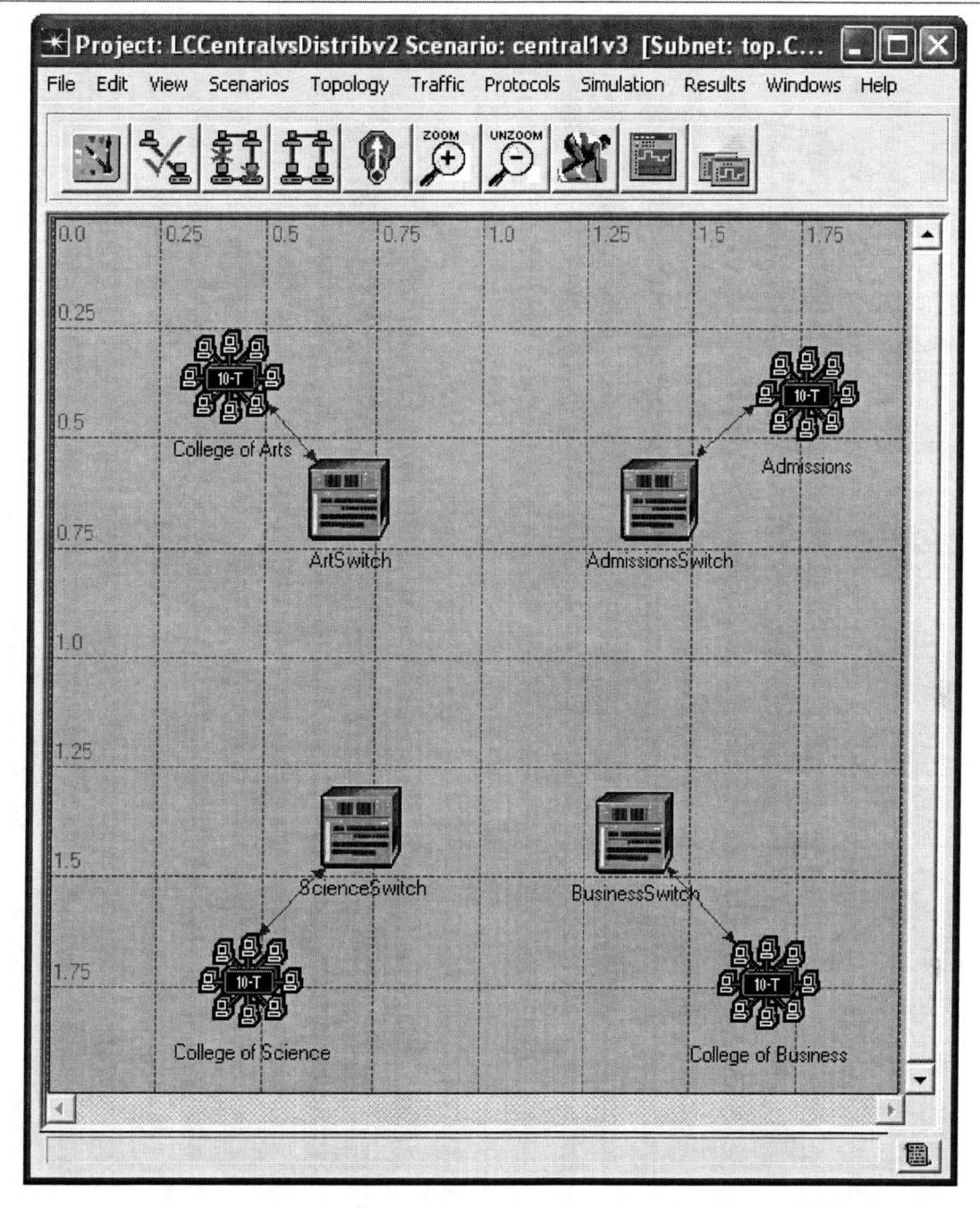

Lab 1 Central versus Distributed Processing

6. Now we will add another switch to the grid and connect the other switches to it via 100BaseT links. 100BaseT links have a speed of 100 Mbps. From the **Object Palette**, click on the **ethernet15_switch** object to select it. Click to paste it in the center of the project grid. Right click to release the object. Right click, and choose **Set Name** from the pull-down menu. Name the switch **ServerSwitch**.

7. From the **Object Palette**, click on the **100BaseT link** object. Click on each college switch and connect it to the **ServerSwitch**. Right click to release the link object.

8. From the **Object Palette**, click on the **ethernet_server** object, and click to paste it underneath the **ServerSwitch** on the project grid. Right click to release the object. Now select a **100BaseT link** object, and connect the **ethernet_server** object to the **serverSwitch**. Right click on the **ethernet_server** and choose **Set Name** from the pull-down menu. Name the server object **Server**.

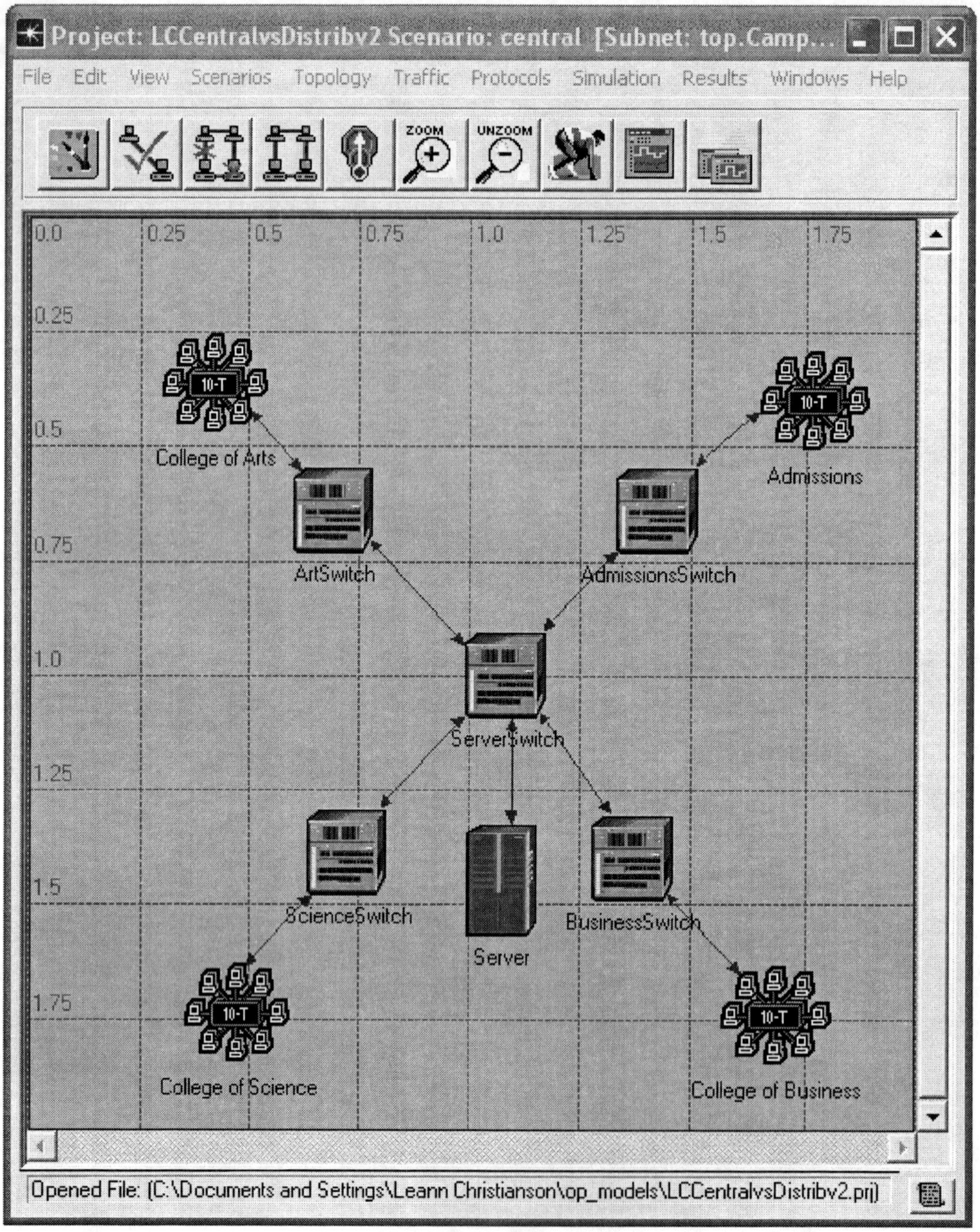
Project: LCCentralvsDistribv2 Scenario: central [Subnet: top.Camp...
File Edit View Scenarios Topology Traffic Protocols Simulation Results Windows Help
ZOOM
UNZOOM
College of Arts
Admissions
ArtSwitch
AdmissionsSwitch
ServerSwitch
ScienceSwitch
Server
BusinessSwitch
College of Science
College of Business
Opened File: (C:\Documents and Settings\Leann Christianson\op_models\LCCentralvsDistribv2.prj)

We will now add an **Application Config** and a **Profile Config** object to the workspace. The application object will allow us to choose and modify applications we want to include in our simulation. We will use Database Access, Email, File Transfer, and Web Browsing. Other applications defined in OPNET IT Guru Academic Edition include: File Print, Telnet, Video Conferencing, and Voice over IP. The Profile Config object allows us to create a traffic profile or choose a predefined profile. We will create our own profile in this project. Predefined profiles include: Engineer, Salesperson, Researcher, E-Commerce, and Multimedia User.

9. From the **Object Palette**, click on the **Profile Config** object, and click to paste it on the project grid. Right click to release the object. Right click on the **Profile Config** object, and choose **Set Name** from the pull-down menu. Name the object **ProfileConfig**.

10. Again, from the **Object Palette**, click on the **Application Config** object and click to paste it on the project grid. Right click to release the object. Right click on the **Application Config** object, and choose **Set**

Name from the pull-down menu. Name the object **AppDef**. Verify that your project looks like the following figure.

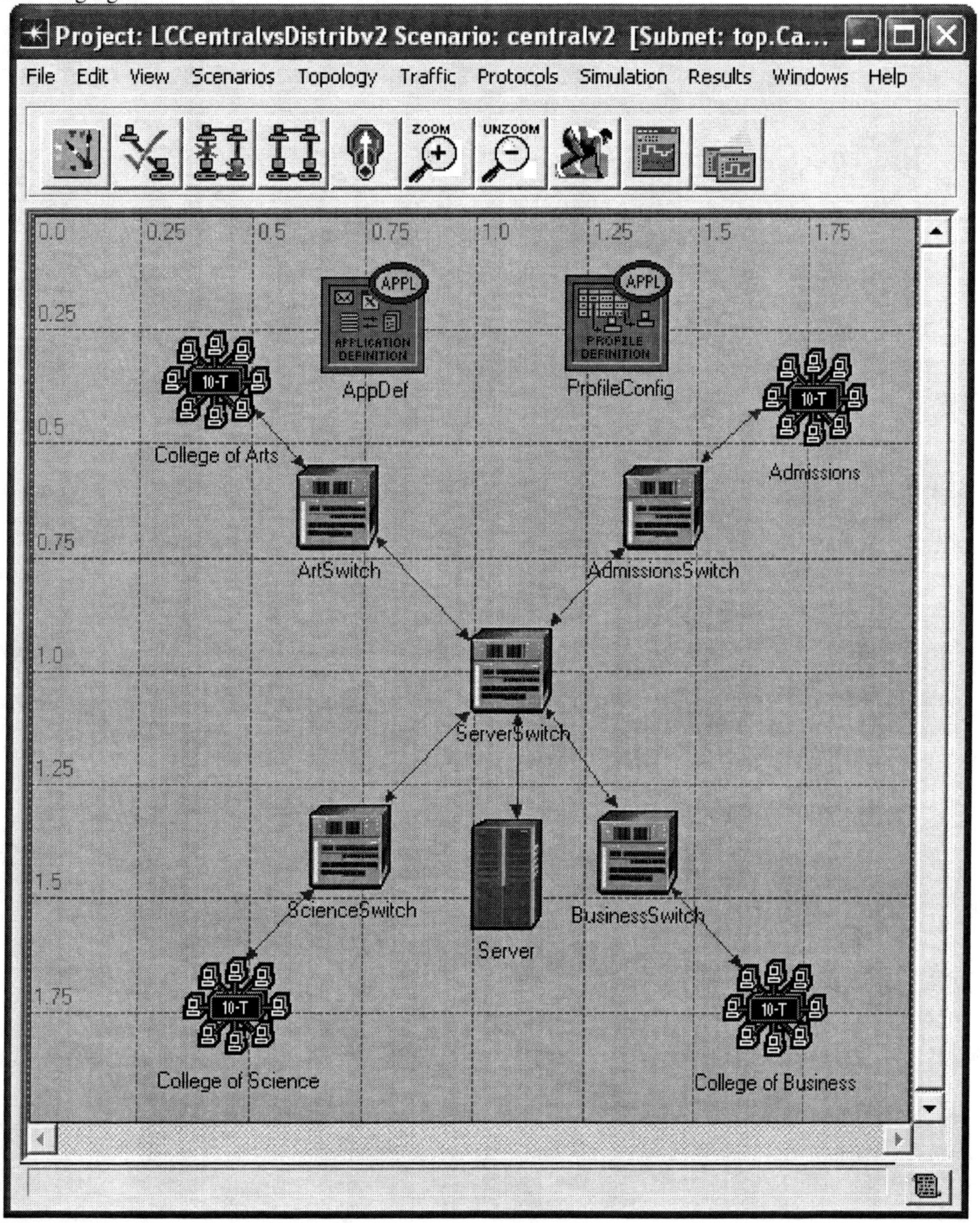
Project: LCCentralvsDistribv2 Scenario: centralv2 [Subnet: top.Ca...
File Edit View Scenarios Topology Traffic Protocols Simulation Results Windows Help
ZOOM
UNZOOM
APPL
APPLICATION DEFINITION
AppDef
PROFILE DEFINITION
ProfileConfig
10-T
College of Arts
Admissions
ArtSwitch
AdmissionsSwitch
ServerSwitch
ScienceSwitch
Server
BusinessSwitch
College of Science
College of Business

11. From the Main Menu, choose **File/Save** to save your project.

Configuring the Network

1. Right click on the **Profile Config** object and choose **Edit Attributes** from the pull-down menu. Edit the **Profile Configuration** attribute and set the value to **Sample Profiles**. Click on the + next to **Profile Configuration** to expand the parameters. Click on the value next to **rows** and choose **Edit**. Change the number of rows to **6**. A new row will appear below the other predefined profiles.

2. Click on the + next to row **5** to expand the parameter list. Click on the value next to **Profile name** and enter the name **Advertisement**. Click on the value next to **Applications** and choose **Edit**. A new window will appear.

3. Click on the value listed next to **Rows** (displayed in the bottom left hand corner), and choose **Edit** from the pull-down menu. Enter the value **4**. You will see four rows added to the table. These rows will specify which applications are associated with our Advertisement profile. Click on the first row under the **Name** column, and choose **Database Access (Heavy)** from the pull-down menu. Repeat this step for each of the other rows choosing **Email (Heavy), File Transfer (Heavy),** and **Web Browsing (Heavy HTTP1.1)**. Click **OK** to close the **Application Table** window.

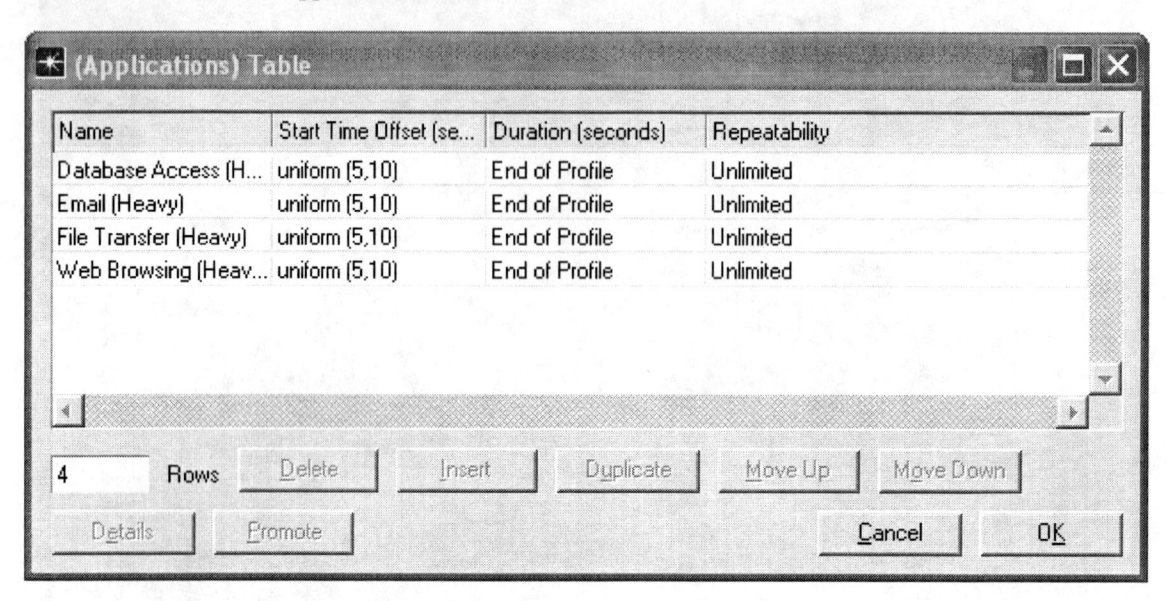

Name	Start Time Offset (se...	Duration (seconds)	Repeatability
Database Access (H...	uniform (5,10)	End of Profile	Unlimited
Email (Heavy)	uniform (5,10)	End of Profile	Unlimited
File Transfer (Heavy)	uniform (5,10)	End of Profile	Unlimited
Web Browsing (Heav...	uniform (5,10)	End of Profile	Unlimited

4. In the **Profile Attribute** window, click on the value next to **Operation Mode** and choose **Simultaneous** from the pull-down menu. Click **OK** to close the **Profile Attribute** window.

Lab 1   Central versus Distributed Processing

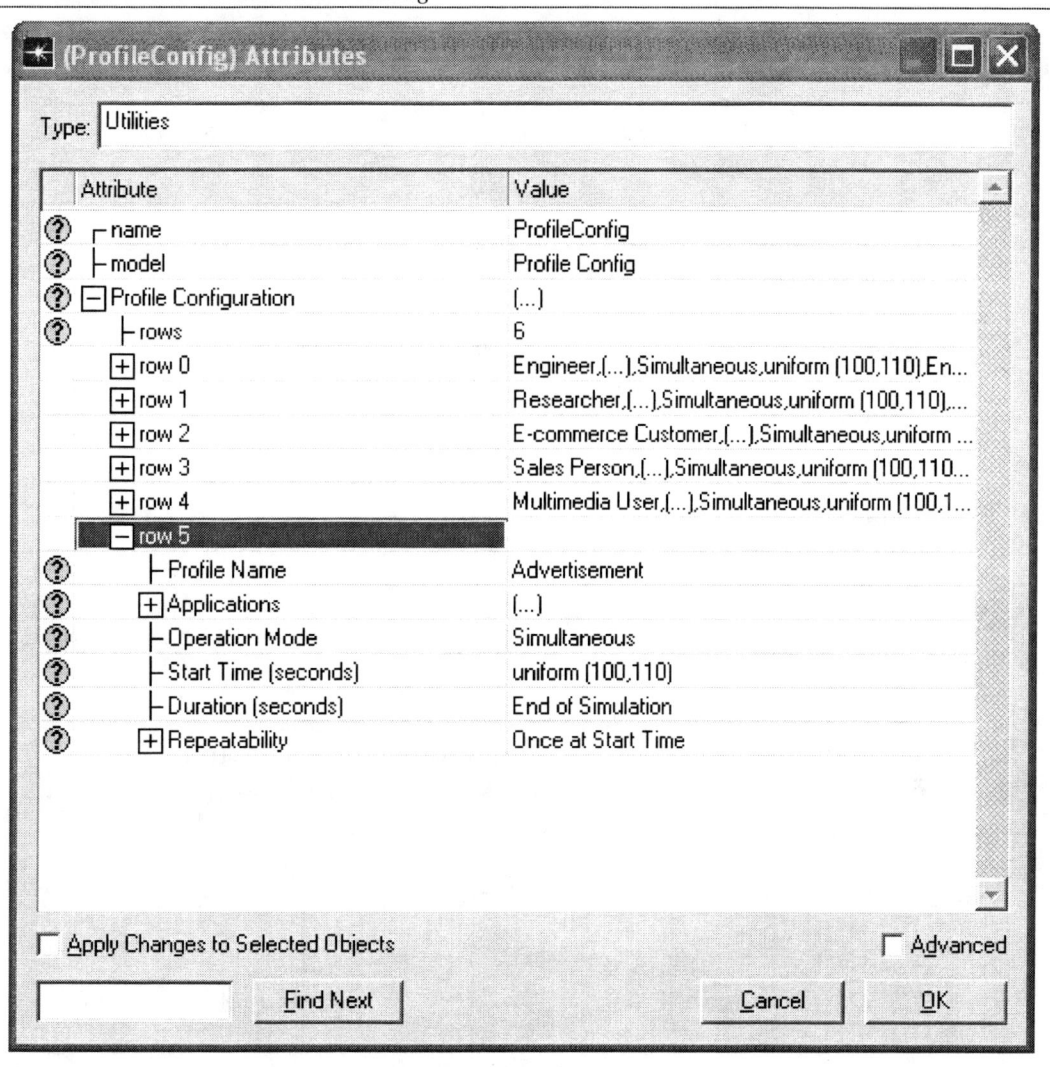

Attribute	Value
name	ProfileConfig
model	Profile Config
⊟ Profile Configuration	(...)
├ rows	6
⊞ row 0	Engineer,(...),Simultaneous,uniform (100,110),En...
⊞ row 1	Researcher,(...),Simultaneous,uniform (100,110),...
⊞ row 2	E-commerce Customer,(...),Simultaneous,uniform ...
⊞ row 3	Sales Person,(...),Simultaneous,uniform (100,110...
⊞ row 4	Multimedia User,(...),Simultaneous,uniform (100,1...
⊟ row 5	
├ Profile Name	Advertisement
⊞ Applications	(...)
├ Operation Mode	Simultaneous
├ Start Time (seconds)	uniform (100,110)
├ Duration (seconds)	End of Simulation
⊞ Repeatability	Once at Start Time

5. Now we will configure the applications that we chose for our new profile. Right click on the **AppDef** object and choose **Edit Attributes** from the pull-down menu. Edit the **Application Definitions** attribute and set the value to **Default**. Click on the + next to **Application Definitions** to expand the parameters. You will see a list of all the applications that OPNET IT Guru Academic Edition supports.

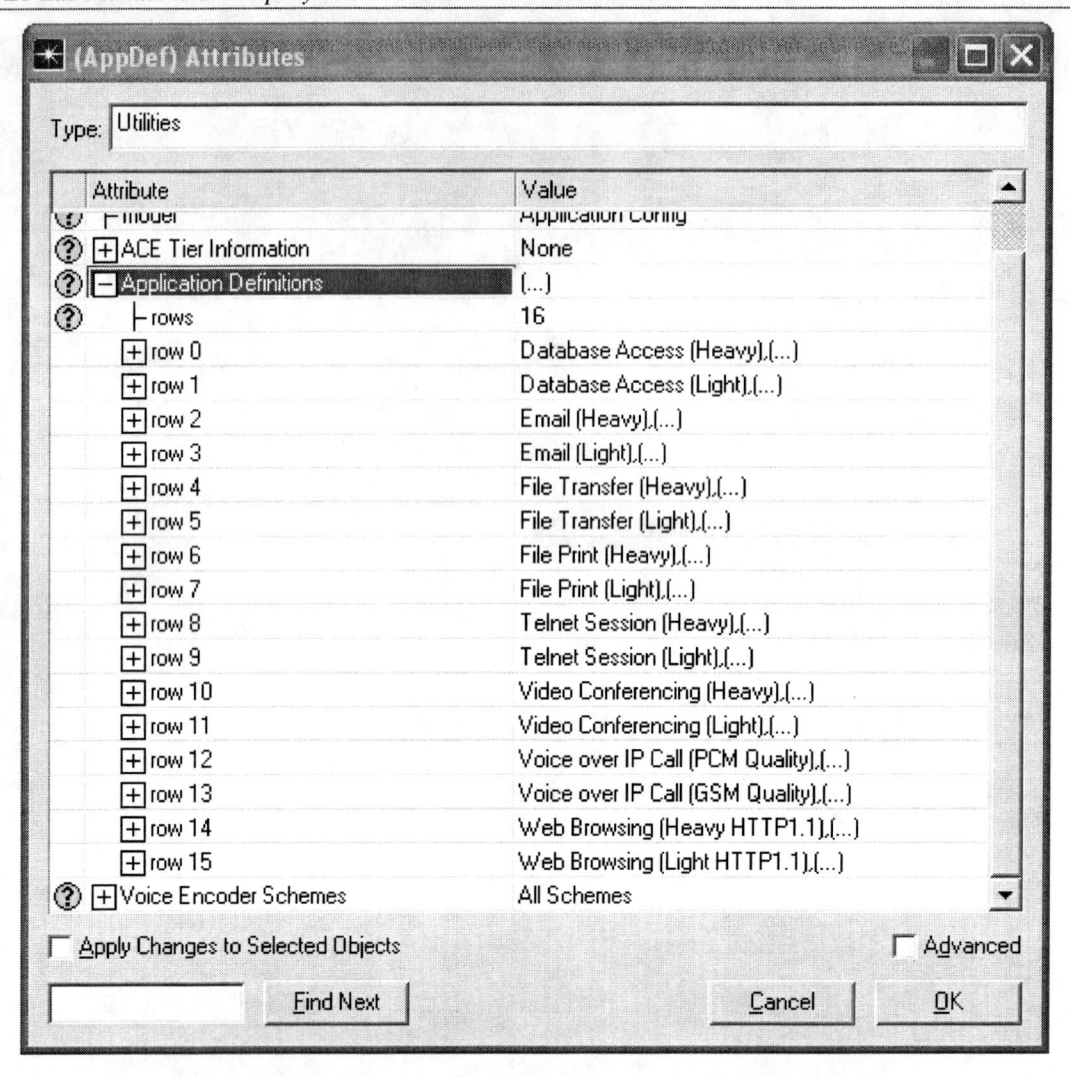

6. Click on the + next to **Database Access (Heavy)**. Click on the + next to **Description**. Next, click on the value next to **Database** and choose **Edit** from the pull-down menu. A new window will appear with the database attribute values. We will change the transaction mix and transaction interarrival time.

7. Click on the value next to **Transaction Mix (Queries/Total Transactions)** and choose **100%** from the pull-down menu. This will make all of our database transactions queries to the server.

8. Click on the value next to **Transaction Interarrival Time**. A new window will appear. Choose **exponential** from the pull-down menu and enter the value **5**. This means that the interarrival time for our queries will have an exponential distribution with an average value of 5 seconds. Click **OK**.

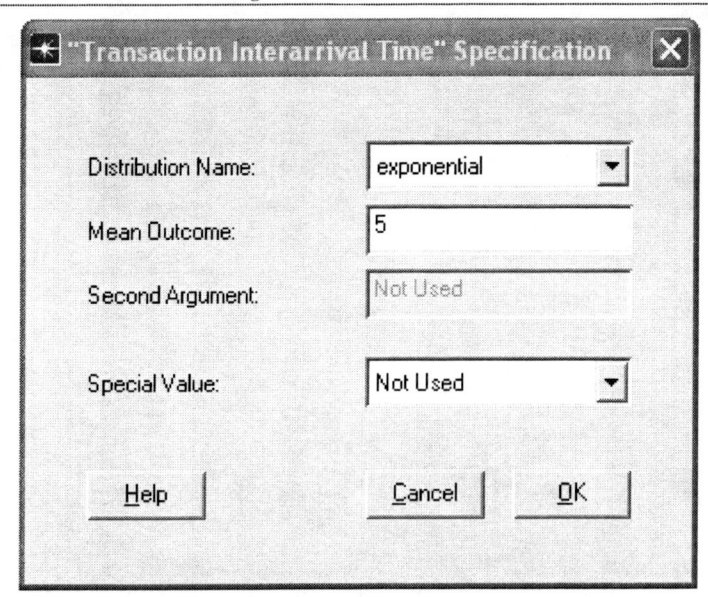

Lab 1   Central versus Distributed Processing

[Transaction Interarrival Time Specification dialog:
- Distribution Name: exponential
- Mean Outcome: 5
- Second Argument: Not Used
- Special Value: Not Used]

9. You should now have the values listed below in the **Database Table** window that follows. Click **OK** to close the window.

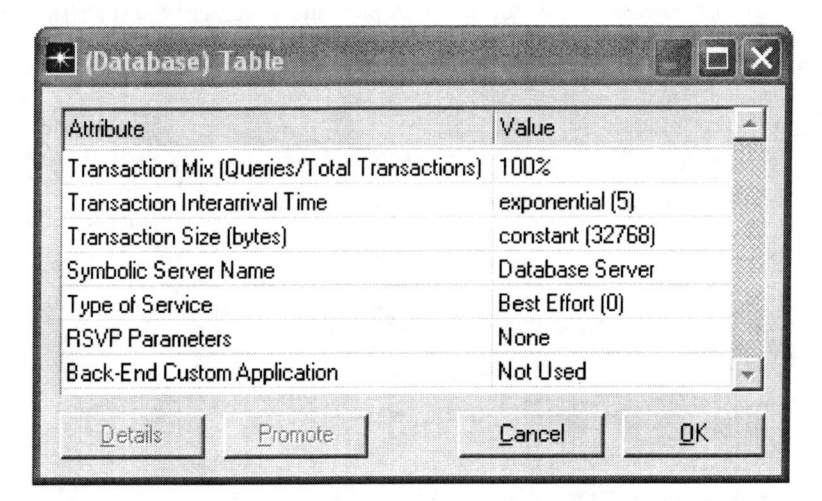

Attribute	Value
Transaction Mix (Queries/Total Transactions)	100%
Transaction Interarrival Time	exponential (5)
Transaction Size (bytes)	constant (32768)
Symbolic Server Name	Database Server
Type of Service	Best Effort (0)
RSVP Parameters	None
Back-End Custom Application	Not Used

10. Now click on the + next to **Email (Heavy).** Click on the + next to **Description**. Next, click on the value next to **Email** and choose **Edit** from the pull-down menu. A new window will appear with the email attribute values. We will change the send and receive interarrival times and the size of email messages.

11. Click on the value next to **Send Interarrival Time (seconds)**. A new window will appear. Choose **exponential** from the pull-down menu and enter the value **10**. This means that the interarrival time for email that is sent will have an exponential distribution with an average value of 10 seconds. Click **OK**.

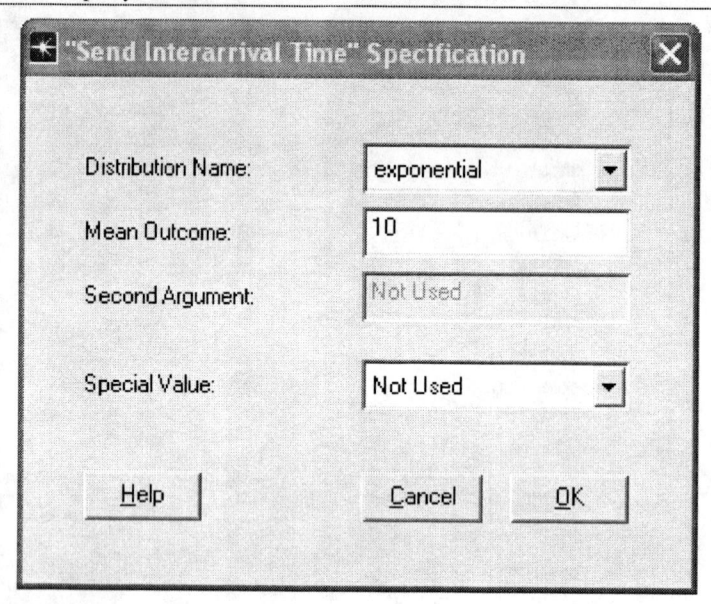

OPNET Lab Manual to accompany Business Data and Communications

12. Click on the value next to **Receive Interarrival Time (seconds)**. A new window will appear. Choose **exponential** from the pull-down menu and enter the value **10**. This means that the interarrival time for email received will also have an exponential distribution with an average value of 10 seconds. Click **OK**.

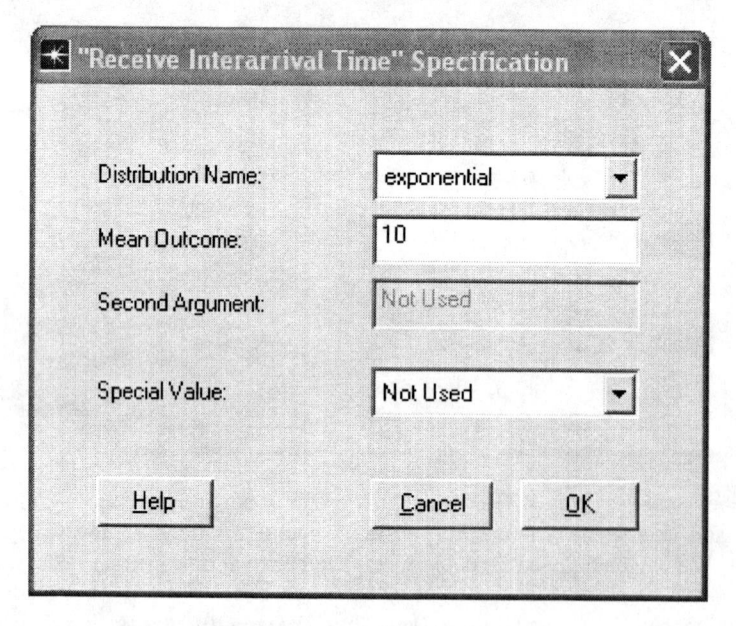

13. Click on the value next to **Email Size (bytes)**. A new window will appear. Choose **constant** from the pull-down menu and enter the value **10000**. Now all email messages will have a size of 10,000 bytes. Click **OK**.

16

Lab 1   Central versus Distributed Processing

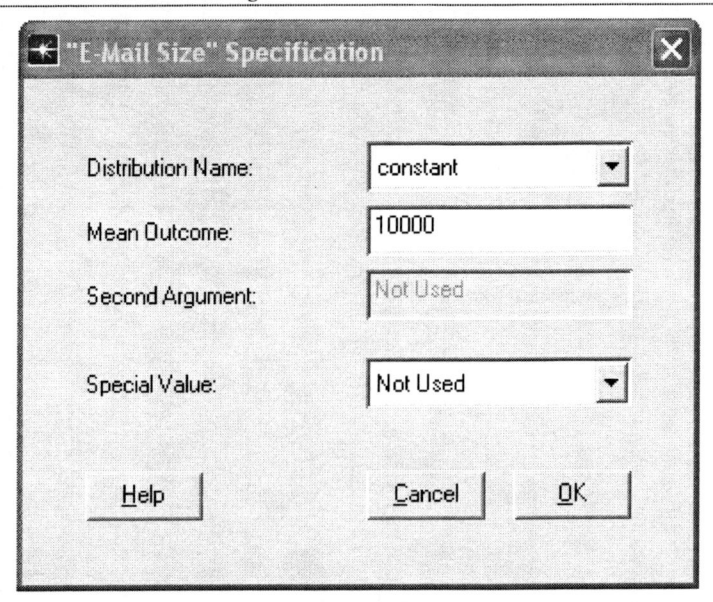

14. You should now have the values listed below in the **Email Table** window that follows. Click **OK** to close the window.

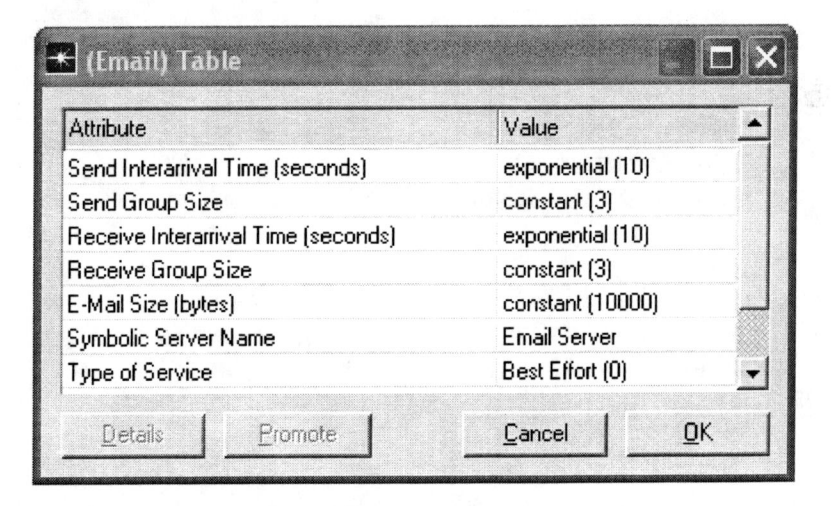

15. Now click on the + next to **File Transfer (Heavy)**. Click on the + next to **Description**. Next, click on the value next to **Ftp** and choose **Edit** from the pull-down menu. A new window will appear with the FTP attribute values. We will change the command mix and the interarrival times of the requests.

16. Click on the value next to **Command Mix (Get/Total)**, and choose **100%** from the pull-down menu. This will make all of our FTP transfers downloads from the server.

17. Click on the value next to **Inter-request Time (seconds)**. A new window will appear. Choose **exponential** from the pull-down menu, and enter the value **10**. This means that the interrequest time for file transfers will have an exponential distribution with an average value of 10 seconds. Click **OK**.

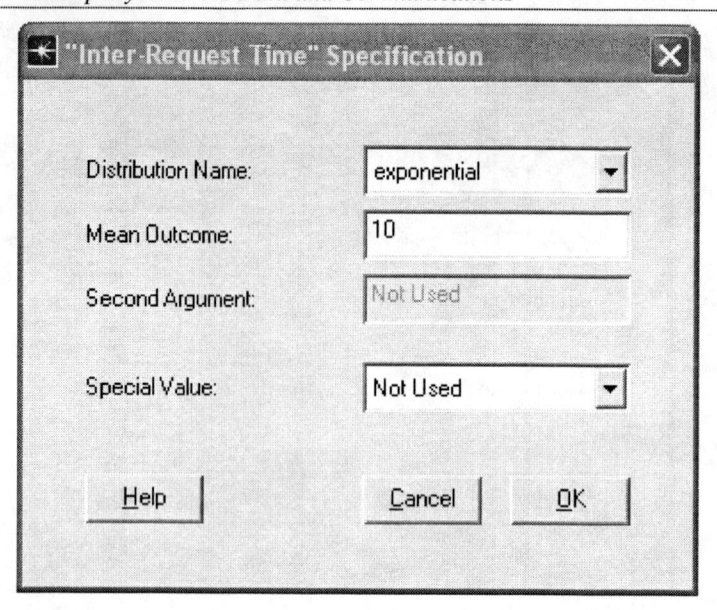

OPNET Lab Manual to accompany Business Data and Communications

[Inter-Request Time Specification dialog: Distribution Name: exponential; Mean Outcome: 10; Second Argument: Not Used; Special Value: Not Used]

18. You should now have the values listed below in the **FTP Table** window that follows. Click **OK** to close the window.

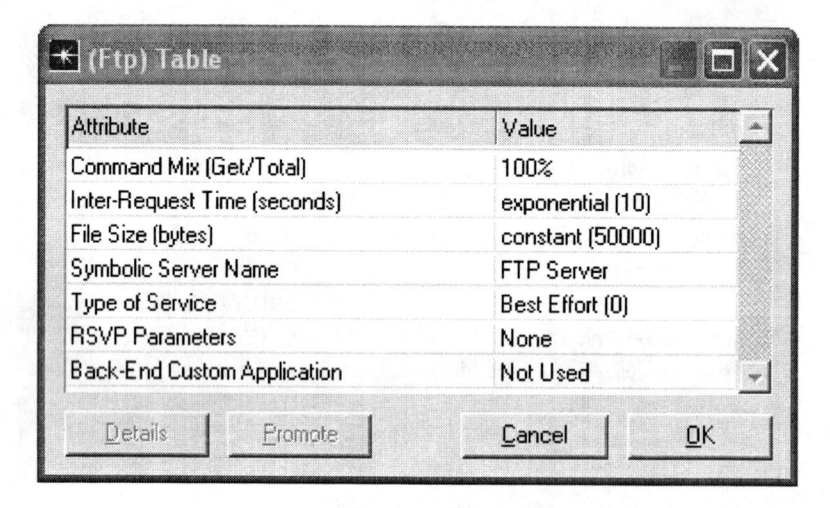

Attribute	Value
Command Mix (Get/Total)	100%
Inter-Request Time (seconds)	exponential (10)
File Size (bytes)	constant (50000)
Symbolic Server Name	FTP Server
Type of Service	Best Effort (0)
RSVP Parameters	None
Back-End Custom Application	Not Used

19. Now click on the + next to **Web Browsing (Heavy HTTP 1.1).** Click on the + next to **Description**. Next click on the value next to **Http,** and choose **Edit** from the pull-down menu. A new window will appear with the HTTP attribute values. We will change the page interarrival time.

20. Click on the value next to **Page Interarrival Time (seconds)**. A new window will appear. Choose **exponential** from the pull-down menu and enter the value **10**. This means that the interrequest time for file transfers will have an exponential distribution with an average value of 10 seconds. Click **OK**.

18

Lab 1   Central versus Distributed Processing

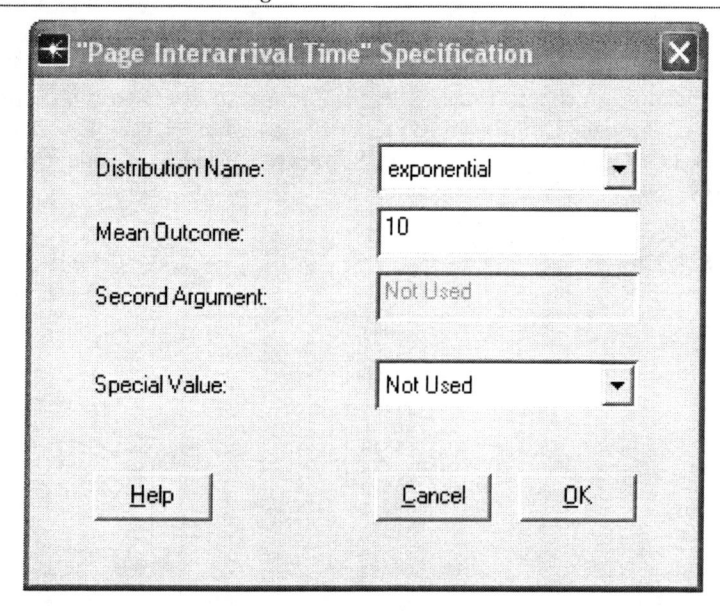

21. You should now have the values listed below in the **HTTP Table** window. Click **OK** to close the window.

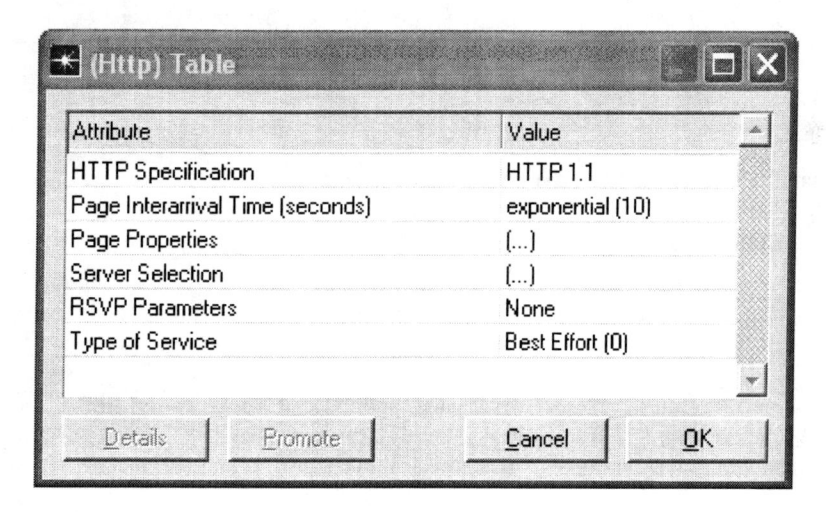

22. From the Main Menu, choose **File/Save** to save your project.

Configuring the LANs and Server

1. Now we will give each LAN the advertisement profile. Right click on one of the **10BaseT_LAN** objects, and choose **Select Similar Nodes** from the pull-down menu. Right click again on one of the selected LAN objects and choose **Edit Attributes.** Click on the box labeled **Apply Changes to Selected Objects.** Next, click on the value next to **Application: Supported Profiles** and choose **Edit** from the pull-down menu. A new window will appear.

2. Click on the value listed next to **Rows** (displayed in the bottom left hand corner), and choose **1** from the pull-down menu. You will see a new row added to the table. Now click on the value under **Profile Name.** You will see a list of profiles to choose from. Choose the **Advertisement** profile that we just created. Click **OK** to close the window. Click **OK** again to close the **Attribute** window.

3. Let us configure the server to provide services for our applications. Right click on the **Server** object and choose **Edit Attributes** from the pull-down menu. Choose the applications that will be used. Click on the value next to **Application: Supported Services,** and choose **Edit.** A new window will appear.

19

4. Click on the value listed next to **Rows** and choose **4** from the pull-down menu. You will see four rows added to the table. Now, click on the first row under the **Name** column. You will see a pull-down menu with a list of applications to choose from. Select **Database Access (Heavy)**. Repeat the same process for each of the rows choosing **Email (Heavy)**, **File Transfer (Heavy)**, and **Web Browsing (Heavy HTTP1.1)**. Click **OK** to close the **Application Supported Services Table**. Click **OK** again to close the **Attribute** window.

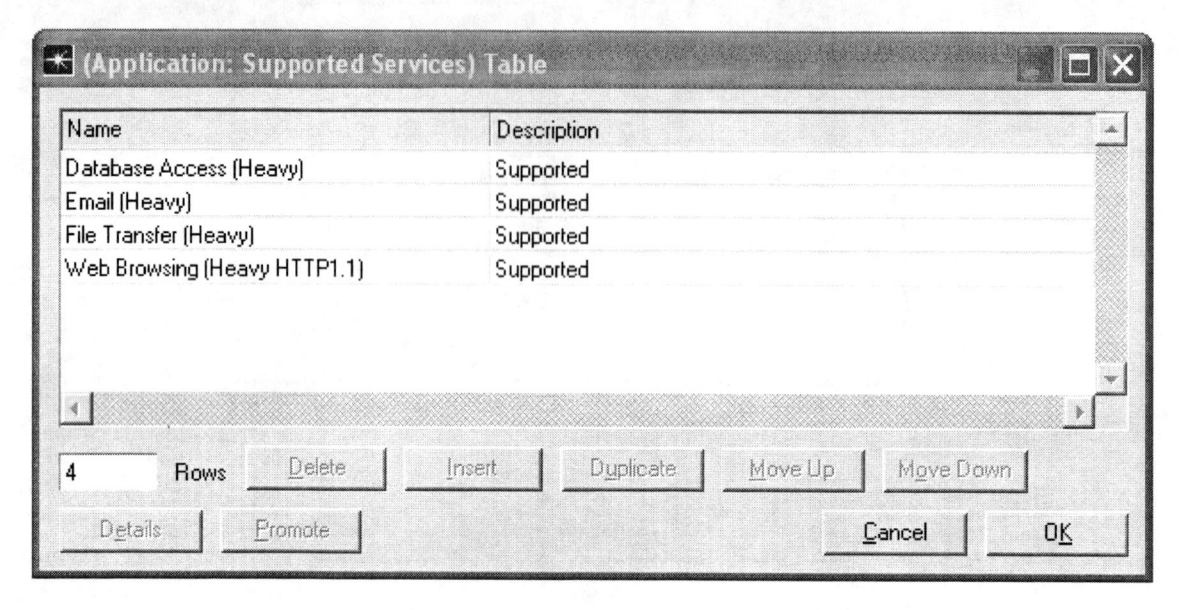

![(Application: Supported Services) Table showing four rows: Database Access (Heavy), Email (Heavy), File Transfer (Heavy), Web Browsing (Heavy HTTP1.1), all marked Supported.]

5. From the Main Menu, choose **File/Save** to save your project.

Configuring the Simulation

At this time, we would like to choose the statistics that we will gather for this simulation. Our goal will be to determine the response times for the applications the server provides as well as to determine the server's CPU load.

1. Right click somewhere on the project workspace, and select **Choose Individual Statistics** from the pull-down menu. Click on the **+** next to **Global** to expand the parameters. Click on the **+** next to **DB Query** to expand the parameters. Click next to **Response Time (sec)**, **Traffic Received (bytes/sec)**, and **Traffic Sent (bytes/sec)**.

2. Now select the statistics for the Email application. Click on the **+** next to **Email** to expand the parameters. Click next to **Download Response Time (sec)**, **Traffic Received (bytes/sec)**, **Traffic Sent (Bytes/Sec)**, and **Upload Response Time (sec)**.

3. Next, select the statistics for the FTP application. Click on the **+** next to **Ftp** to expand the parameters. Click next to **Download Response Time (sec)**, **Traffic Received (bytes/sec)**, **Traffic Sent (bytes/sec)**, and **Upload Response Time (sec)**.

4. Finally, select the statistics for the Web Browsing application. Click on the **+** next to **HTTP** to expand the parameters. Click next to **Object Response Time (seconds)**, **Page Response Time (seconds)**, **Traffic Received (bytes/sec)**, and **Traffic Sent (bytes/sec)**.

Lab 1 Central versus Distributed Processing

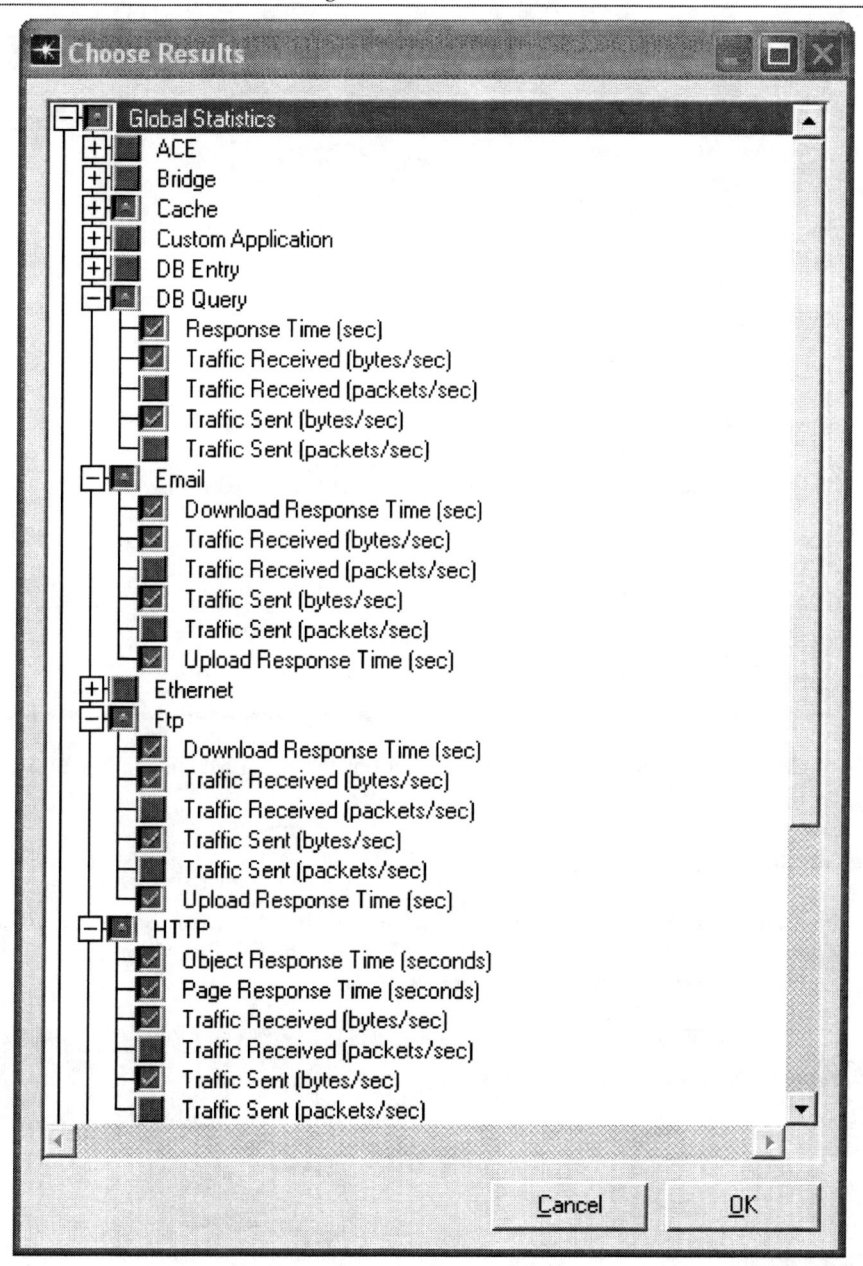

5. We would also like to gather statistics regarding CPU utilization of the server. This will allow us to measure how hard the centralized server is working. Click on the + next to **Node** to expand the parameters. Click on the + next to **CPU,** and click next to **Utilization (%).**

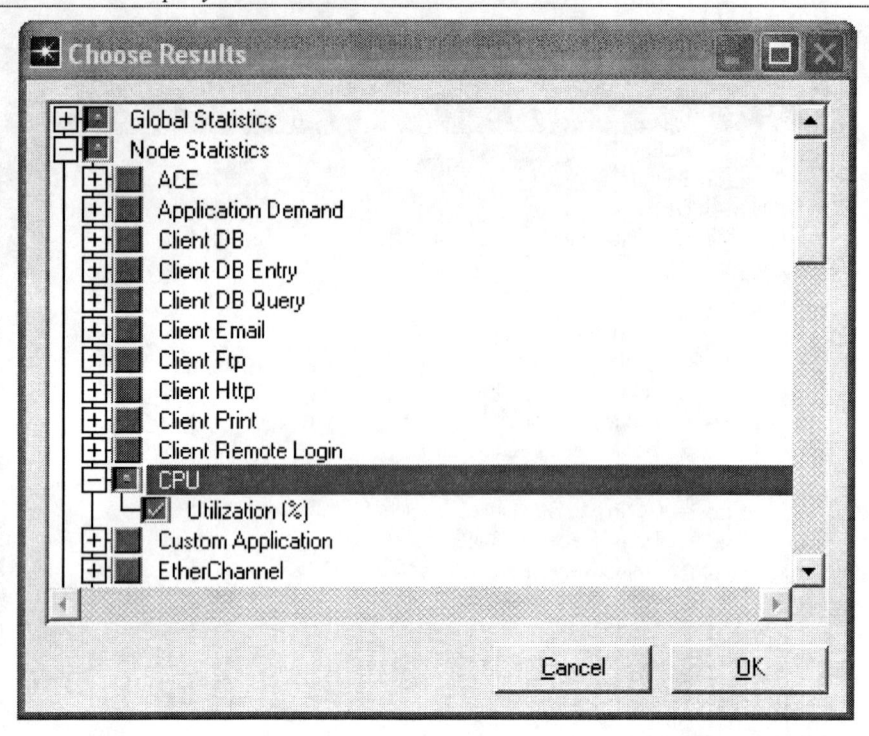

OPNET Lab Manual to accompany *Business Data and Communications*

6. Click **OK** to close the **Choose Results** window. From the Main Menu, choose **File/Save** to save your project.

Running the Simulation

1. Click on the **Configure and Run** Button. Change the **Duration** of the simulation to **5**. Choose **minute(s)** from the pull-down menu. Click on the **Run** button to run the simulation. Depending on your processor, the simulation should take several minutes to run.

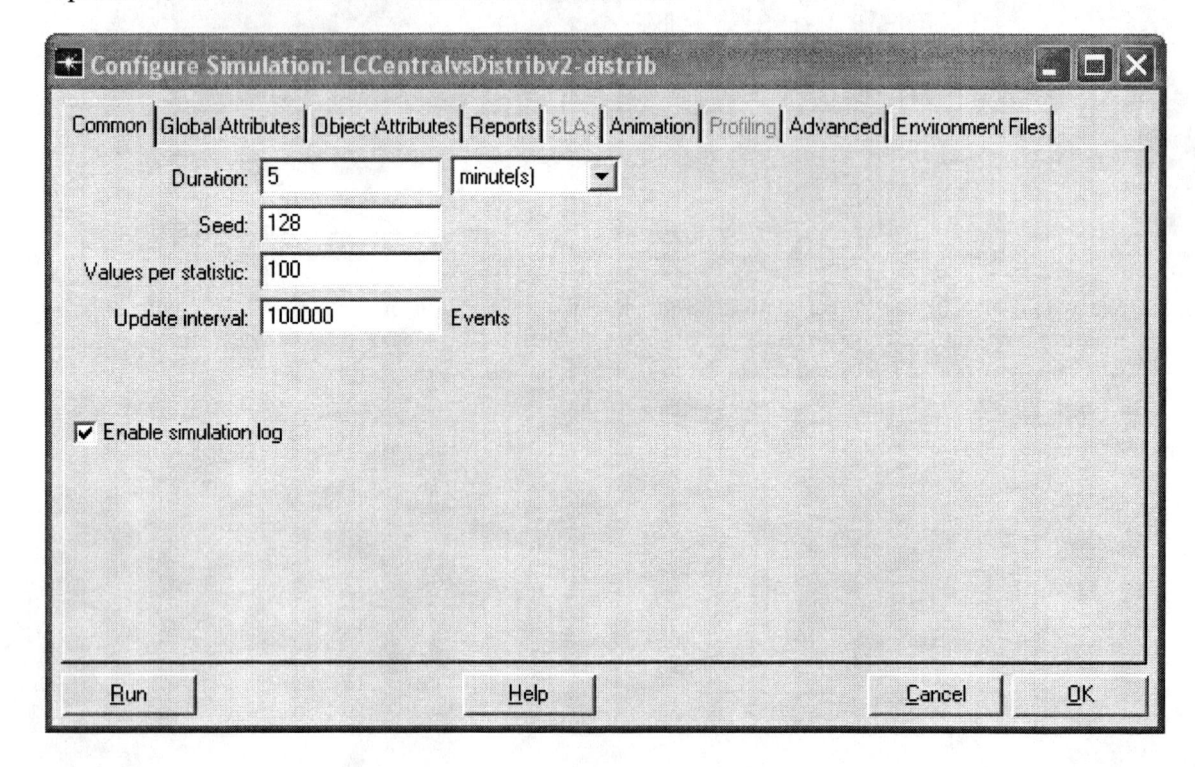

2. We will view our results after completing our second distributed scenario.

22

Lab 1   Central versus Distributed Processing

Creating the Distributed Services Scenario

1. We will now create a duplicate scenario with four servers, each providing services for one type of application. We will then compare application response times and server CPU utilization for both scenarios. We expect that response times and CPU utilization will decrease when services are distributed. From the Main Menu, choose **Scenarios/Duplicate Scenario**. Give the new scenario a name such as **Distributed**.

2. Click on the **Server** object to select it. From the Main Menu, choose **Edit/Copy** and then **Paste** three copies of the server on the project grid. Each time you paste a copy of the server, click on the **Switch** object to connect it to the switch. Right click to release the link object. For each server, right click and choose **Set Name** from the pull-down menu. Name the servers, **DatabaseServer, EmailServer, FTPServer,** and **HTTPServer.** Your project should resemble the following figure:

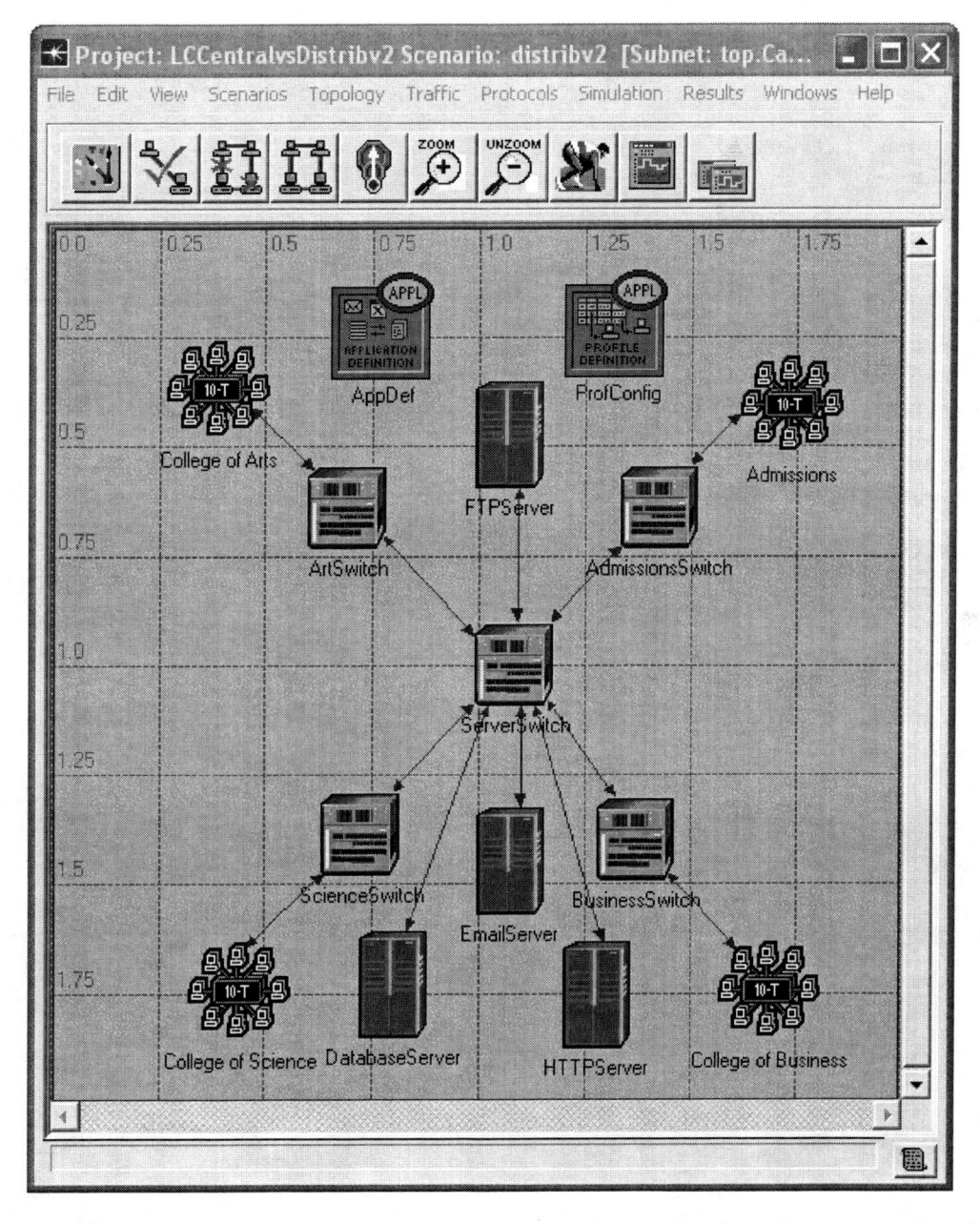

3. Now we need to change the server objects so that they provide services for only one type of application. Click on the **DatabaseServer** object to select it and choose **Edit Attributes** from the pull-down menu.

23

Click on the value next to **Application Supported Services** and choose **Edit** from the pull-down menu. Click on **Email (Heavy)** to select it, and then click on the **Delete** button to delete this application. Repeat this step to delete **File Transfer (Heavy)** and **Web Browsing (Heavy HTTP1.1)** The DatabaseServer object should now be configured to provide only Database service. Click **OK** to close the **Attribute** window.

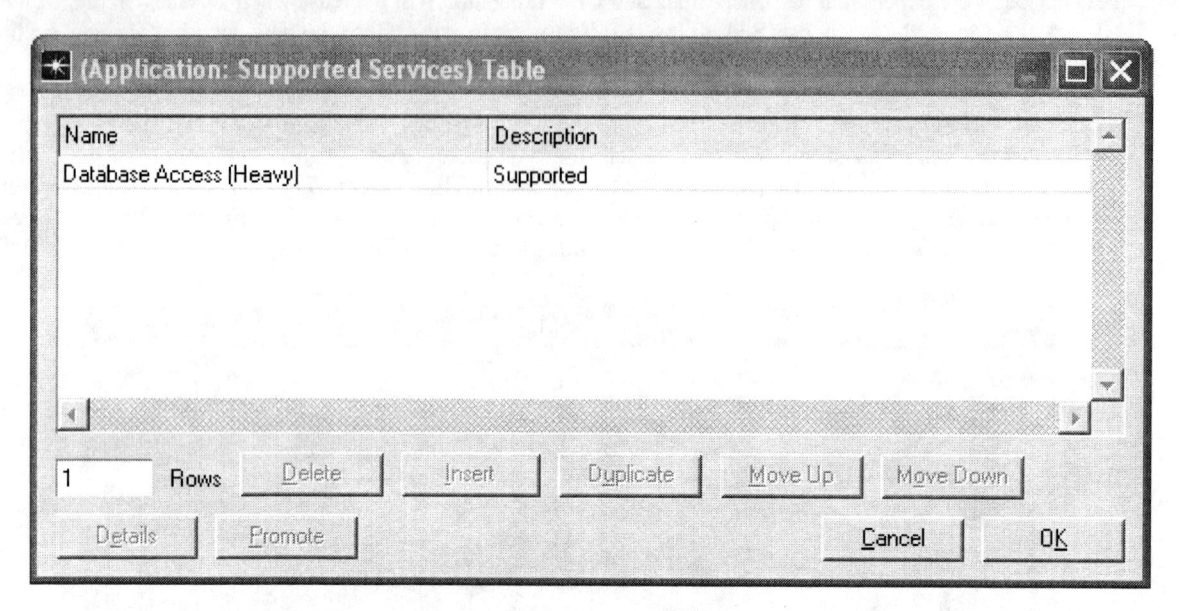

4. Repeat the preceding step to configure each server to provide services for one application. The EmailServer should provide Email (Heavy) services, the FTPServer should provide File Transfer (Heavy) service, and the WebServer should provide Web Browsing (Heavy HTTP1.1) service.

5. We would also like to gather statistics regarding load at each of the servers. Right click on the project workspace and select Choose Individual Statistics. Click on the + next to Node to expand the parameters. Scroll down, click on the + next to Server DB Query, and click next to Load (Requests/Sec). Repeat the step to select Load statistics for the Email Server, FTP Server, and Web Server.

Lab 1 Central versus Distributed Processing

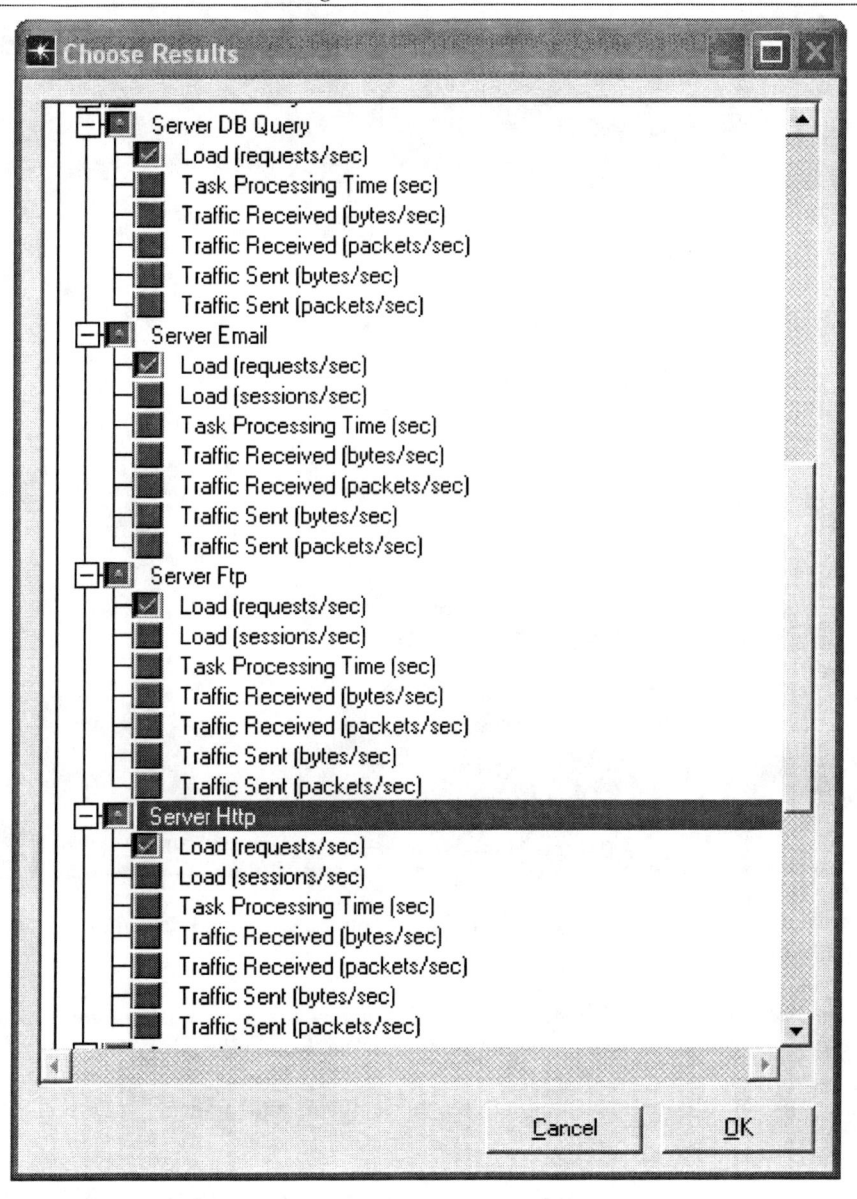

Running the Simulation

1. From the Main Menu, choose **Simulation/Run Discrete Event Simulation**. Depending on your processor, the simulation should take several minutes to run.

2. We can now compare results for both the centralized and distributed scenarios. From the Main Menu, choose **Results/Compare Results**. To compare results for **Database Query Response Time**, click on the + next to **DB Query**, and click next to **Response Time (Seconds)**. Choose **average** from the pull-down menu and click **Show**. This will give you a graph with data from both scenarios as well as a legend for the graph. Your graph should resemble the one shown next. Notice that the response time for the distributed scenario is less than that of the centralized scenario.

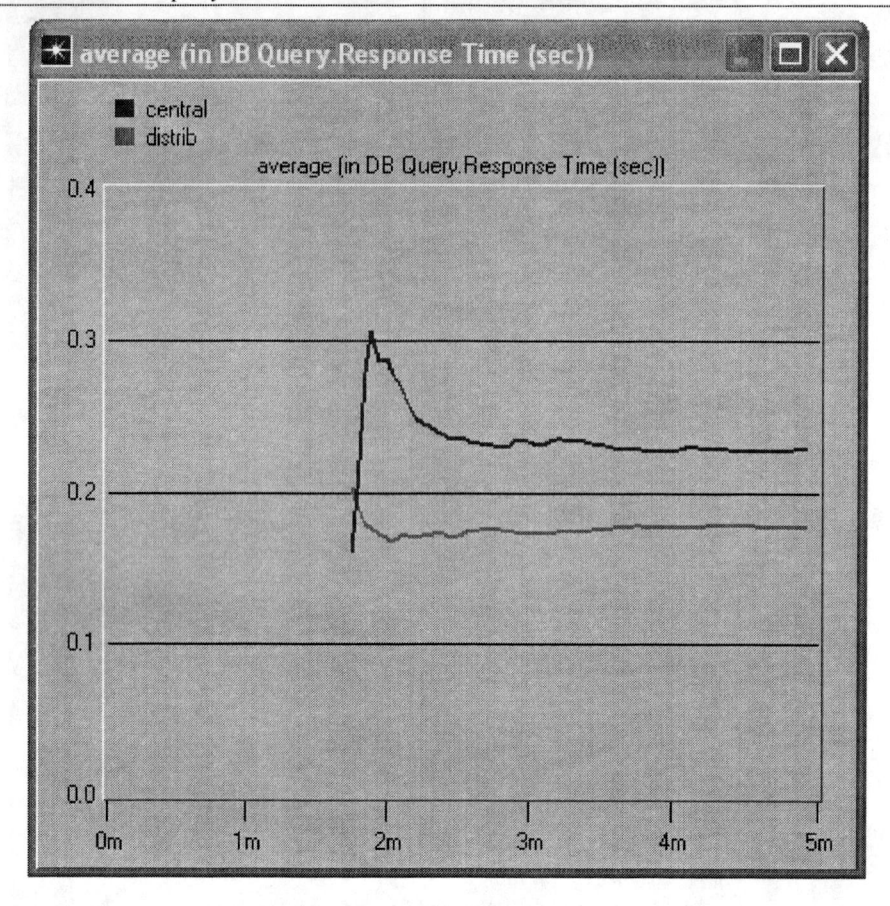

You can **Zoom In** on an area of the graph by left clicking on the graph and dragging a square around the area that you would like to enlarge.

Click on the X to close the window. The **Close Analysis** window will appear asking you if you want to hide the panel or permanently delete. Click on the **Delete** button to delete the window.

You may discard the results of your simulations, rerun them, and manage your scenarios by doing the following:

Choose **Scenarios/Manage Scenarios** from the Main Menu. A window will appear with the names of your scenarios, the length of time that the simulation ran, and the status of the data.

Lab 1   Central versus Distributed Processing

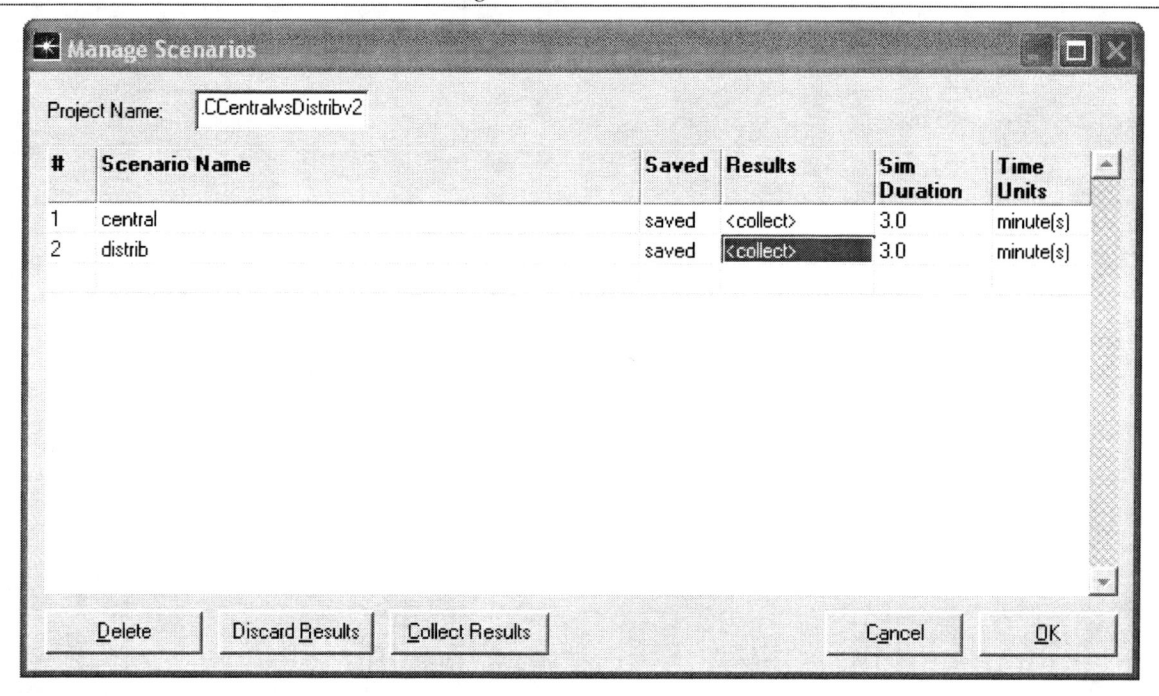

By clicking on the values under the **Results** column, you may choose to **collect** or **recollect** results. By clicking on a scenario name, you may choose to **Delete** the scenario or **Discard Results**. Click **OK** to close the window.

From the Main Menu, choose **File/Save** to save your project.

Use the results you obtained in your simulations to answer the following questions.

Questions

1. What is the average Database query response time in both the centralized and the distributed scenarios? Which scenario has the shortest response time?
2. What is the average download response time for the FTP application in both the centralized and the distributed scenarios? How does the download time compare to the upload time? Which one is longer?
3. What is the average page response time for the HTTP Application in both the centralized and the distributed scenarios? Which scenario has the longest page response time? How does the object response time compare to the page response time?
4. Look at the number of bytes sent and received for the Email application in the centralized system. Are these numbers equivalent? Explain your results.
5. What is the CPU utilization for the server in the Centralized Server Scenario? What is the CPU utilization for each of the servers in the distributed scenario? Has utilization decreased in the distributed scenario?
6. In the distributed scenario, which server had the highest load? Which server had the lowest load?

Summary Question

1. Based on what you have observed, would you advise the Computer Services Department to modify their network to a distributed architecture? Write a paragraph summarizing your results and stating your recommendations. What other considerations should the department make before implementing their choice?

Lab 2 TCP and UDP

Objectives
1. To simulate client/server communication over an IP network.
2. To observe the effect of using TCP and UDP transport layer protocols for file transfer.

Motivation
EmpireStateStock is a stock exchange company located in New York, NY. They have recently opened a regional office in San Francisco, CA. The main service provided by the San Francisco office will be to gather and distribute monthly and annual reports describing stocks handled by the main office. Instead of setting up a private network between the two offices, EmpireStateStock would like to use the Internet. The San Francisco office will use the File Transfer Protocol (FTP) to download reports as needed. There are two transport protocols that the company may use with their FTP application. These are the Transmission Control Protocol (TCP) or the User Datagram Protocol (UDP). EmpireStateStock would like to run tests to discover how these two transport protocols affect file transfer.

Description
The Transmission Control Protocol (TCP) is a reliable protocol. This means that error checking is performed to ensure that there are no lost or dropped packets, that all packets are placed in order at the destination, and that any bit errors are corrected. TCP is connection oriented. It requires a three-way handshake for connection setup and a four-way handshake for connection teardown. The User Datagram Protocol (UDP) is unreliable, meaning that it is a best effort service, and no error checking is performed. UDP is also connectionless; thus, packets can be sent immediately without waiting for connection set up. UDP packets, therefore, are sent with less delay than TCP packets. However, there is no guarantee that UDP packets will reach their destination, arrive in order, and be error free.

In this lab, we will create a Wide Area Network (WAN) that uses the Internet Protocol (IP) to route traffic between San Francisco and New York. The two sites will communicate in client/server fashion. The San Francisco site will request files from the New York site, and the New York site will respond by providing the requested file. We will examine performance statistics, such as throughput and delay for file transfers using TCP and UDP. We will also examine the effect of packet discards on throughput and delay.

Follow the directions below to create an OPNET model, run simulations, and view results. Questions regarding the lab are listed at the end.

Create a New Project

1. Start the **OPNET IT Guru Academic Edition Application.** Create a **New Project** by selecting **File/New/Project** from the Main Menu. Click **OK.**

2. Give the project a name such as **your_initials_TCPvsUDP.** Give the scenario a name such as **TCP.** Click **OK.** You should see the **Startup Wizard Initial Topology** dialog box.

3. Verify that **Create Empty Scenario** is selected in the **Initial Topology Dialog Box.** Click **Next.** You will now see the **Network Scale** dialog box.

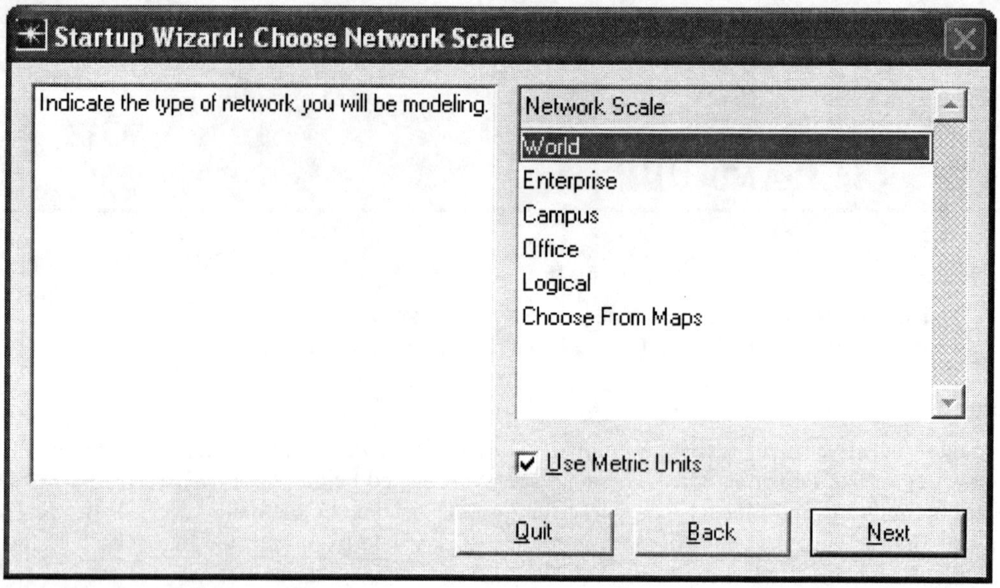
Startup Wizard: Choose Network Scale
Indicate the type of network you will be modeling.
Network Scale
World
Enterprise
Campus
Office
Logical
Choose From Maps
Use Metric Units
Quit
Back
Next

4. Choose **World** from the **Network Scale** list. Click **Next.** Choose **USA** as your map and click **Next.** You should see the **Select Technologies** dialog box.

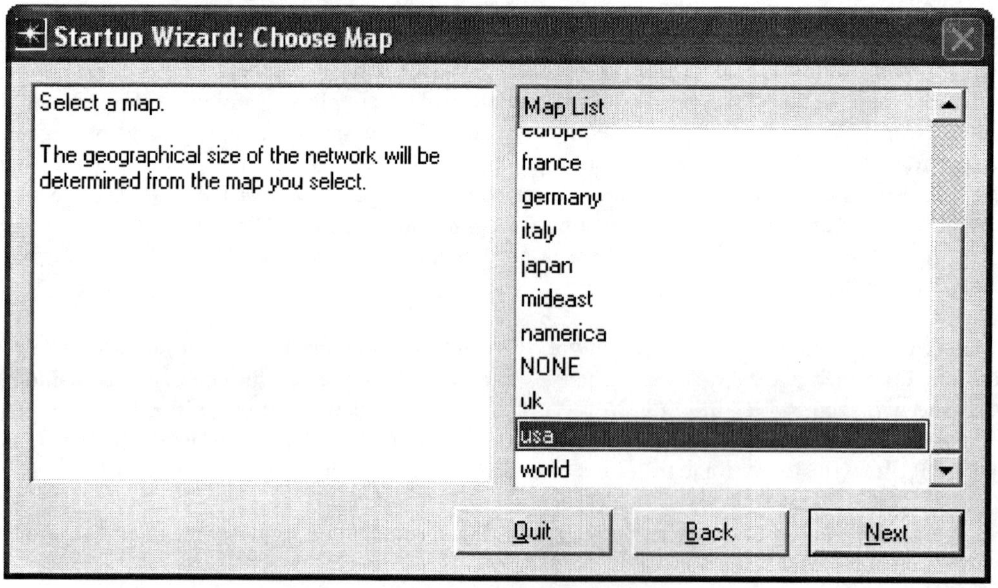
Startup Wizard: Choose Map
Select a map.
The geographical size of the network will be determined from the map you select.
Map List
france
germany
italy
japan
mideast
namerica
NONE
uk
usa
world
Quit
Back
Next

5. Scroll down until you see **internet_toolbox** under the **Model Family** column. Under the **Include column,** click on the box next to **internet_toolbox**. This will change the "no" to "yes" indicating that internet technologies will be included. Click **Next.**

Lab 2   TCP and UDP

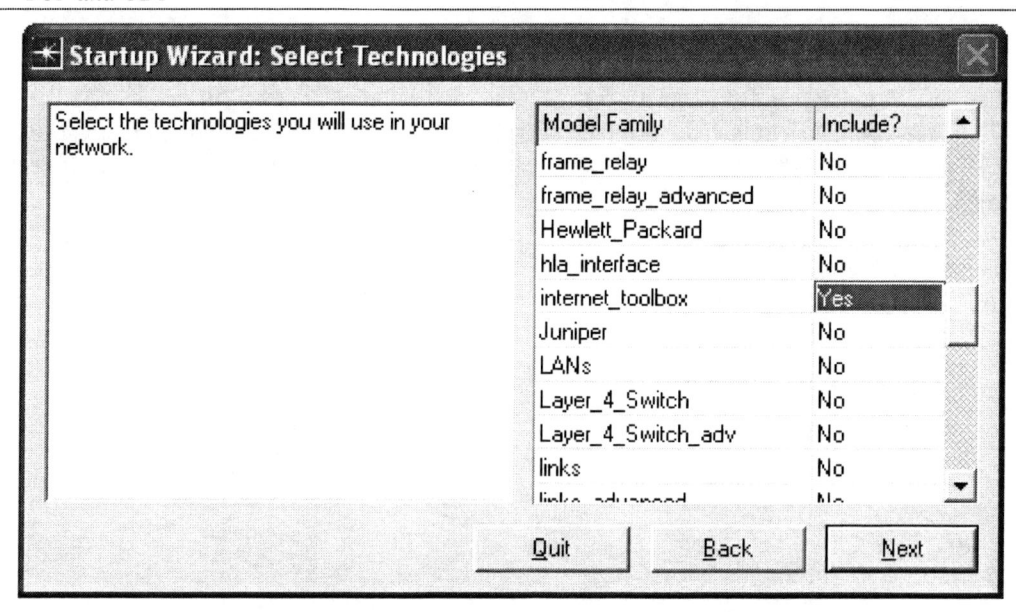

6. In the **Startup Wizard Review** dialog box, verify that the internet_toolbox technologies are chosen, that the scale is World, and that the map is USA. Click **OK**. You should see the **Object Palette** window.

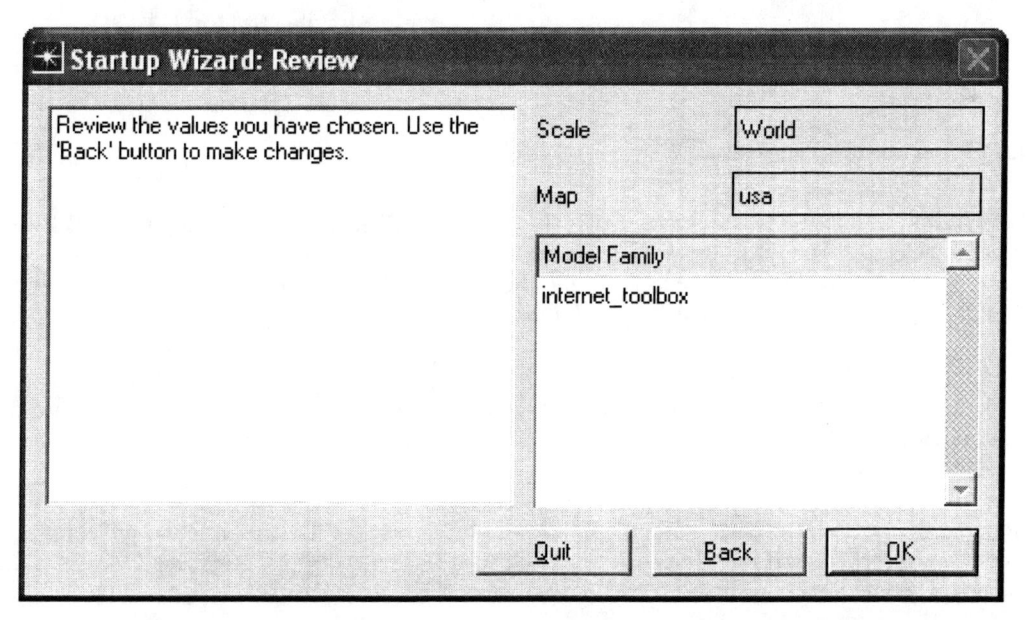

7. We would like to create a network where the server is a Point-to-Point (PPP) server and the client is a PPP advanced workstation. To include these objects, we need to configure the object palette. In the **Object Palette Window**, click on the **Configure Palette Button.** A new window will appear.

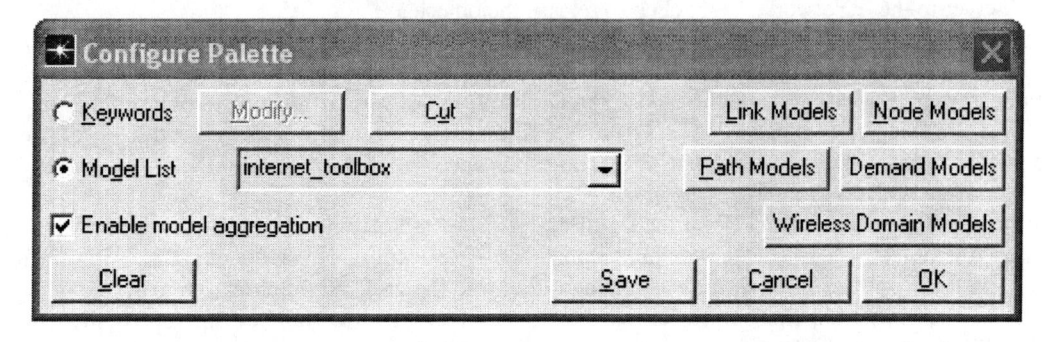

8. Click on the **Node Models Button.** This will bring up a window with an alphabetical list of all the possible object nodes that you can use in your projects. Scroll down until you find the PPP objects. Click on the value next to **PPP_server_adv** to include it in the project. Click on the value next to **PPP_workstation_adv** to include it in the project. We will use the advanced objects because these allow us to select from a variety of transport layer protocols. Click **OK** to save the Object Toolbox Configuration.

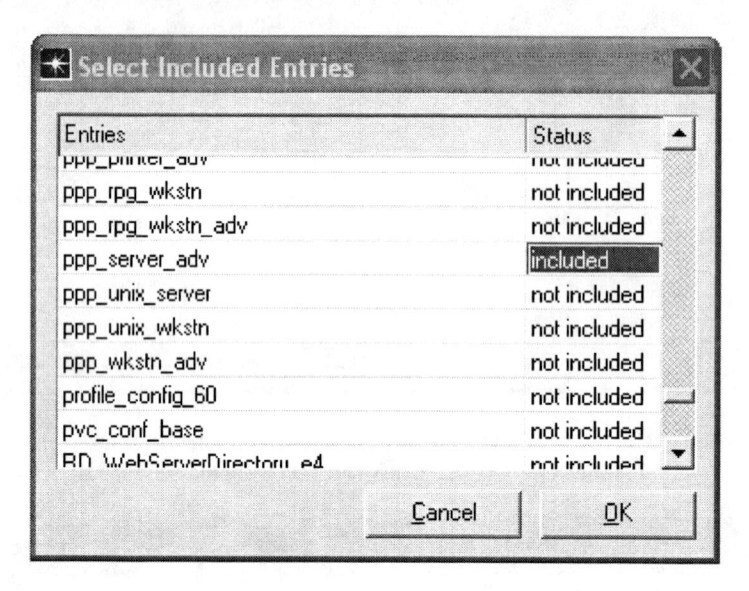

Building the Client/Server Network

1. We will now create the Client/Server network. From the **Object Palette**, click on the **PPP_sever_adv** object to select it. Drag it to the project grid in the area of New York. Click to paste a copy of it on the map. Right click to release the object. Right click on the **PPP_server_adv** object and choose **Set Name** from the pull-down menu. Name the object **FTPserver**.

2. From the **Object Palette**, click on the **PPP_workstation_adv** object to select it. Drag it to the project grid in the area of San Francisco. Click to paste a copy of it on the map. Right click to release the object. Right click on the **PPP_workstation_adv** object and choose **Set Name** from the pull-down menu. Name the object **FTPClient**.

3. From the **Object Palette,** click on the **IP32_Cloud** object to select it. Drag it to the project grid and paste it between the client and the server. Right click to release the object. Right click on the **IP32_Cloud** object and choose **Set Name** from the pull-down menu. Name the object **IPNetwork**.

4. We will use a DS1 link to connect our client and server to the IP Network. DS1 speed is the combination of 24 DS0 (64Kbps); therefore, they have a link speed of 1.544Mbps. From the **Object Palette**, click on the **DS1 link** object to select it. Connect the FTPClient to the IPNetwork object. Next, connect the FTPServer to the IPNetwork. Right click to release the object.

5. Now, we will add an **Application Definition** and **Profile Configuration** object to the workspace. The application object will allow us to choose the applications we want to include in our simulation. We will use the file transfer application. The Profile Configuration object allows us to use default traffic profiles for different types of users or to create new profiles. We will create our own profile for this project.

    From the **Object Palette**, click on the **Profile Definition** object and drag it to the grid. Right click to release the object. Right click on the object and choose **Set Name**. Name the object **ProfileConfig**. Next, click on the **Application Definition** object and drag it to the grid. Right click to release the object. Right click on the object and choose **Set Name**. Name the object **ApplicationConfig**. Verify that your project looks like the figure shown next.

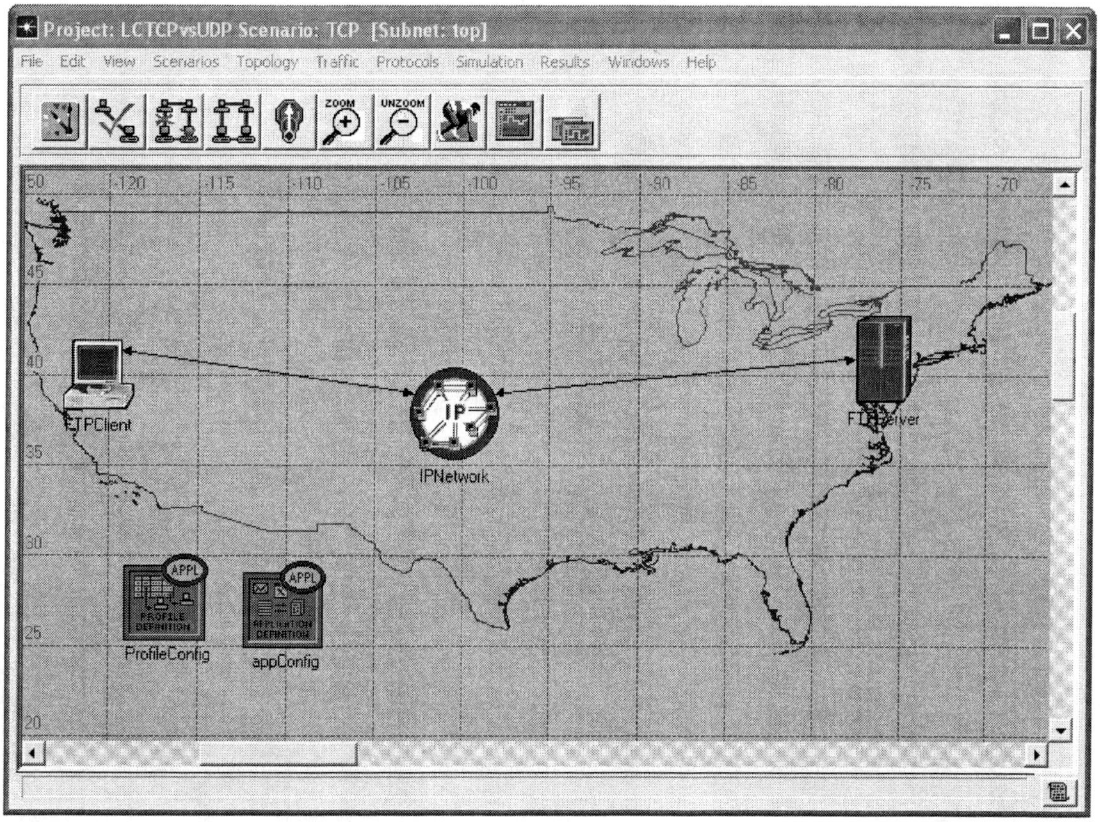
Project: LCTCPvsUDP Scenario: TCP [Subnet: top]
File Edit View Scenarios Topology Traffic Protocols Simulation Results Windows Help
ZOOM
UNZOOM
FTPClient
IPNetwork
FTPServer
ProfileConfig
appConfig
APPL

From the main menu, choose **File/Save** to save your project.

Configuring the Network

1. We will start by configuring the application object to provide the application we are interested in using in our simulation. OPNET IT Guru allows you to define new applications or use default descriptions for standard applications. We are interested in the file transfer application, that is predefined; however, we will modify the size of the file that is transferred as well as how often file transfers occur. Right click on the **Application Configuration** object and choose **Edit Attributes.** Click on the value next to **Application Definitions.** Choose **Default** from the pull-down menu.

2. Click on the + next to **Application Definitions** to expand the parameters. Click on the + next to **File Transfer (Heavy)**. Click on the field next to **Description** and choose **Edit** from the pull-down menu.

(appConfig) Attributes

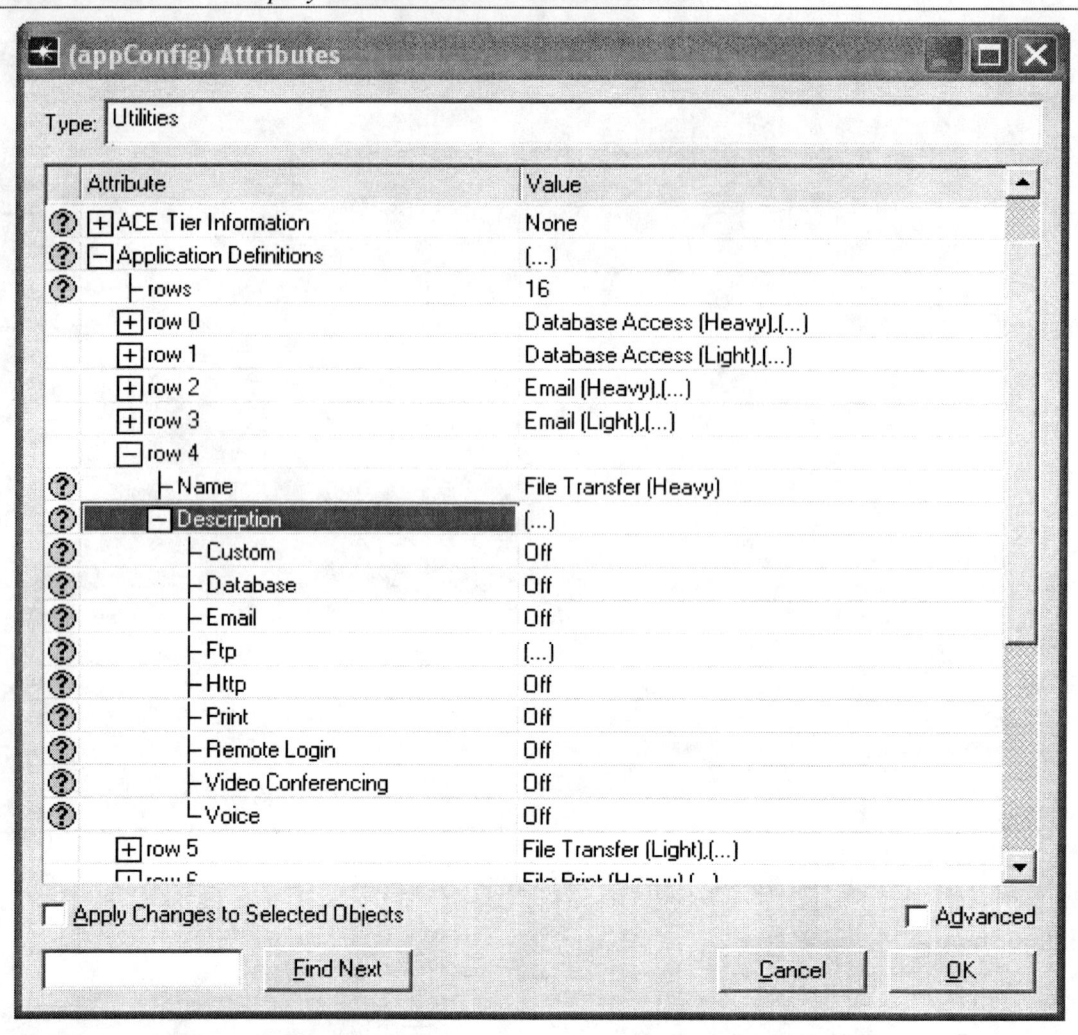

Type: Utilities

Attribute	Value
⊞ ACE Tier Information	None
⊟ Application Definitions	(...)
├ rows	16
⊞ row 0	Database Access (Heavy),(...)
⊞ row 1	Database Access (Light),(...)
⊞ row 2	Email (Heavy),(...)
⊞ row 3	Email (Light),(...)
⊟ row 4	
├ Name	File Transfer (Heavy)
⊟ Description	(...)
├ Custom	Off
├ Database	Off
├ Email	Off
├ Ftp	(...)
├ Http	Off
├ Print	Off
├ Remote Login	Off
├ Video Conferencing	Off
└ Voice	Off
⊞ row 5	File Transfer (Light),(...)

☐ Apply Changes to Selected Objects   ☐ Advanced

[Find Next]   [Cancel]   [OK]

3. Click on the value next to **Ftp** and choose **Edit**. First, we will modify the command mix. It is currently sent to 50%, meaning that 50% of the commands will be get-file commands. Click on the value next to **Command Mix (Get/Total)** and change it to 100%. Now all of our commands will be get-file commands.

4. We will now modify the value which controls how often file transfers occur. Click on the value **exponential(360)** for the **Inter-Request Time** attribute. A new window will appear. Change the value to **exponential(10)** and click **OK**. This will create file transfer requests every 10 minutes.

5. Next we will change the size of the file that is transferred. We will set it to be 5000 bytes. Click on the value **constant(50000)**. A new window will appear. Change the value to **constant(5000)** and click **OK**. You should now have the values displayed on the following screen:

Lab 2   TCP and UDP

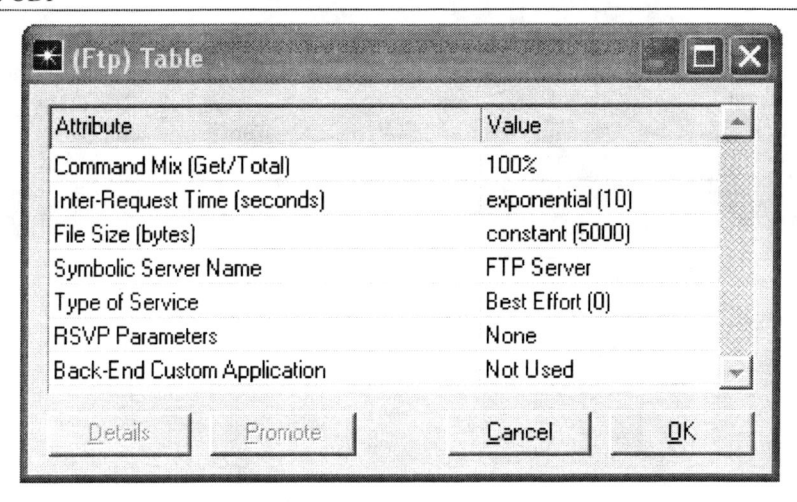

6. Click **OK** to close the Application Configure Window.

7. Now we will create a new traffic profile and call the profile **FTP**. The profile will include the **File Transfer (Heavy)** application. Right Click on the **Profile Definition** object and choose **Edit Attributes**. Click on the + next to **Profile Configuration**. Click on the value next to **rows** and choose **Edit**. Change the number of rows to **1**. Click on the value next to the new added row and choose **Expand Row**. Next to **Profile Name** type **FTP**. Choose **Simultaneous** under **Operation Mode**.

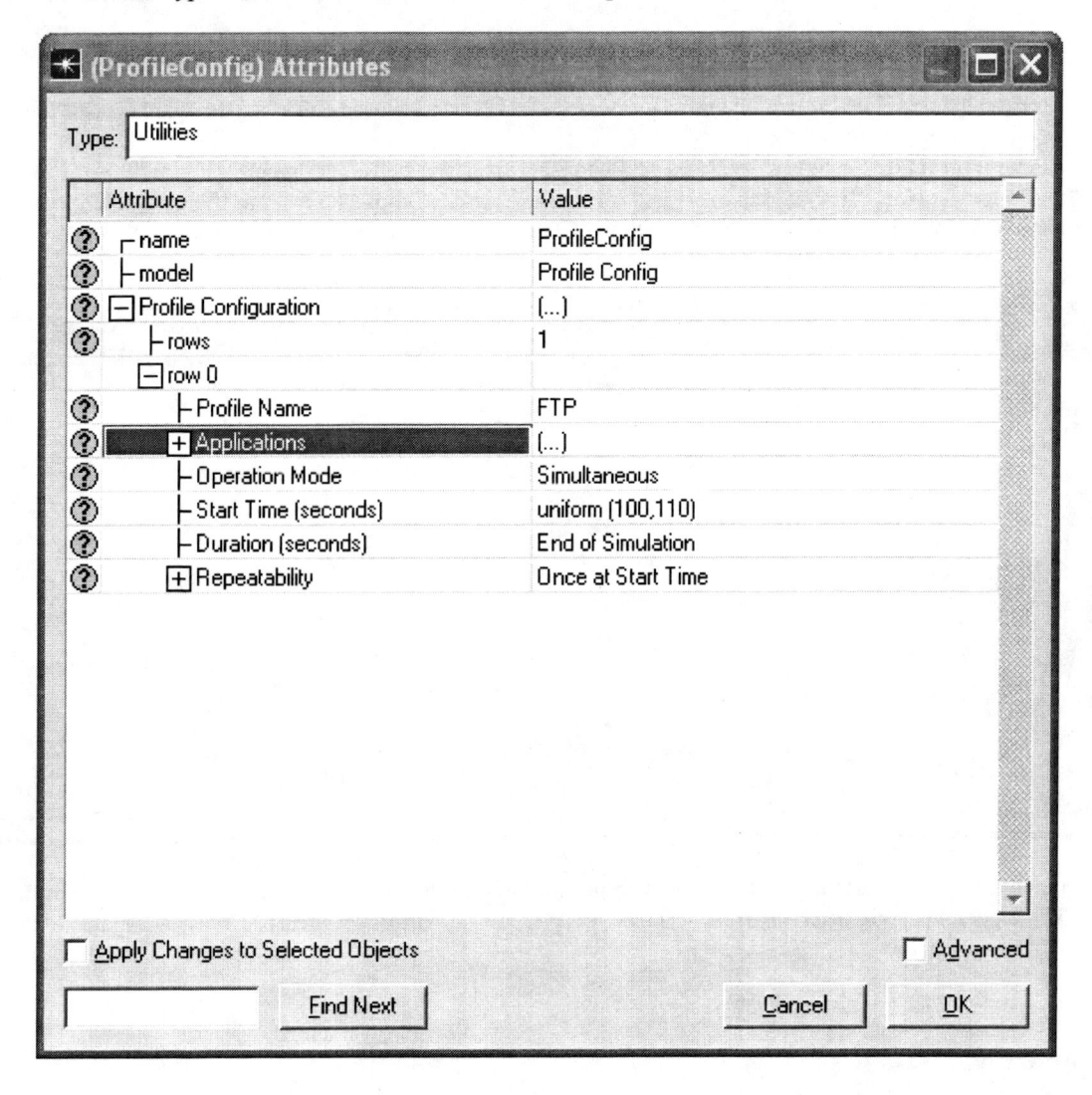

8. Click on the value next to **Applications** and choose **Edit** from the pull-down menu. A new window will appear. Next to **Rows** at the bottom of the window, choose **1** from the pull-down menu. A new row will appear. Click on the value under the **Name** column, choose **File Transfer (Heavy)** from the pull-down menu, and click **OK**. Click **OK** again to close the **Profile Configuration** window.

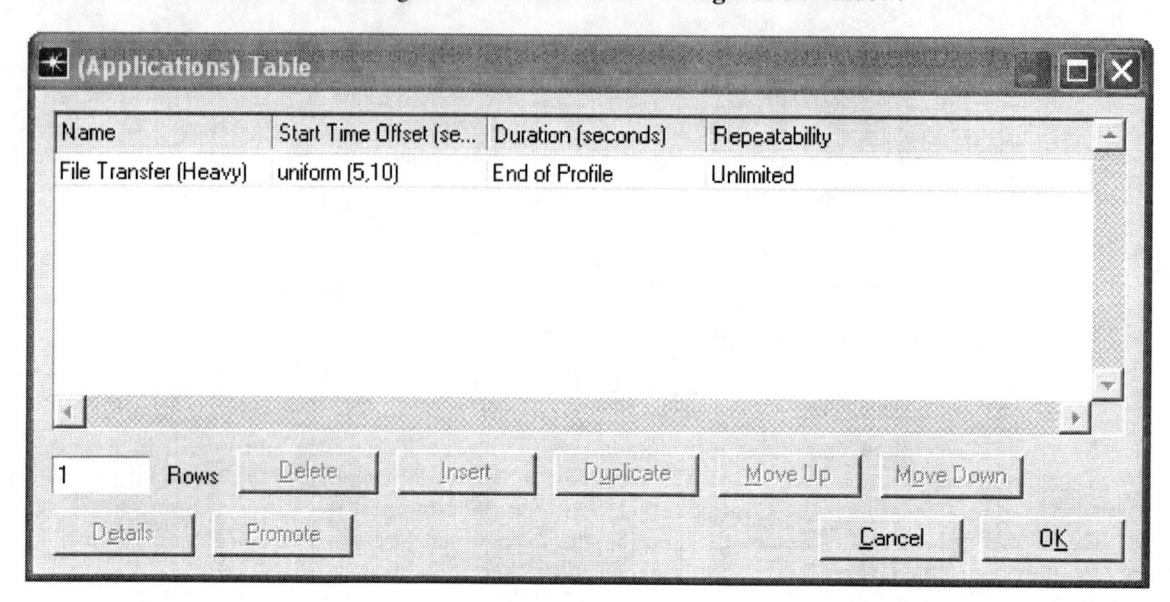

9. Now we will give the San Francisco site the FTP profile. Right click on the **FTPclient** object and choose **Edit Attributes**. Next, click on the **+** next to **Application Supported Profiles** to expand the parameters. Click on the value next to **Application Supported Profiles** and choose **Edit**. A new window will appear.

10. Click on the value listed next to **Rows** and choose **1** from the pull-down menu. You will see a row added to the table. Now click on the value under **Profile Name**. Choose **FTP** from the pull-down menu. This configures the client to generate traffic with the FTP profile. Click **OK** to close the window.

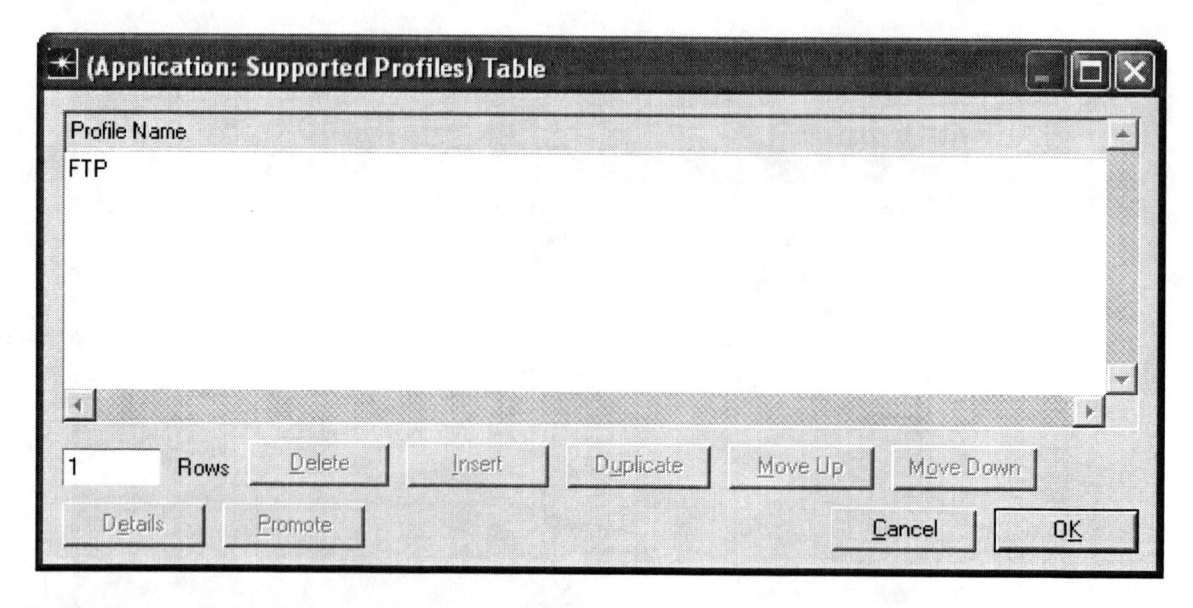

11. Finally, we will configure the **FTPClient** to use the TCP protocol. Click on the value next to **Application: Transport Protocol Specification**. Choose **TCP** from the pull-down menu. Click **OK** again to close the **Attribute** window.

36

Lab 2   TCP and UDP

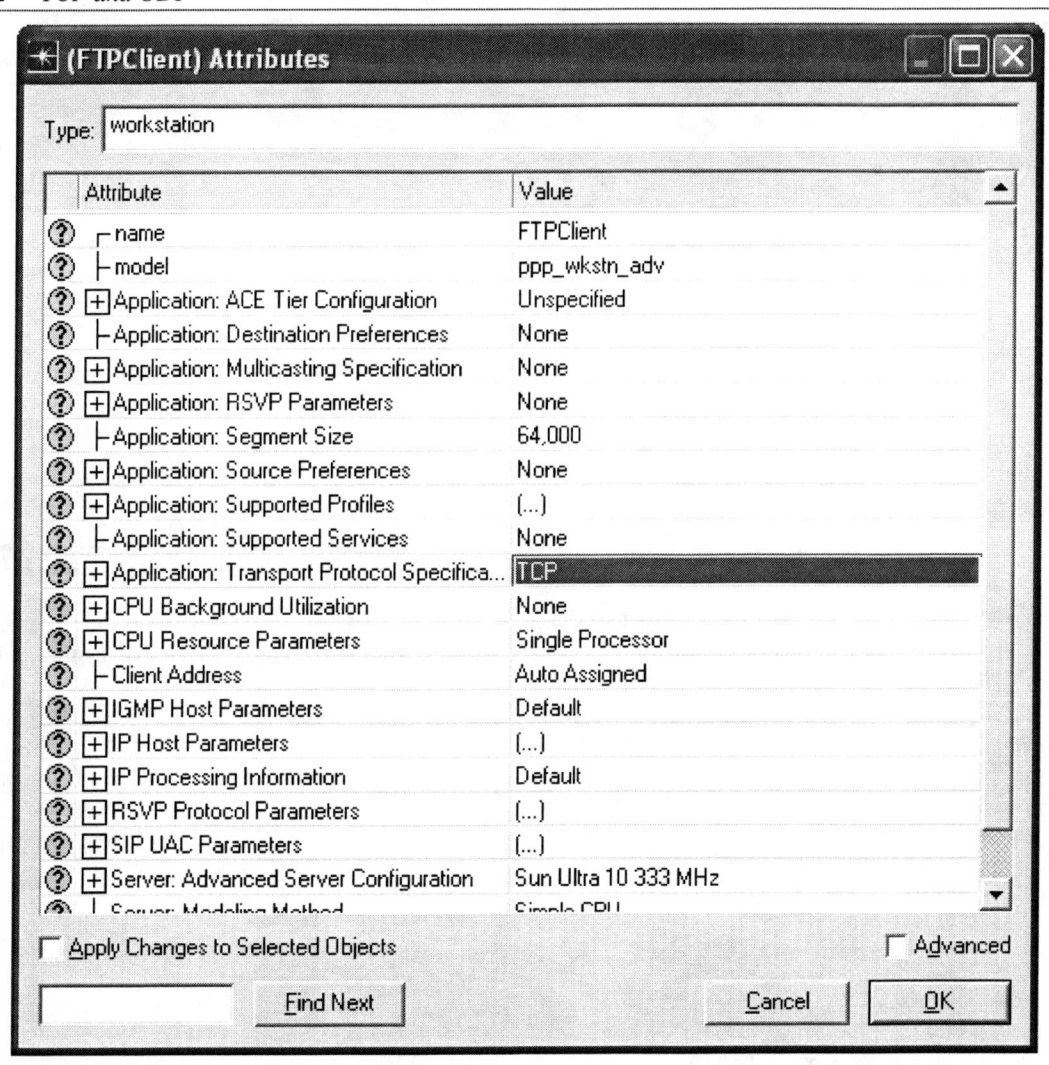

12. Now we will configure the New York server so that it will provide file transfer services. Right click on the **FTPServer** object and choose **Edit Attributes**. Next, click on the value next to **Application Supported Services** and choose **Edit**. A new window will appear.

13. Click on the value listed next to **Rows** and choose **1** from the pull-down menu. You will see a row added to the table. Click on the value under **Application Name**. Choose **File Transfer (Heavy)** from the pull-down menu. Now the **FTPServer** will provide support to this application. Click **OK** to close the window.

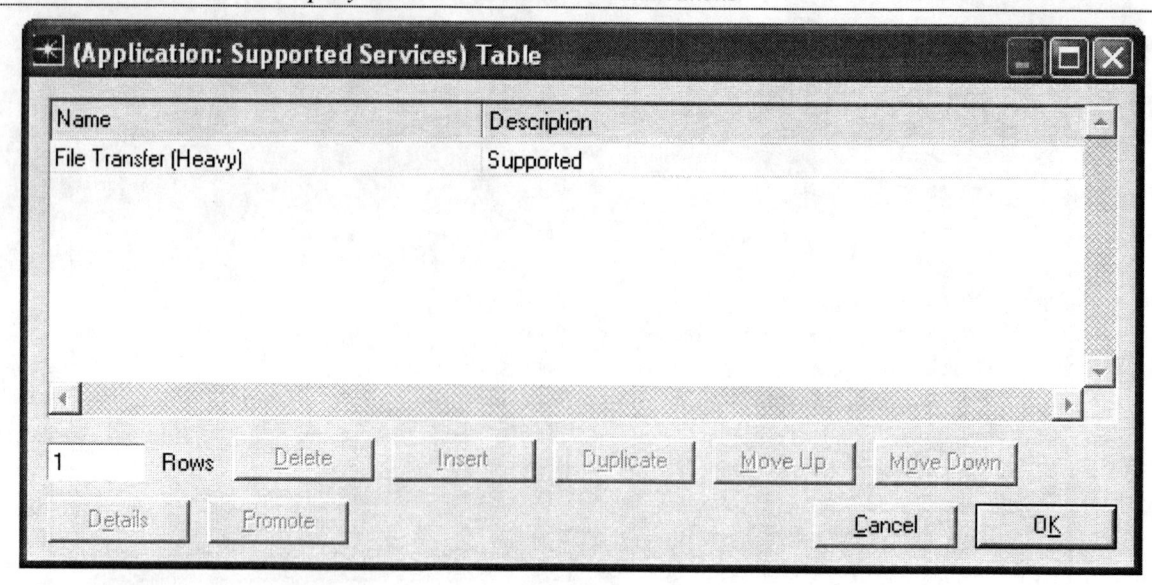

Name	Description
File Transfer (Heavy)	Supported

14. We will configure the **FTPServer** to use the TCP protocol. Click on the value next to **Application: Transport Protocol Specification.** Choose **TCP** from the pull-down menu. Click **OK** again to close the **Attribute** window.

Lab 2   TCP and UDP

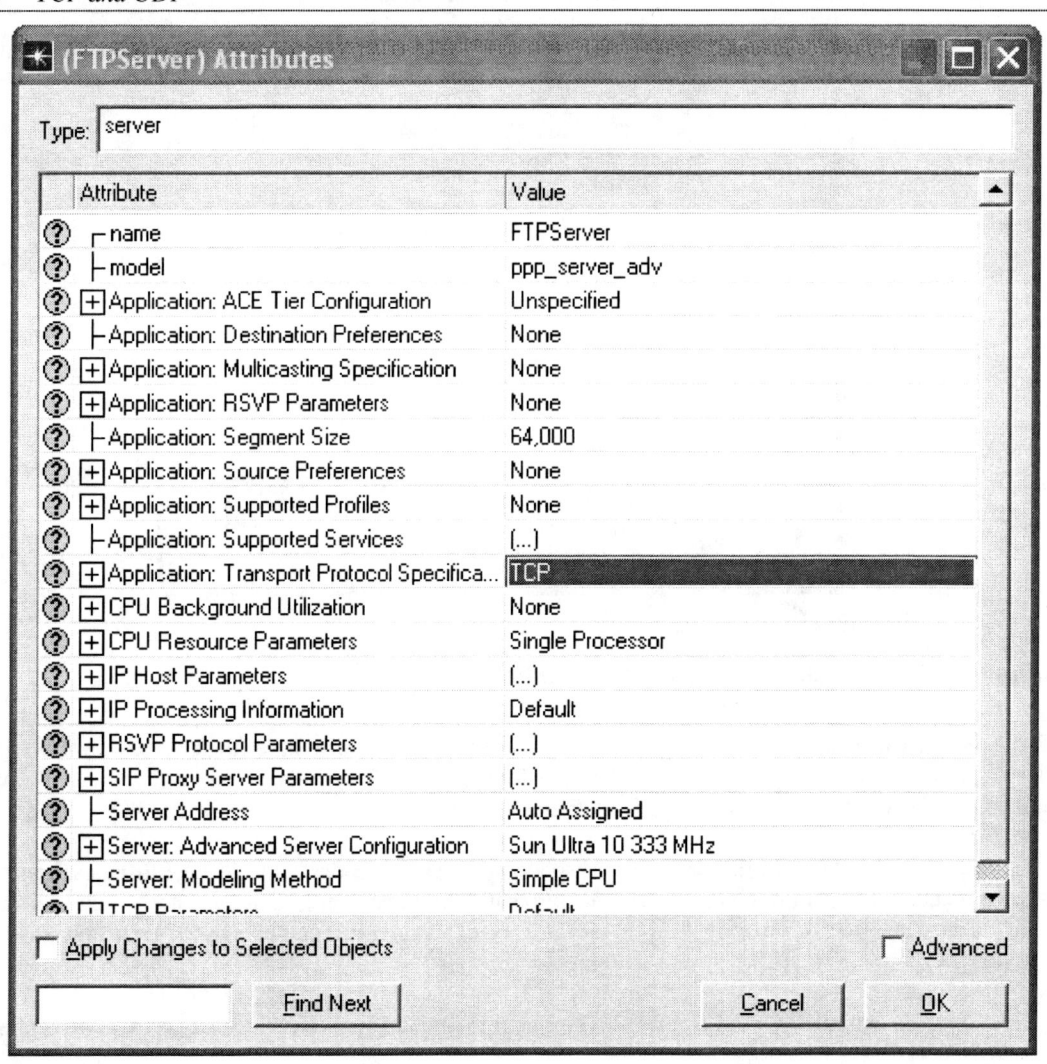

Attribute	Value
name	FTPServer
model	ppp_server_adv
Application: ACE Tier Configuration	Unspecified
Application: Destination Preferences	None
Application: Multicasting Specification	None
Application: RSVP Parameters	None
Application: Segment Size	64,000
Application: Source Preferences	None
Application: Supported Profiles	None
Application: Supported Services	(...)
Application: Transport Protocol Specifica...	TCP
CPU Background Utilization	None
CPU Resource Parameters	Single Processor
IP Host Parameters	(...)
IP Processing Information	Default
RSVP Protocol Parameters	(...)
SIP Proxy Server Parameters	(...)
Server Address	Auto Assigned
Server: Advanced Server Configuration	Sun Ultra 10 333 MHz
Server: Modeling Method	Simple CPU
TCP Parameters	Default

15. Finally, we will configure the IPNetwork. Right click on the **IPNetwork** object and choose **Edit Attributes.** Click on the value next to **Packet Latency**. A new window will appear. Choose **Not Used** from the pull-down menu next to the **Special Value** field. For the **Distribution Name** choose **constant**, and for the **Mean Outcome** enter **.25**. The latency for our packets in this simulation will always be .25 second.

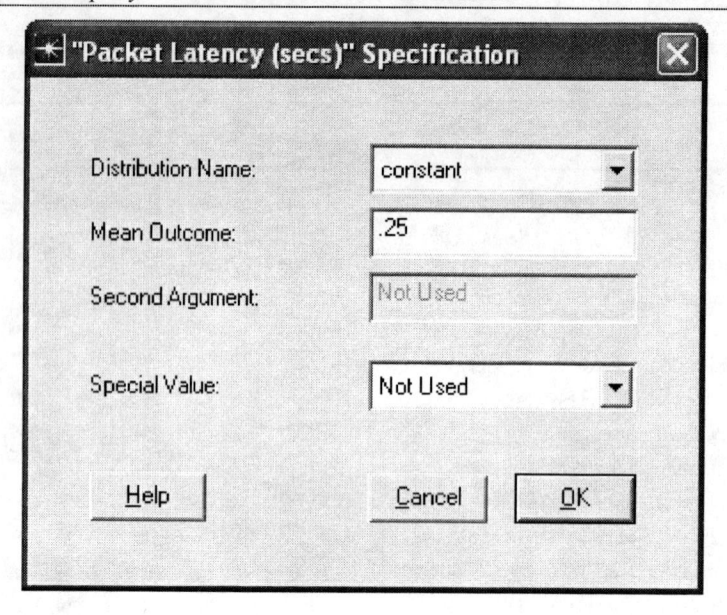

Configuring the Simulation and Choosing Statistics

At this time we would like to choose the statistics that will help EmpireStateStock decide on the appropriate transport protocol for their file transfers. We will look at throughput, delay, and utilization. We will also examine the effect of packet discards on these statistics.

1. Right click somewhere on the project workspace and select **Choose Individual Statistics** from the pull-down menu. Click on the + next to **Global** to expand the parameters. Click on the + next to **FTP**. Click on the box next to **Download Response Time (sec), Traffic Received (bytes/sec),** and **Traffic Sent (bytes/sec).**

2. Click on the + next to **TCP**. Click on the box next to **Delay (Sec), Retransmission Count,** and **Segment Delay (sec).**

3. Click on the + next to **Link Statistics** to expand the parameters. Click on the + next to **point-to-point**. Click on the box next to **Throughput (bits/sec)** →, **Throughput (bits/sec)** ←, **utilization** →, and **utilization** ←. Your values should match those shown next. Click **OK** to close the window.

Lab 2   TCP and UDP

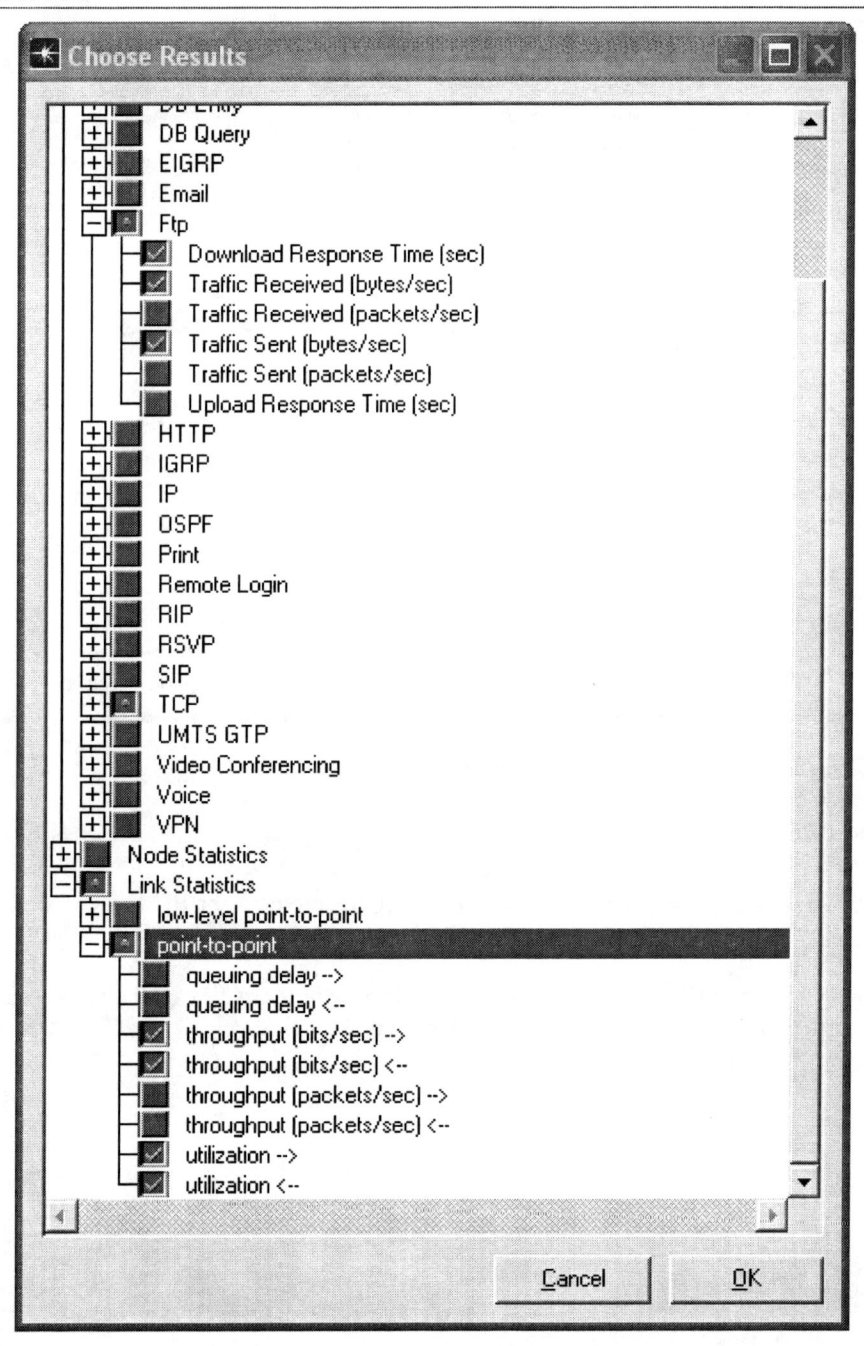

[Screenshot of "Choose Results" dialog showing a tree of statistics. Under Ftp: Download Response Time (sec) ✓, Traffic Received (bytes/sec) ✓, Traffic Received (packets/sec), Traffic Sent (bytes/sec) ✓, Traffic Sent (packets/sec), Upload Response Time (sec). Other categories listed: HTTP, IGRP, IP, OSPF, Print, Remote Login, RIP, RSVP, SIP, TCP, UMTS GTP, Video Conferencing, Voice, VPN. Under Link Statistics > point-to-point: queuing delay -->, queuing delay <--, throughput (bits/sec) --> ✓, throughput (bits/sec) <-- ✓, throughput (packets/sec) -->, throughput (packets/sec) <--, utilization --> ✓, utilization <-- ✓.]

Running the Simulation

1. Click on the **Configure and Run** Button. Verify that the **Duration** value is **1**. Verify that **Hour(s)** is chosen for the time unit. Click the **Run** button to run the simulation. Depending on your processor, the simulation should take several minutes to run.

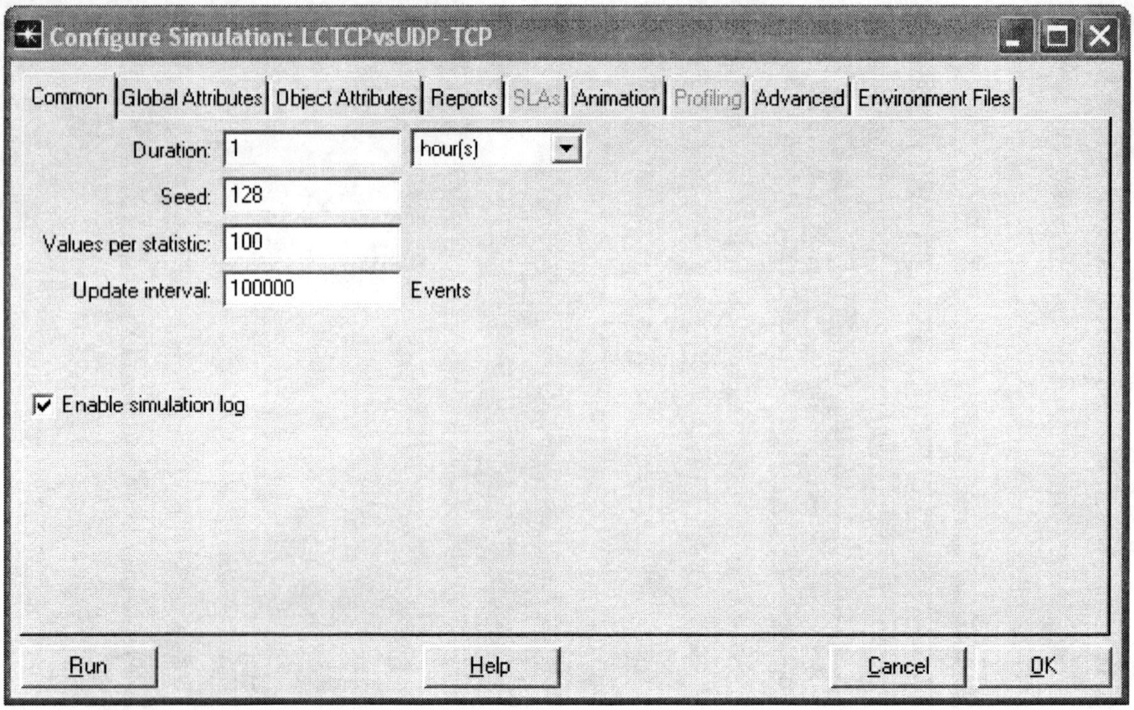
Configure Simulation: LCTCPvsUDP-TCP
Common
Global Attributes
Object Attributes
Reports
SLAs
Animation
Profiling
Advanced
Environment Files
Duration: 1
hour(s)
Seed: 128
Values per statistic: 100
Update interval: 100000
Events
Enable simulation log
Run
Help
Cancel
OK

2. Choose **File/Save** from the Main Menu to save your project.

Creating the UDP Scenario

1. Now we will duplicate the scenario and choose the UDP protocol for file transmission. From the Main Menu choose **Scenarios/Duplicate Scenario**. Give the new scenario a name such as **UDP.**

2. Right click on the **FTPClient** object and choose **Edit Attributes.** Click on the value next to **Application: Transport Protocol Specification.** Choose **UDP** from the pull-down menu. Click **OK** again to close the **Attribute** window.

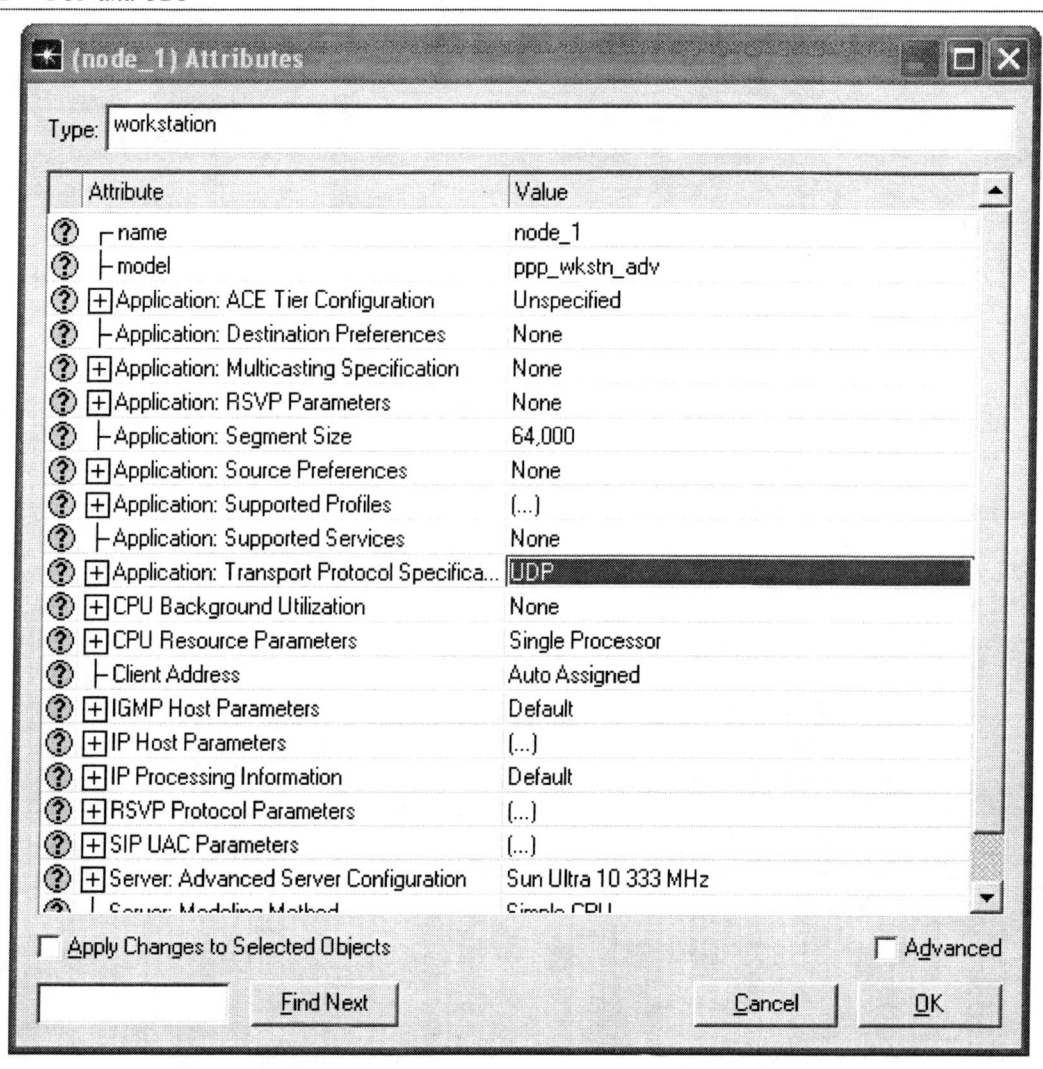

Lab 2 TCP and UDP

Attribute	Value
─ name	node_1
├ model	ppp_wkstn_adv
⊞ Application: ACE Tier Configuration	Unspecified
├ Application: Destination Preferences	None
⊞ Application: Multicasting Specification	None
⊞ Application: RSVP Parameters	None
├ Application: Segment Size	64,000
⊞ Application: Source Preferences	None
⊞ Application: Supported Profiles	(...)
├ Application: Supported Services	None
⊞ Application: Transport Protocol Specifica...	**UDP**
⊞ CPU Background Utilization	None
⊞ CPU Resource Parameters	Single Processor
├ Client Address	Auto Assigned
⊞ IGMP Host Parameters	Default
⊞ IP Host Parameters	(...)
⊞ IP Processing Information	Default
⊞ RSVP Protocol Parameters	(...)
⊞ SIP UAC Parameters	(...)
⊞ Server: Advanced Server Configuration	Sun Ultra 10 333 MHz
Server: Modeling Method	Simple CPU

3. Next, right click on the **FTPServer** and choose **Edit Attributes**. Click on the value next to **Application: Transport Protocol Specification**. Choose **UDP** from the pull-down menu. Click **OK** again to close the **Attribute** window.

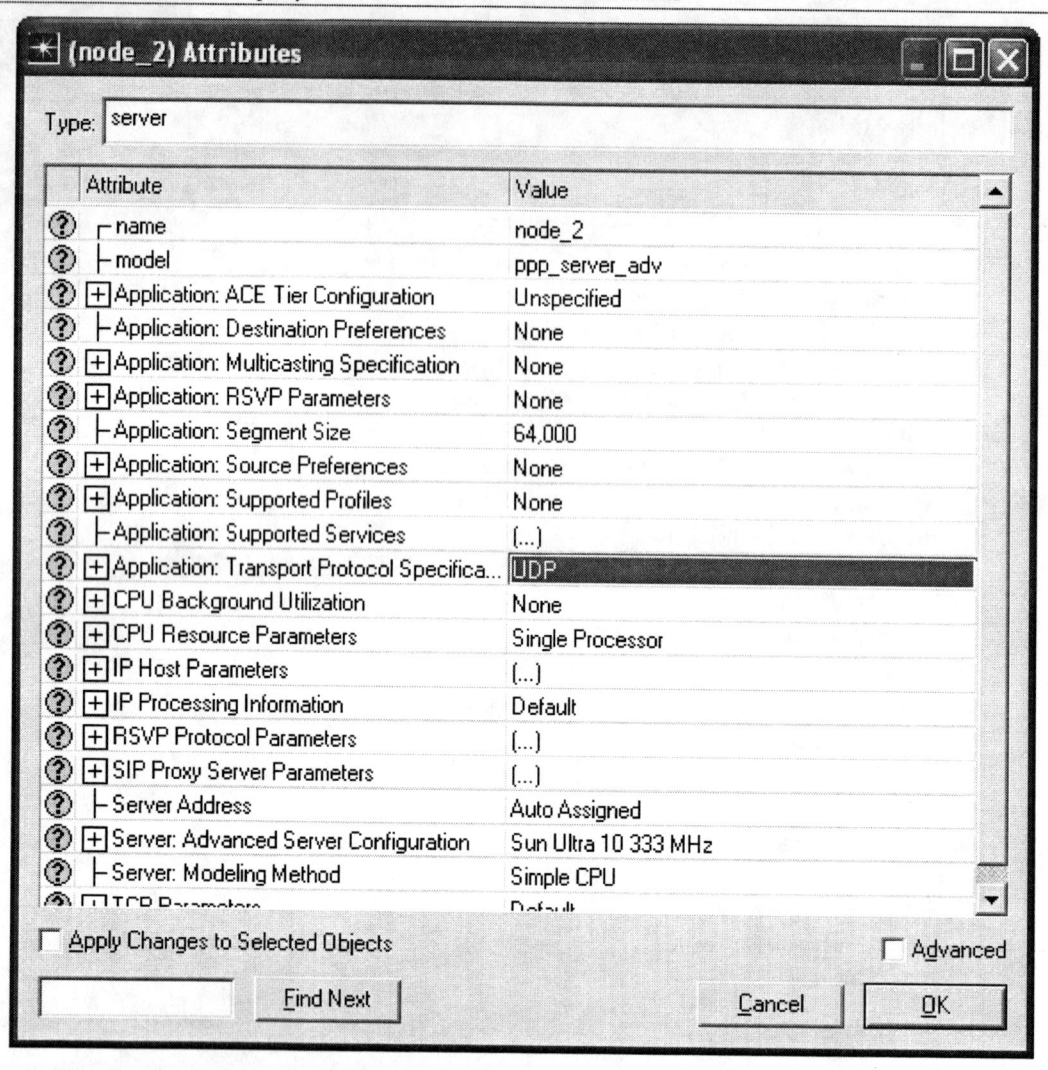

OPNET Lab Manual to accompany Business Data and Communications

3. Choose **File/Save** from the Main Menu to save your project.

4. Again from the Main Menu, choose **Simulation/Run Discrete Time Simulation**.

Comparing Results

5. To view results, choose **Results/ Compare Results** from the Main Menu. To view results for FTP download response time, click on the **+** next to **FTP,** and click next to **Download Response Time (seconds).** You can view the results **As Is** or select **Time Average** from the pull-down menu. Choose **Time_Average.** Click on the **Show** button to display the graph in a new window. Your graph should resemble the figure below. Notice that the response time for TCP is longer than that of UDP.

Lab 2   TCP and UDP

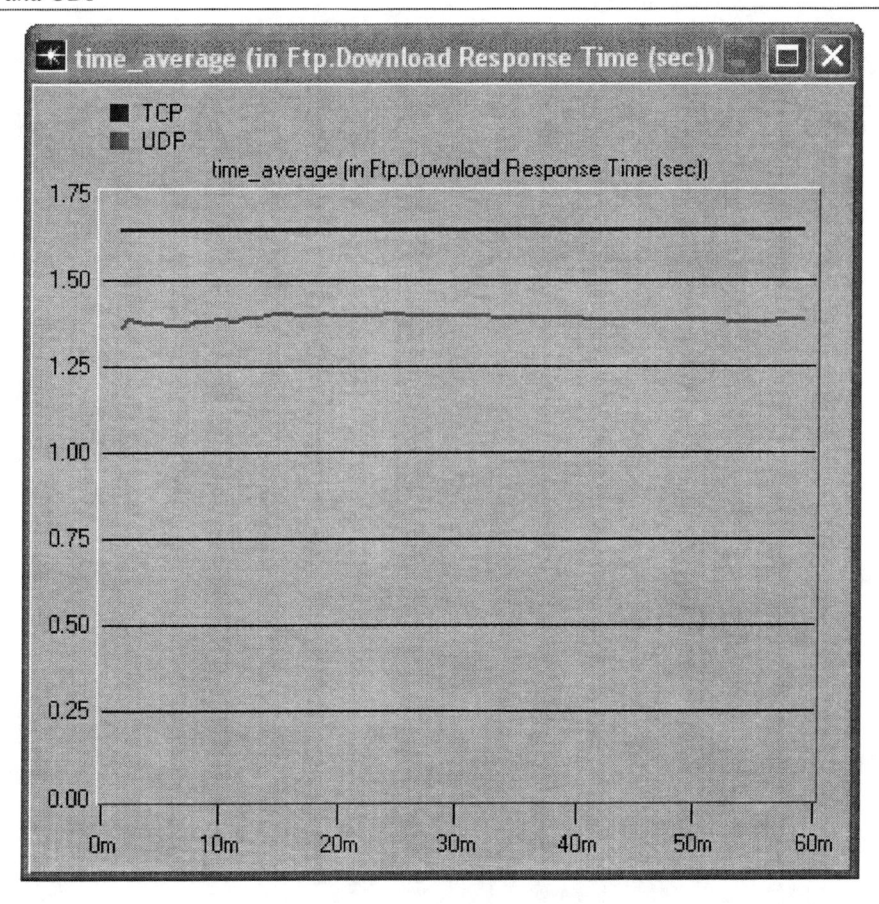

[Figure: time_average (in Ftp.Download Response Time (sec)) — graph showing TCP (upper line near 1.65) and UDP (lower line near 1.40) from 0m to 60m]

If you would like to **Zoom In** on an area of the graph, you may left click on the graph and drag a square around the area that you would like to enlarge.

Click on the **X** to close the window. The **Close Analysis** window will appear asking you if you want to hide the panel or permanently delete it. Click on the **Delete** button to delete the window.

You may discard the results of your simulations, rerun them, and manage your scenarios by doing the following:

Choose **Scenarios/Manage Scenarios** from the Main Menu. A window will appear with the names of your scenarios, the length of time that the simulation ran, and the status of the data.

By clicking on a scenario name, you may choose to **Discard Results** or **Collect Results**. By clicking on the values under the **Results** column, you may choose **recollect** or **collect** from the pull-down menu. Click **OK** to close the window.

Use the results you obtained in your simulations to answer the following questions.

Questions

1. What is the average FTP download response time for TCP? What is it for UDP? Is this what you expected based on what you know about these two protocols? Explain your answer.
2. Based on what you observed, which protocol do you think EmpireStateStock should use for their file transfers? Explain your reasoning.
3. Look at Traffic Sent (bytes/sec). On average, is the number of bytes sent the same for TCP and UDP?
4. Look at Link Throughput, sent and received. What is the average throughput for TCP? How does this compare to average throughput for UDP?
5. What is the link utilization for the TCP and UDP scenarios?

6. When we configured our application we set the FTP file size to be a constant size of 5000 bytes. How does file size affect the simulation? In particular, what do you think the results would be if we chose a larger file size? Comment on TCP and UDP delay, as well as link utilization.
7. Based on what you observed regarding delay, which protocol would be best for transferring multimedia traffic? Explain your answer.
8. To evaluate how packet discard rate affects results, duplicate the TCP scenario. Change the packet discard rate of the IPNetwork object to be 5%. This should cause TCP to retransmit 5% of the sent packets. Now duplicate the UDP scenario. Again, change the packet discard rate of the IPNetwork object to 5%. Compare FTP download response times for all scenarios. How does the packet discard rate affect the UDP and TCP results? Explain what you observed.

LAB 3   RIP AND ROUTING TABLE CONVERGENCE

Objectives
1. To create a network of routers configured to use the Routing Information Protocol (RIP).
2. To evaluate the routing tables created by the RIP routing algorithm and to use the tables to trace a path between two routers.
3. To determine the routing table convergence time after a link is broken.

Motivation
Tony Nguyen has been assigned the task of monitoring a group of routers for SpeedPath, a network service provider. The routers have been configured to use the Routing Information Protocol (RIP). Tony's job is to evaluate the routing tables created by this protocol to ensure that they are accurate. He has also been asked to discover how long it takes for the routing tables to be created, and how long it takes for an alternative route to be found when a path becomes unavailable.

Description
The Routing Information Protocol (RIP) is a popular intradomain routing protocol that was originally distributed with BSD Unix. RIP creates routing tables based on a distance vector algorithm. Using this algorithm, each router will build a table of "vectors" which consist of other router node names and link costs. Each router will then distribute its table to its directly connected neighbors. Table exchanges take place every 30 seconds. We say that the tables have "converged" when exchanges no longer cause table updates.

In this lab, we will design a network of routers configured to use the RIP protocol. The cost associated for each link connecting a pair of routers will be 0; therefore, the routes in the routing table should correspond to the path with the fewest hops or links. We will configure a simulation to create routing tables. After the simulation is finished, the routing tables will be available for you to view. We will look at the routing tables and find the paths that were selected.

Finally, we will break a link in the router network and rerun the simulation. We will then be able to evaluate how long it takes for the routing tables to converge. Using the new routing tables, we will also be able to see which new path was chosen.

Note that in this lab we will concentrate on the routing tables and will not run application traffic. If we wanted our simulation to create routing tables and run application traffic, we would need to delay the application traffic until the tables were created in order for the simulation to work correctly.

Follow the next set of directions to create an OPNET model, run simulations, and view routing tables. Questions regarding the lab are listed at the end.

Create a New Project

1. Start the **OPNET IT Guru Academic Edition Application.** Create a **New Project** by selecting **File/New/Project** from the Main Menu. Click **OK**.

2. Give the project a name such as **your_initials_RIPConvergence.** Give the scenario a name such as **RIPStartConfig.** Click **OK.** You should see the **Startup Wizard Initial Topology** dialog box.

3. Verify that **Create Empty Scenario** is selected in the **Initial Topology Dialog Box.** Click **Next.** You will now see the **Network Scale** dialog box.

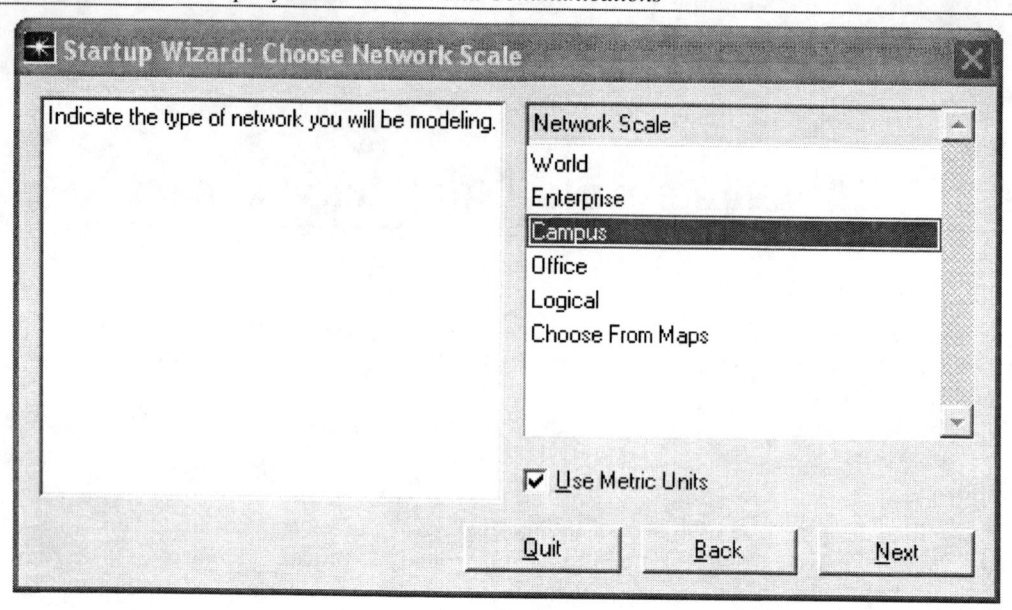

OPNET Lab Manual to accompany Business Data and Communications

> **Startup Wizard: Choose Network Scale**
>
> Indicate the type of network you will be modeling.
>
> Network Scale:
> - World
> - Enterprise
> - **Campus**
> - Office
> - Logical
> - Choose From Maps
>
> ☑ Use Metric Units
>
> [Quit] [Back] [Next]

4. Choose **Campus** from the **Network Scale** list. Click **Next**. Choose **Miles** in the **size** drop-down box. Type **5** for the **X span** and 5 for the **Y span.** Click **Next**. You should see the **Select Technologies** dialog box.

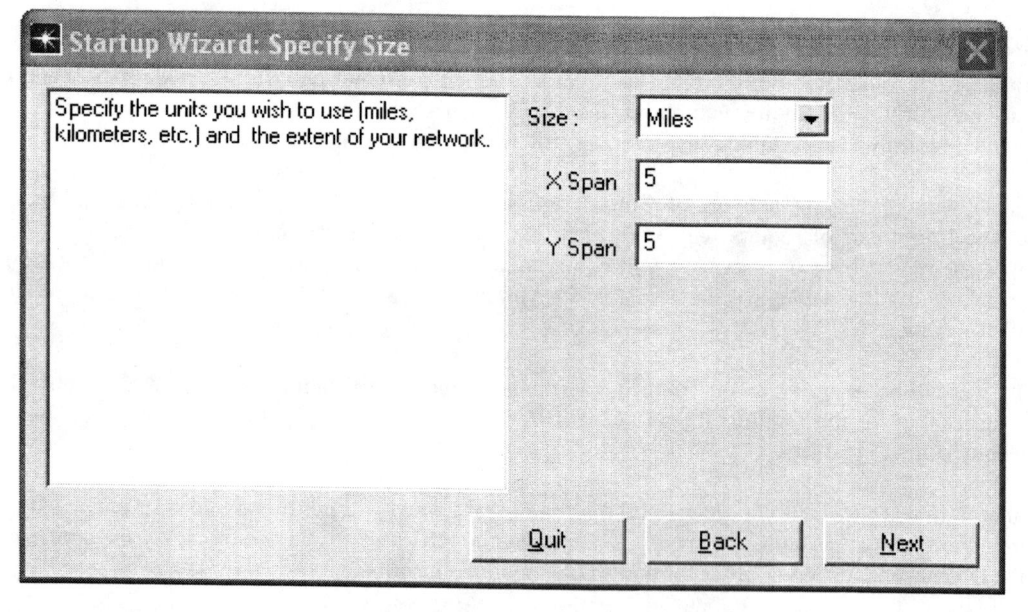

> **Startup Wizard: Specify Size**
>
> Specify the units you wish to use (miles, kilometers, etc.) and the extent of your network.
>
> Size: Miles
> X Span: 5
> Y Span: 5
>
> [Quit] [Back] [Next]

5. Scroll down until you see **internet_toolbox** under the **Model Family** column. Under the **Include column**, click on the box next to **Internet_toolbox**. This will change the "no" to "yes" indicating that **internet_toolbox** technologies will be included. Scroll down a bit more until you see **routers** under the **Model Family** column. Click on the box next to **routers** to also include these technologies in your project. Click **Next**.

48

Lab 3   RIP and Routing Table Convergence

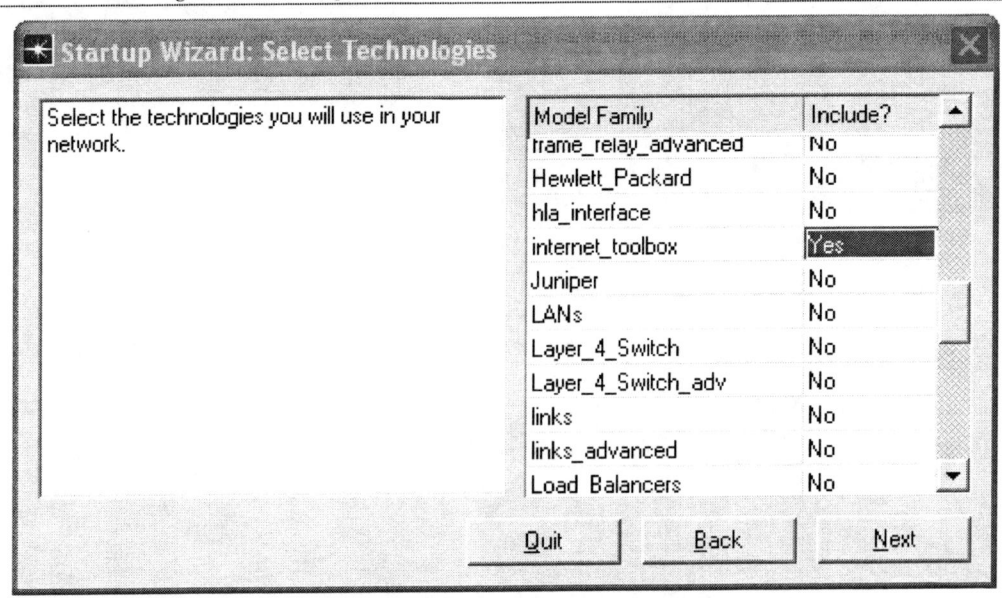

6. In the **Startup Wizard Review** dialog Box, verify that the **internet_toolbox** and **routers** technologies are chosen and that the scale is Campus 5 miles x 5 miles. Click **OK**. You should see the **Object Palette** window.

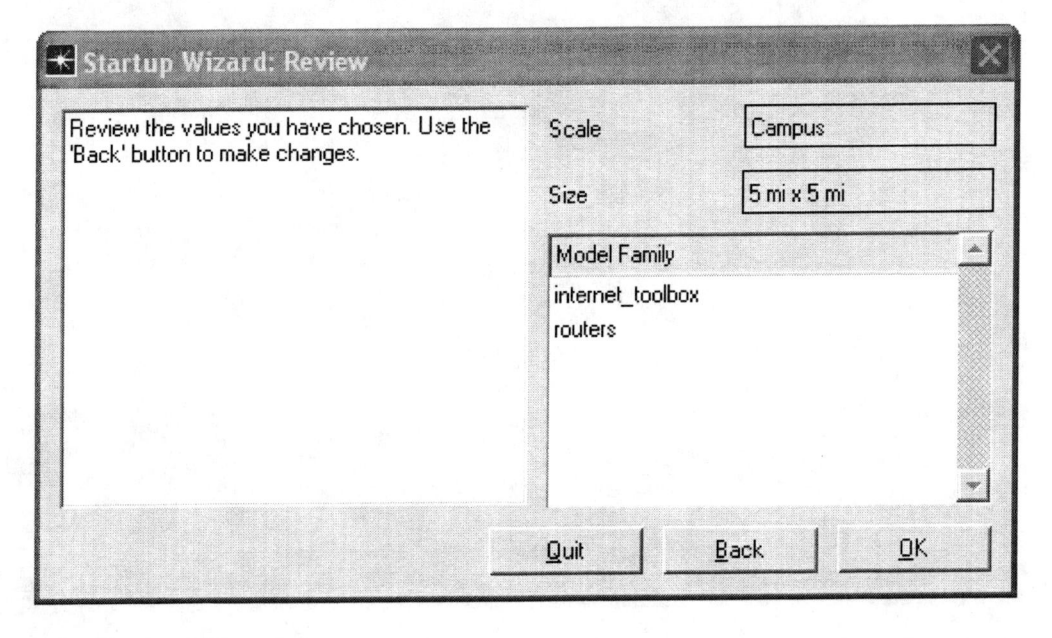

Building the Router Network

1. We will now create the router network. Look at the next diagram. It shows where the routers should be placed. We will use five routers named node 0 through 4. From the **Object Palette,** click on the **ethernet2_slip8_gtwy** object to select it. Drag it to the middle of the project grid and click to paste five copies of the router on the grid Right click to release the **ethernet2_slip_gtwy** object when you are through. Verify that your routers are in the correct place on the grid and that the names also correspond to those in the next picture. You may click to select a router object and drag it to a different position if you like.

    The **ethernet2_slip8_gtwy** supports two Ethernet connections and eight point-to-point connections.

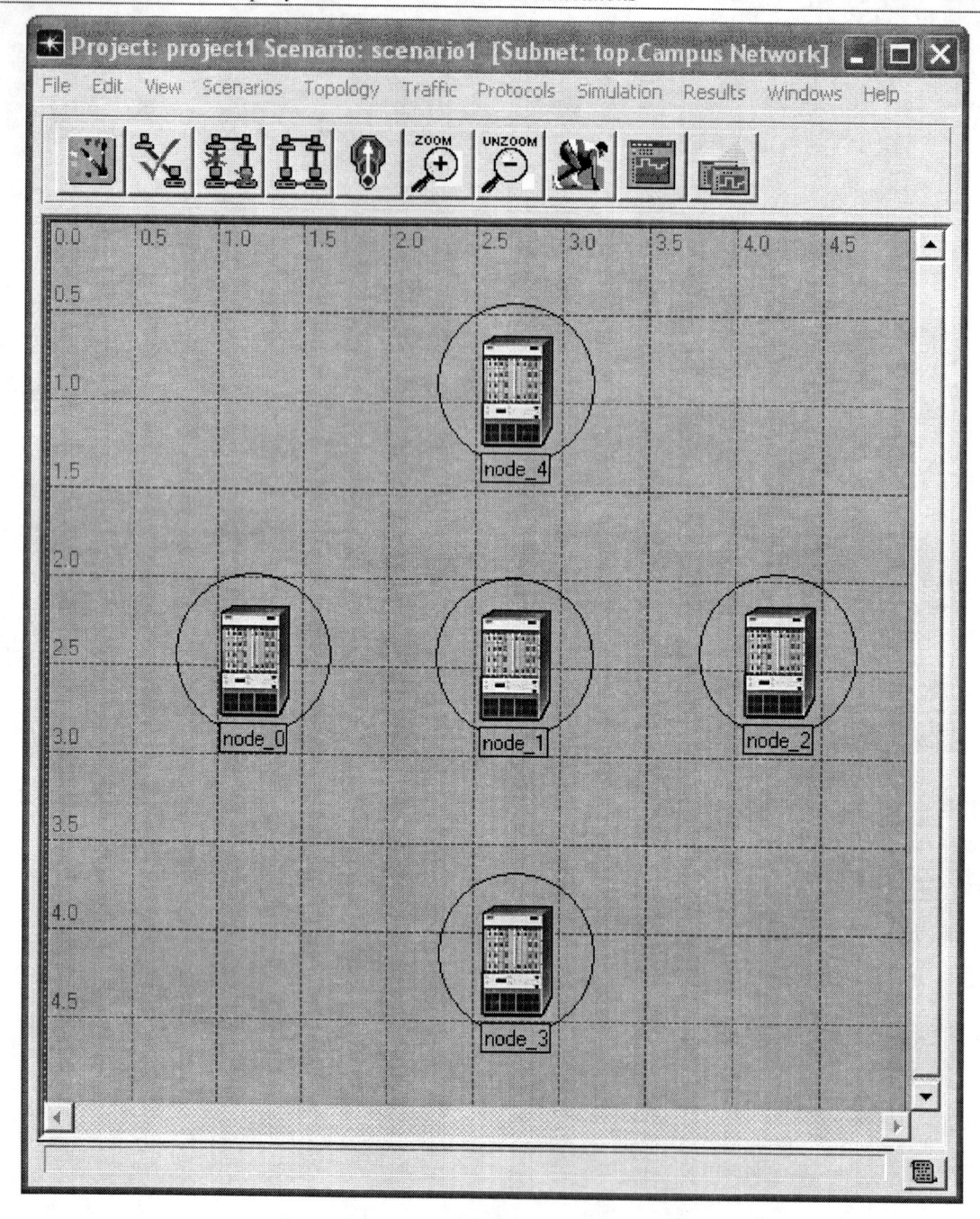

OPNET Lab Manual to accompany Business Data and Communications

2. We will now connect the routers using point-to-point DS1 links. A DS1 is 1.544Mbps. From the **Object Palette**, click on the **PPP_DS1 link** object. Click on each node to connect them in the configuration shown next. Right click to release the **PPP_DS1 link object**.

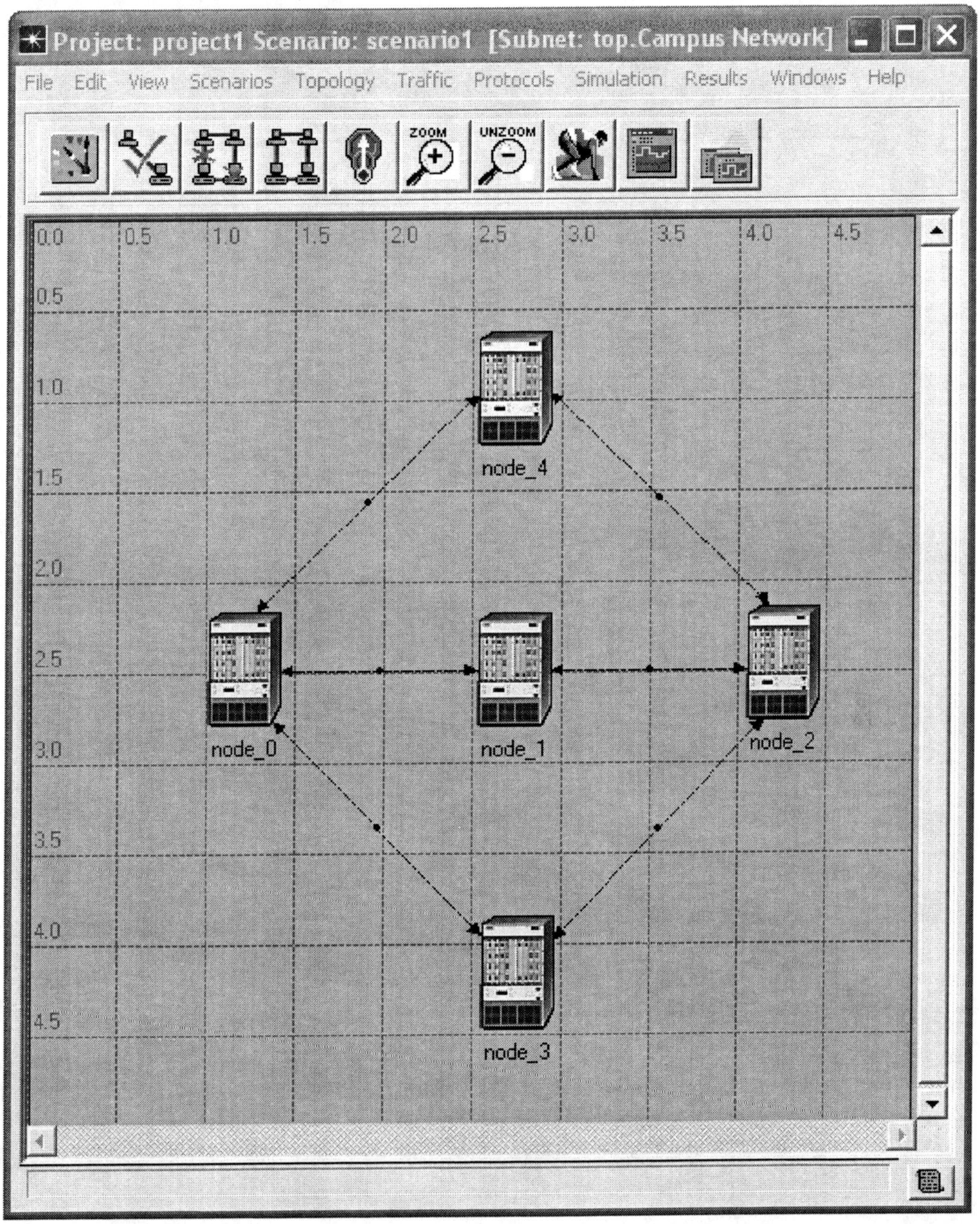

We now have a campus intranet of five routers. Note that there are multiple paths between each node. For instance, if a packet is to be sent from node_0 to node_2, it may take a route through node_4, node_1, or node_3. Each path is currently of equal length or cost. We will now configure each node to use RIP to build routing tables. We will then be able to trace a path from node_0 to node_1 and discover which path the distance vector algorithm found.

3. Click on one of the **ethernet2_slip8_gtwy** nodes. You will see a circle around the object indicating that it is selected. Right click and choose **Select Similar Nodes** from the pull-down menu. Right click again on one of the **ethernet2_slip8_gtwy** objects and choose **Edit Attributes** from the pull-down menu.

4. Click on the box next to **Apply Changes to Selected Objects** at the bottom of the window. This will ensure that the changes are applied to all the other router nodes. Click on the + next to **IP Routing Parameters** to expand the parameter list. Now Click on the + next to **Routing Table Export.** Choose

Once at End of Simulation from the drop-down box as shown on the next screen display. This will allow us to view the routing tables after the simulation has run.

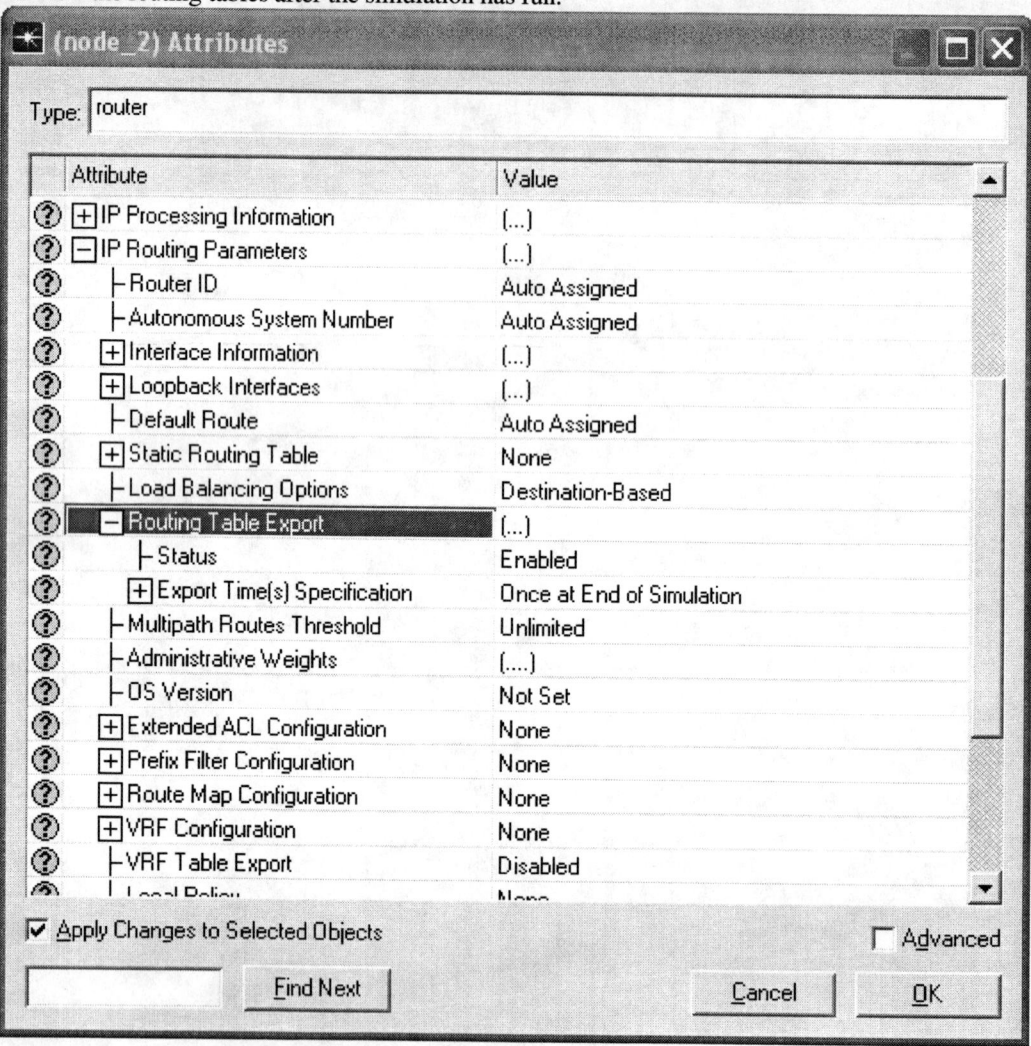

Attribute	Value
⊞ IP Processing Information	(...)
⊟ IP Routing Parameters	(...)
Router ID	Auto Assigned
Autonomous System Number	Auto Assigned
⊞ Interface Information	(...)
⊞ Loopback Interfaces	(...)
Default Route	Auto Assigned
⊞ Static Routing Table	None
Load Balancing Options	Destination-Based
⊟ Routing Table Export	(...)
Status	Enabled
⊞ Export Time(s) Specification	Once at End of Simulation
Multipath Routes Threshold	Unlimited
Administrative Weights	(....)
OS Version	Not Set
⊞ Extended ACL Configuration	None
⊞ Prefix Filter Configuration	None
⊞ Route Map Configuration	None
⊞ VRF Configuration	None
VRF Table Export	Disabled

5. Now scroll down and click on the **+** next to **RIP Parameters** to expand the parameter list. Notice that the **Start Time** has a value of **constant(5)**. This means that the RIP protocol will begin after 5 seconds. Edit the **Stop Time** and change its value to **1,000 seconds** as indicated in the next screen image. This will ensure that the protocol will not stop before the routing tables have converged.

Lab 3   RIP and Routing Table Convergence

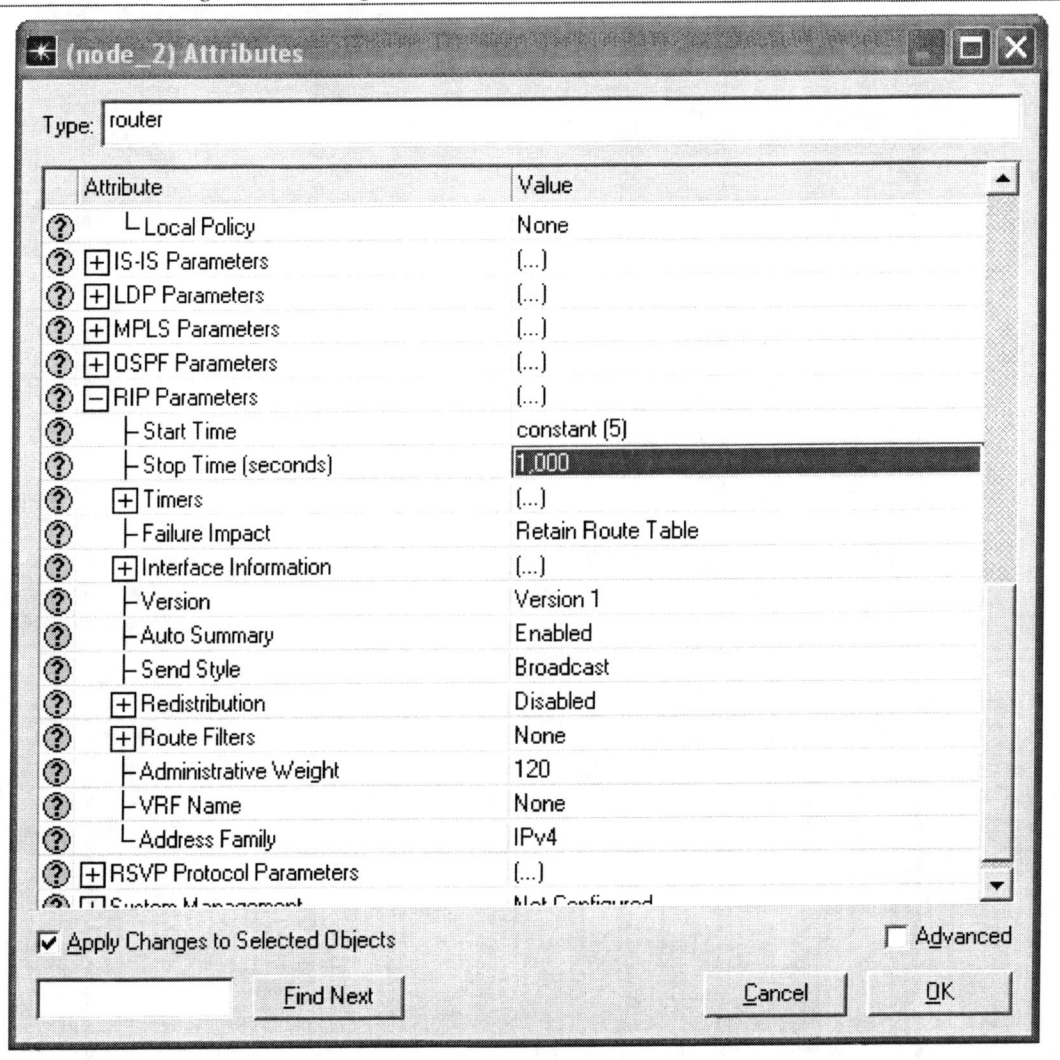

Attribute	Value
└ Local Policy	None
⊞ IS-IS Parameters	(...)
⊞ LDP Parameters	(...)
⊞ MPLS Parameters	(...)
⊞ OSPF Parameters	(...)
⊟ RIP Parameters	(...)
├ Start Time	constant (5)
├ Stop Time (seconds)	1,000
⊞ Timers	(...)
├ Failure Impact	Retain Route Table
⊞ Interface Information	(...)
├ Version	Version 1
├ Auto Summary	Enabled
├ Send Style	Broadcast
⊞ Redistribution	Disabled
⊞ Route Filters	None
├ Administrative Weight	120
├ VRF Name	None
└ Address Family	IPv4
⊞ RSVP Protocol Parameters	(...)

6. Under the **RIP Parameters,** click on the + next to **Timers** to expand the parameter list. Notice that the **Update Interval** is **30 seconds**. This means that table exchanges should occur every 30 seconds. OPNET IT Guru also uses "triggered updates." If a table exchange causes a table to be modified, a new exchange will be triggered immediately between 1 and 5 seconds afterwards.

7. From the main menu choose **File/Save** to save your project.

Configuring the Routing Simulation

1. Right click anywhere on the grid workspace and select **Choose Individual Statistics** from the menu. Click on the + next **Node** to expand the parameters. Click on the + next to **Route Table** to expand the parameters. Click on **Total number of updates.** This will allow us to keep track of the number of updates for each simulation.

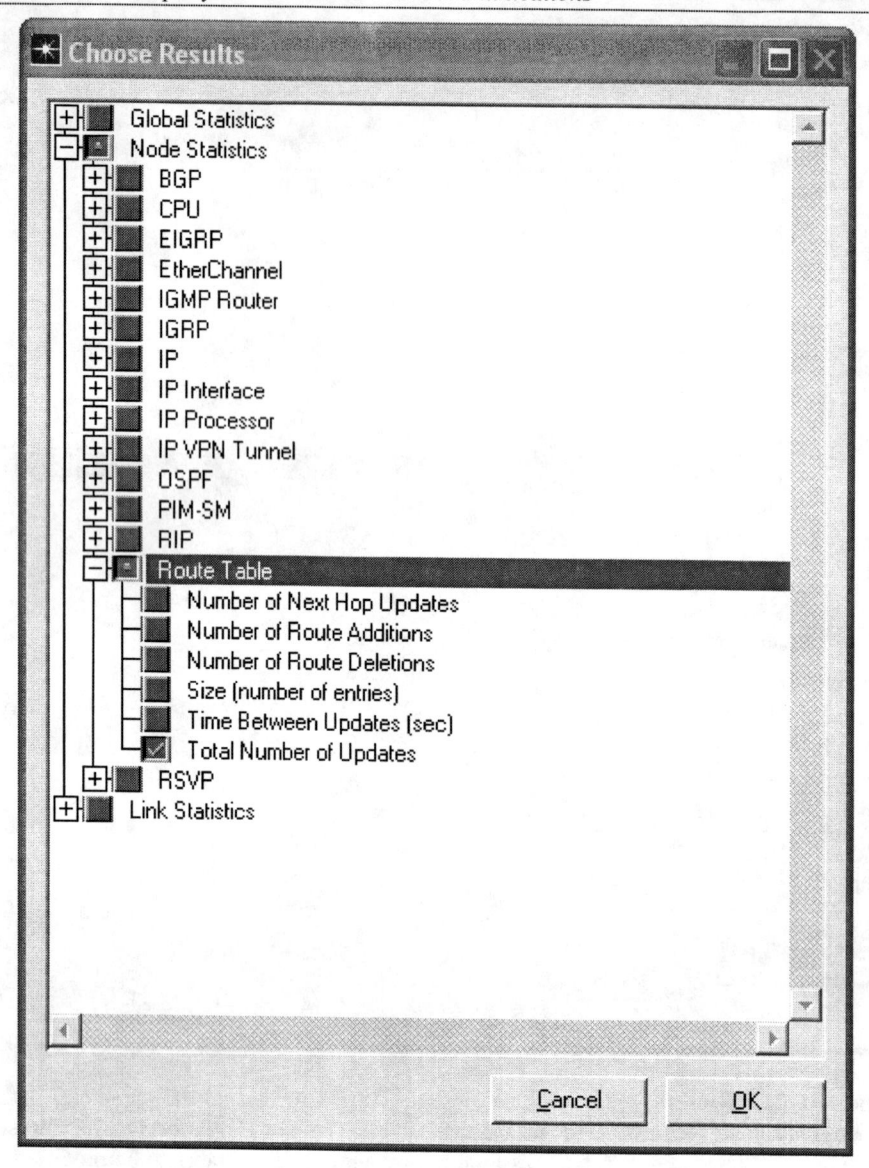

OPNET Lab Manual to accompany Business Data and Communications

[Screenshot of "Choose Results" dialog showing expanded Node Statistics tree with BGP, CPU, EIGRP, EtherChannel, IGMP Router, IGRP, IP, IP Interface, IP Processor, IP VPN Tunnel, OSPF, PIM-SM, RIP, Route Table (expanded with Number of Next Hop Updates, Number of Route Additions, Number of Route Deletions, Size (number of entries), Time Between Updates (sec), Total Number of Updates checked), RSVP, and Link Statistics.]

2. Click on the **Configure and Run** button. Click on the **Global Attributes** Tab. Verify that the **IP Dynamic Routing Protocol** is **RIP**. We will now change four different attribute values in this window by clicking on the value and selecting that option from the pull-down menu.

   Change **IP Routing Interface Addressing Mode** to **Auto Addressed/Export.**
   Change **IP Routing Table Export/Import** to **Export.**
   Change **RIP Sim Efficiency** to **Disabled.**
   Change **RIP Stop Time** to **1000.** (Note: Do not use a comma when typing this value.)

Lab 3   RIP and Routing Table Convergence

Configure Simulation: LCsmRoute-scenario1

Common | Global Attributes | Object Attributes | Reports | SLAs | Animation | Profiling | Advanced | Environment Files

Attribute	Value
IP Dynamic Routing Protocol	RIP
IP Interface Addressing Mode	Auto Addressed/Export
IP Routing Table Export/Import	Export
LDP Discovery End Time	250
LDP Discovery Start Time	100
LSP Signaling Protocol	RSVP
OSPF Sim Efficiency	Enabled
OSPF Stop Time	260
RIP Sim Efficiency	Disabled
RIP Stop Time	1000
RSVP Sim Efficiency	Enabled
Tracer Packet Redundancy	Enabled

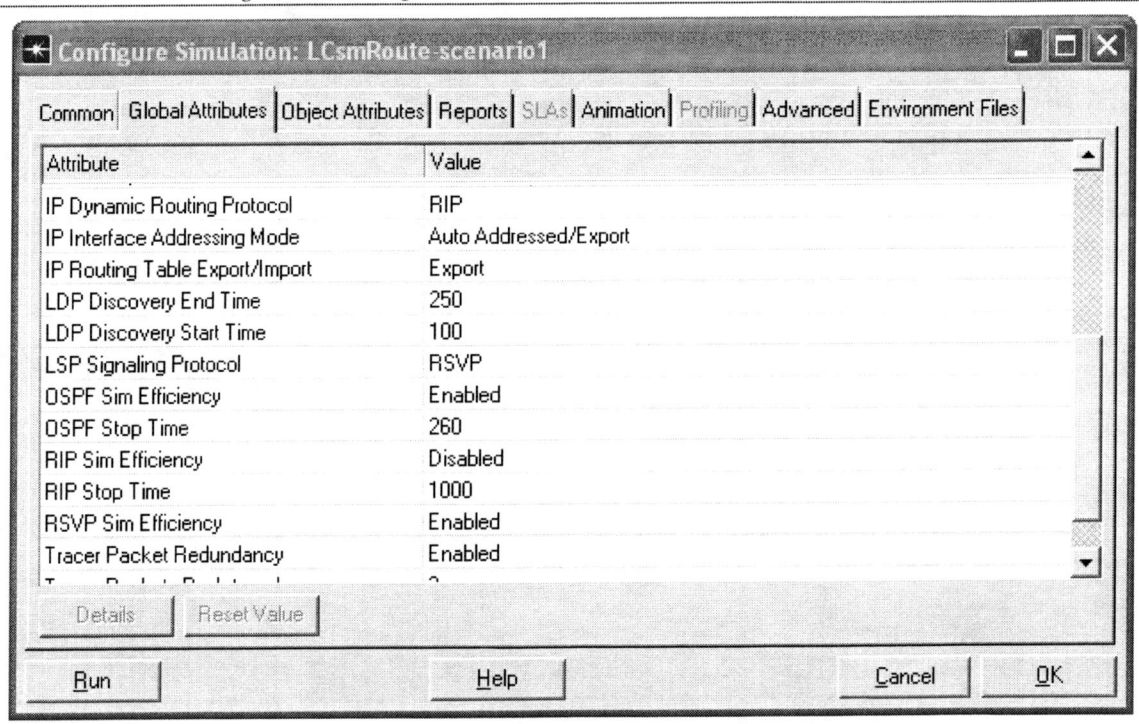

These changes will export the address table for each of the routing interfaces to a file and will also export the routing tables at the end of the simulation to a file. Disabling the simulation efficiency will cause it to run for the expected amount of time, assuming that triggered updates occur in 1 to 5 seconds and updates occur every 30 seconds. The **RIP Stop Time** will ensure our simulation will run long enough for the tables to converge.

Now click on the **Common** tab at the top of the window. Set the duration of the simulation to **180 second(s)**. With triggered updates and our simple intranet, we can expect that the routing tables will have converged by this time. Click on the **Run** button to run the simulation.

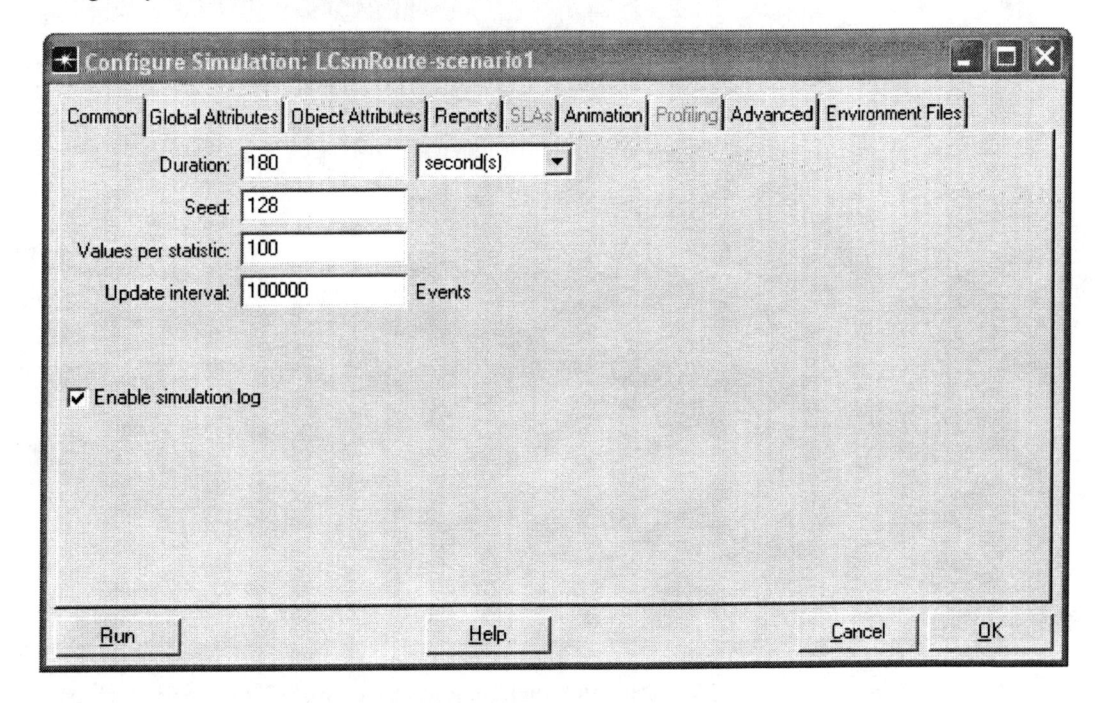

Configure Simulation: LCsmRoute-scenario1

Common | Global Attributes | Object Attributes | Reports | SLAs | Animation | Profiling | Advanced | Environment Files

Duration: 180   second(s)
Seed: 128
Values per statistic: 100
Update interval: 100000   Events

☑ Enable simulation log

3. From the main menu, choose **File/Save** to save your project.

Gathering Results

1. Now we would like to look at the address file associated with our intranet. From the main menu, choose **File/Model Files/Refresh Model Directories**. This will update the model files with our last simulation run.

2. From the main menu, choose **File/Open/Generic Data File**. You will see many files associated with your project and other OPNET models. Select the file named **your_initialsRIPConvergenceRIPStartConfig-ip_addr** The contents of the file will look similar to the following:

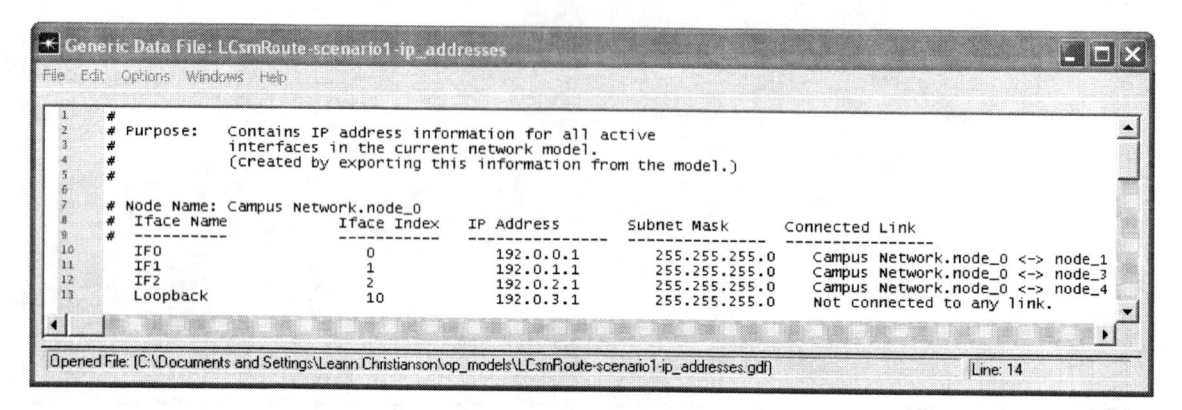

```
Generic Data File: LCsmRoute-scenario1-ip_addresses
File Edit Options Windows Help

# Purpose:   Contains IP address information for all active
#            interfaces in the current network model.
#            (created by exporting this information from the model.)
#
# Node Name: Campus Network.node_0
# Iface Name    Iface Index    IP Address      Subnet Mask       Connected Link
# ----------    -----------    ----------      -----------       --------------
  IF0           0              192.0.0.1       255.255.255.0     Campus Network.node_0 <-> node_1
  IF1           1              192.0.1.1       255.255.255.0     Campus Network.node_0 <-> node_3
  IF2           2              192.0.2.1       255.255.255.0     Campus Network.node_0 <-> node_4
  Loopback      10             192.0.3.1       255.255.255.0     Not connected to any link.

Opened File: [C:\Documents and Settings\Leann Christianson\op_models\LCsmRoute-scenario1-ip_addresses.gdf]    Line: 14
```

Looking at the interface addresses for each router node, draw a picture of your intranet topology, label each node, and link it with its associated interfaces. This will help you trace routes between the nodes and answer the questions at the end of the lab. Note that each node will have a "loopback interface." Loopback interfaces are often used for testing. They allow routers to send packets to themselves.

3. From the main menu, choose **Results/Open Simulation Log**. Click on the + next to **Simulation Log**. Click on the + next to **Categories**. Click on the + next to **Classes**. Click on the + next to **IP**. Finally, click on **Route Table** as shown on the following screen:

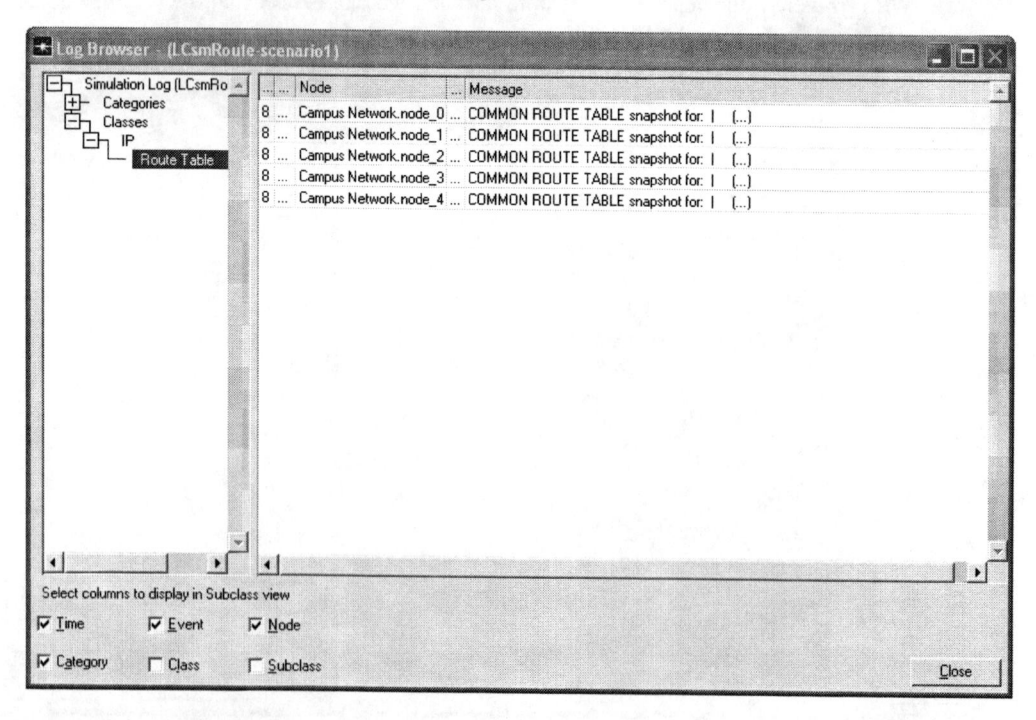

```
Log Browser - (LCsmRoute-scenario1)

Simulation Log (LCsmRo...     Node                           Message
  Categories              8 ... Campus Network.node_0 ... COMMON ROUTE TABLE snapshot for: |  (...)
  Classes                 8 ... Campus Network.node_1 ... COMMON ROUTE TABLE snapshot for: |  (...)
    IP                    8 ... Campus Network.node_2 ... COMMON ROUTE TABLE snapshot for: |  (...)
      Route Table         8 ... Campus Network.node_3 ... COMMON ROUTE TABLE snapshot for: |  (...)
                          8 ... Campus Network.node_4 ... COMMON ROUTE TABLE snapshot for: |  (...)

Select columns to display in Subclass view
  ☑ Time        ☑ Event       ☑ Node
  ☑ Category    ☐ Class       ☐ Subclass                                           Close
```

You will see the names of each node. You can click under the message box column beside the node name to view its routing table. Following is a sample routing table:

56

Lab 3 RIP and Routing Table Convergence

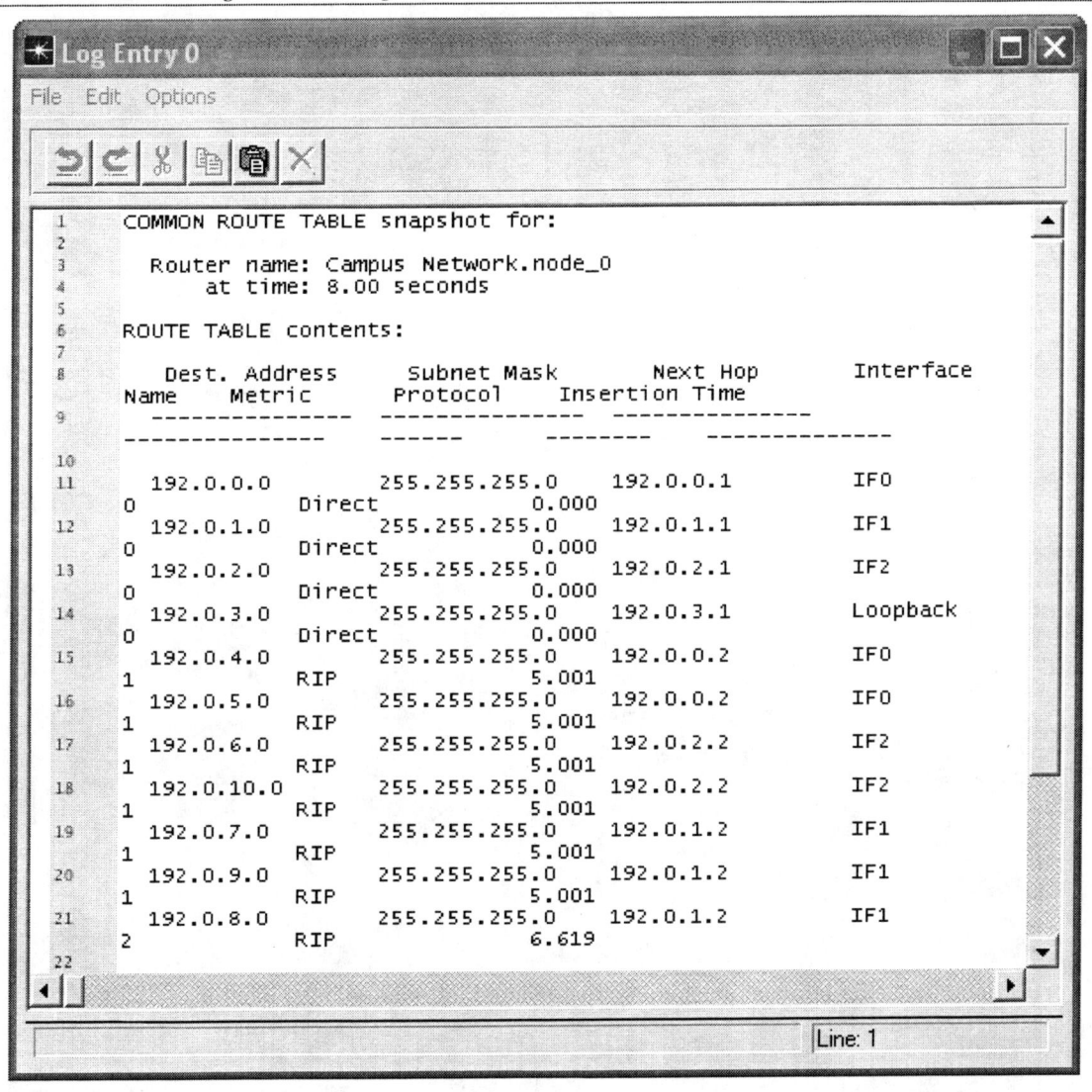

```
COMMON ROUTE TABLE snapshot for:

   Router name: Campus Network.node_0
       at time: 8.00 seconds

ROUTE TABLE contents:

     Dest. Address      Subnet Mask       Next Hop        Interface
   Name    Metric        Protocol      Insertion Time
   ----------------    ---------------  ---------------   ---------------
   -----------------   --------         ---------         ---------------

   192.0.0.0           255.255.255.0    192.0.0.1         IF0
 0                     Direct             0.000
   192.0.1.0           255.255.255.0    192.0.1.1         IF1
 0                     Direct             0.000
   192.0.2.0           255.255.255.0    192.0.2.1         IF2
 0                     Direct             0.000
   192.0.3.0           255.255.255.0    192.0.3.1         Loopback
 0                     Direct             0.000
   192.0.4.0           255.255.255.0    192.0.0.2         IF0
 1                     RIP                5.001
   192.0.5.0           255.255.255.0    192.0.0.2         IF0
 1                     RIP                5.001
   192.0.6.0           255.255.255.0    192.0.2.2         IF2
 1                     RIP                5.001
   192.0.10.0          255.255.255.0    192.0.2.2         IF2
 1                     RIP                5.001
   192.0.7.0           255.255.255.0    192.0.1.2         IF1
 1                     RIP                5.001
   192.0.9.0           255.255.255.0    192.0.1.2         IF1
 1                     RIP                5.001
   192.0.8.0           255.255.255.0    192.0.1.2         IF1
 2                     RIP                6.619
```

We know that a routing table has converged when each interface address of the network is listed in the routing table of each node in the network. You can use these routing tables to find the path chosen by the distance vector RIP algorithm between each node.

Scenario 2   Convergence after a Link Break

1. From the main menu, choose **Scenarios/Duplicate Scenario**. This will bring up a dialog box. Give your scenario a name such as **RIPFailure** and click **OK**.

2. Click on the **Object Palette** Button. Select **utilities** from the pull-down menu. Find the **Failure Recovery** object and click on it to select it. Place it on the workspace grid. Right click to release the object. Close the **Object Palette**. The scenario should look like the following:

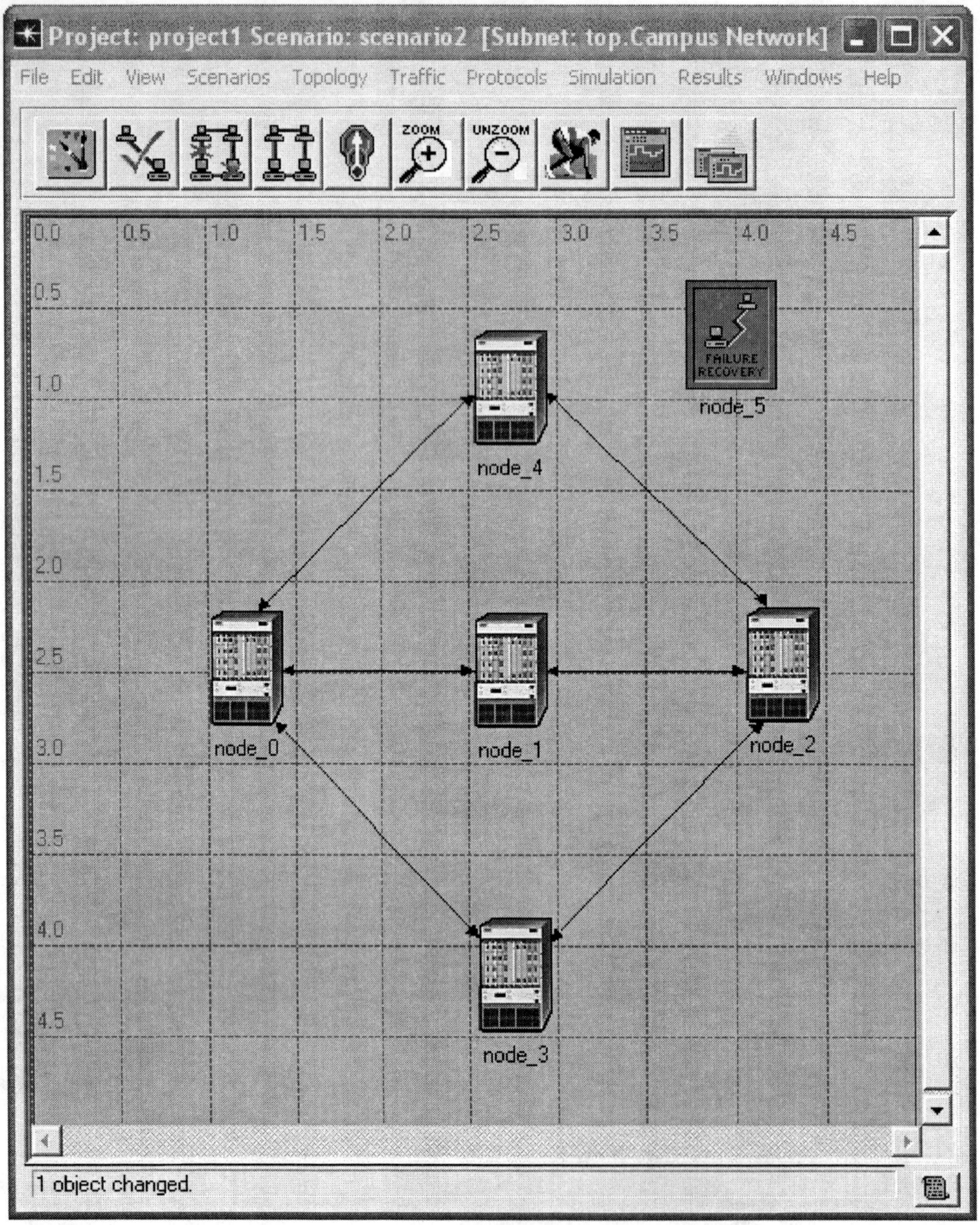
Project: project1 Scenario: scenario2 [Subnet: top.Campus Network]
File Edit View Scenarios Topology Traffic Protocols Simulation Results Windows Help
ZOOM
UNZOOM
FAILURE RECOVERY
node_5
node_4
node_0
node_1
node_2
node_3
1 object changed.

3. Right click on the **Failure Recovery** object and choose **Edit Attributes** from the pull-down menu. We will now introduce a link failure to this scenario. Click on the + next to **Link Failure/Recovery Specification** to expand the parameters. Click on the value next to **row** and choose **1** from the pull-down menu. This will add one row.

4. Click on the + next to **Row 0** to expand the attributes. We will edit the name and the time attributes. Choose **CampusNetwork_node_0 <-> node_3** from the pull-down menu for the **name** attribute. Click on the value next to the time attribute. Edit the field and set the value to **60**. Compare your values to those listed in the following figure:

Lab 3 RIP and Routing Table Convergence

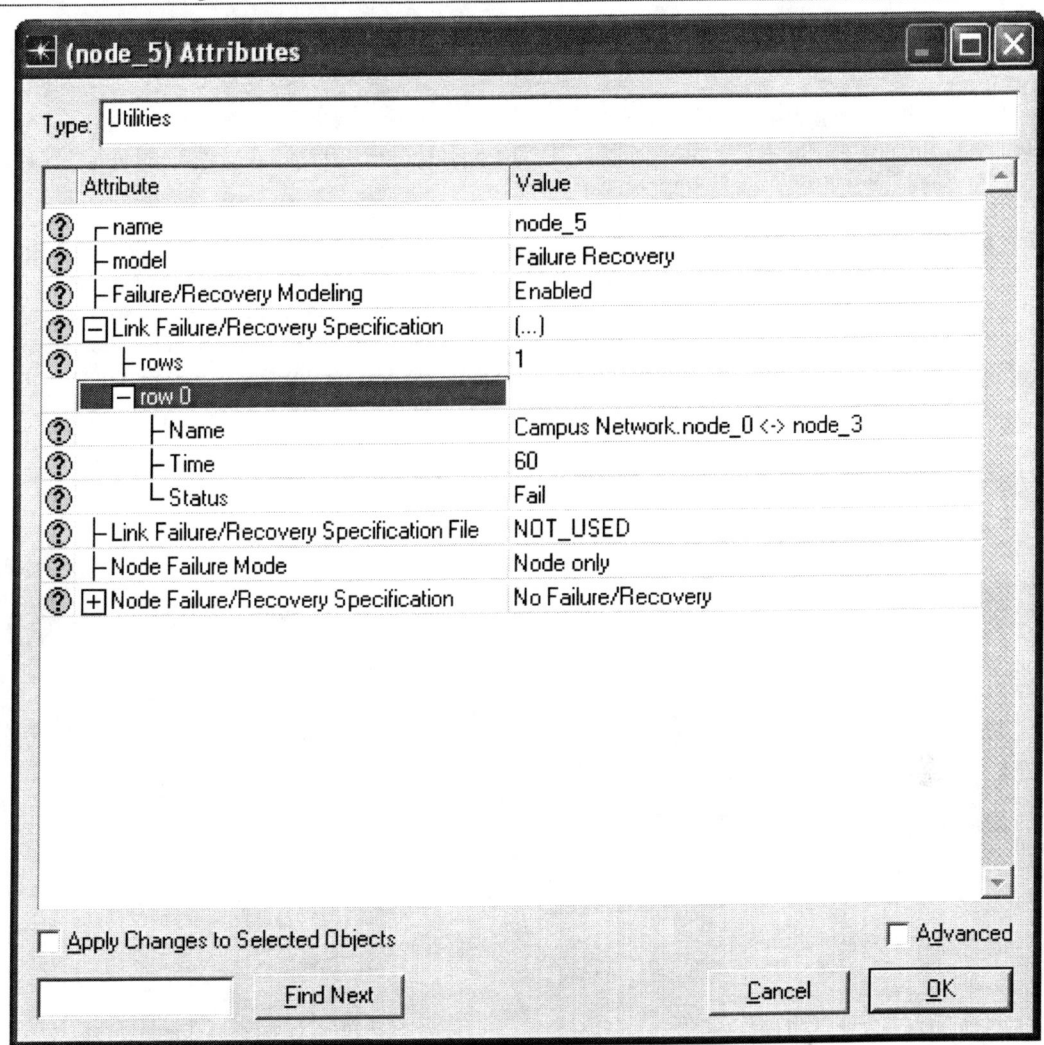

5. Click on the **Configure and Run** button. Set the **Duration** of the simulation to **180 seconds**. After 60 seconds, the link from node_0 to node_3 will fail, table updates will be exchanged, and the tables will again converge within 180 seconds. We leave it as an exercise for you to discover exactly how long it takes for the tables to converge. Click on the **Run** button to run the simulation.

6. From the main menu choose **File/Save** to save your project.

Questions

1. Using the routing tables created after convergence for the **RIPStart** Scenario, find a path from one of the interface addresses of node_0 to one of the interface addresses of node_3. To do this, you will need to look at the routing table for node_0.

    Under the destination addresses, find the node_3 interface address to which you want to find a path. Examine the next hop address for this address. Now find another routing table that has a directly connected interface address with this same next hop address. Look at this routing table and again look for the destination address and next hop.

    Repeat until you find that the destination address next hop is the same as the destination address itself. This means that it is directly connected. Record the nodes you travel through on the route.

59

Interface addresses have the format 192.0.2.1. You can think of this address as subnet 2. You can also ignore the last digit when matching interface addresses to networks. For example, 192.0.2.0 matches 192.0.2.1.

2. Repeat problem 1 for the **RIPFailure** scenario. The path should differ. What is the new path? Is this what you expected? What other paths might have been chosen?

3. Our simulations start after 5 seconds. Triggered updates occur between 1 and 5 seconds. With distance vector routing, nodes exchange tables until updates no longer occur. We call this convergence. Based on this information, try to find the exact time of convergence for the **RIPFailure** simulation. (Remember that the link will not fail until 60 seconds into the simulation.) It may help to run the simulation for a very short time to see what the tables look like before convergence. Try to find the exact time for convergence. Is the time for convergence what you expected? Explain your results.

LAB 4 TCP PARAMETERS FOR FLOW CONTROL

Objective
1. To create a Wide Area Network (WAN) that connects an email client to an email server.
2. To evaluate the effect of TCP segment size and TCP receive window sizes on delay and throughput.

Motivation
SilicoSolutions is a software development company located in Mt. View, CA. Recently, SilcoSolutions has outsourced some of its development projects to an affiliate company located in Bombay, India. The two companies collaborate regularly via extensive email messages. Smitha Johnson, a network administrator for SilicoSolutions, is interested in tuning the network in order to reduce upload delay to the email server in India. SilicoSolution's email is sent using the transport protocol TCP. Smitha believes that changing the TCP segment size and TCP receive window size will significantly improve response time. She would like to run tests to evaluate her hypothesis in order to choose the correct TCP parameters for SilicoSolution's email application.

Description
The Transmission Control Protocol (TCP) is a reliable protocol that provides flow control between senders and receivers. This ensures that a fast sender will not overrun a slow receiver. Flow control is implemented via send and receive windows. The receiver advertises how much space is available for buffering incoming segments in its receive window, and the sender restricts its sending rate to match the advertised value. Throughput is affected by the size of the receive window. If the window is too small, only a few segments will be transmitted per time unit. TCP segment size also affects throughput.

In this lab, we will create a Wide Area Network (WAN) that uses the Internet Protocol (IP) to route traffic between California and India. The California site will transmit email messages to an email server located in India. We will examine performance statistics such as throughput and delay for email uploads using various receive window sizes and TCP segment sizes.

Follow the next set of directions to create an OPNET model, run simulations, and view results. Questions regarding the lab are listed at the end.

Create a New Project

1. Start the **OPNET IT Guru Academic Edition Application.** Create a **New Project** by selecting **File/New/Project** from the Main Menu. Click **OK**.

2. Give the project a name such as **your_initials_TCPFlow.** Give the scenario a name such as **8k.** Click **OK.** You should see the **Startup Wizard Initial Topology** dialog box.

3. Verify that **Create Empty Scenario** is selected in the **Initial Topology Dialog Box**. Click **Next**. You will now see the **Network Scale** dialog box.

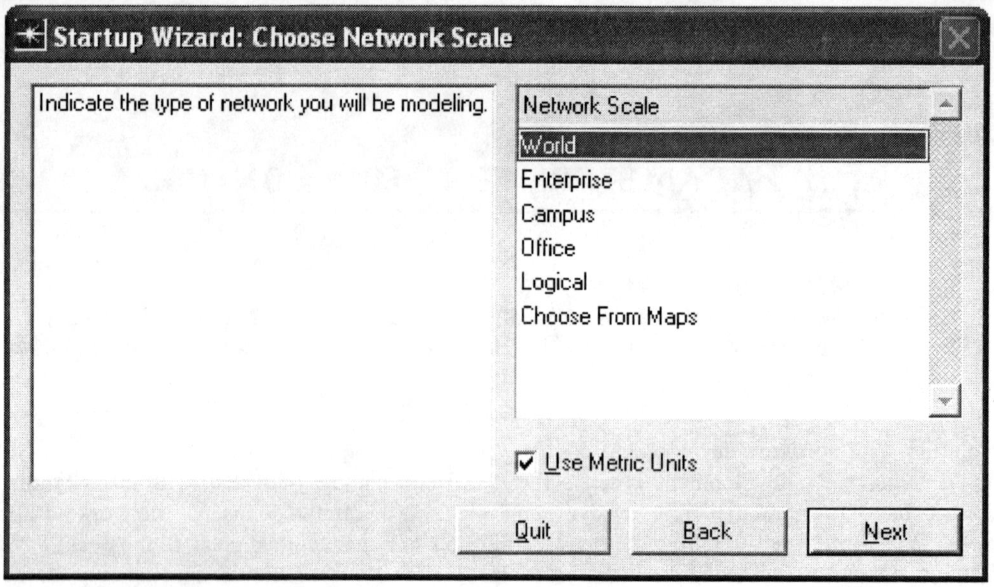
Startup Wizard: Choose Network Scale
Indicate the type of network you will be modeling.
Network Scale
World
Enterprise
Campus
Office
Logical
Choose From Maps
Use Metric Units
Quit
Back
Next

4. Choose **World** from the **Network Scale** list. Click **Next.** Choose **World** as your map and click **Next.** You should see the **Select Technologies** dialog box.

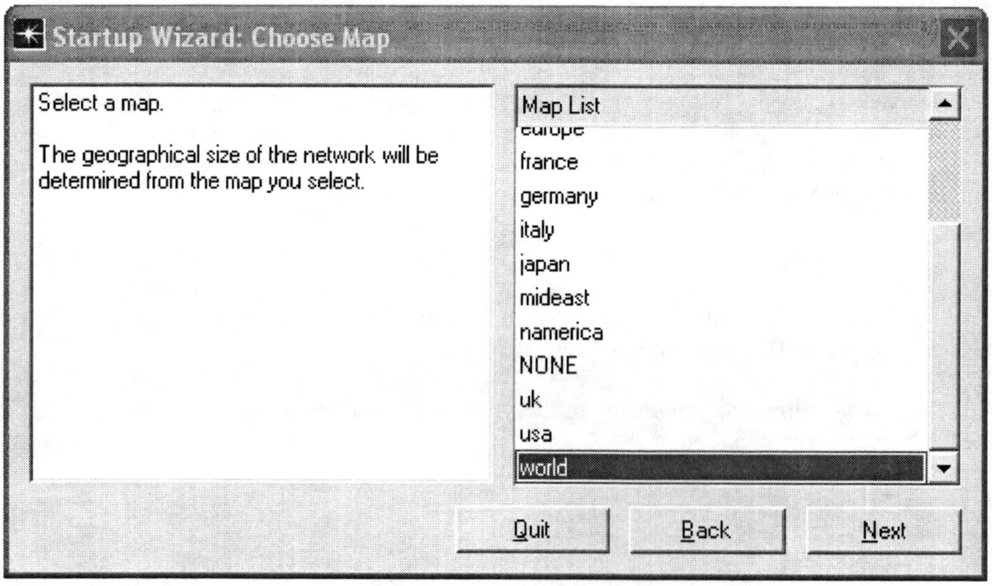
Startup Wizard: Choose Map
Select a map.
The geographical size of the network will be determined from the map you select.
Map List
france
germany
italy
japan
mideast
namerica
NONE
uk
usa
world
Quit
Back
Next

5. Scroll down until you see **internet_toolbox** under the **Model Family** column. Under the **Include column,** click on the box next to **internet_toolbox**. This will change the "no" to "yes" indicating that Internet technologies will be included. Click **Next.**

Lab 4   *TCP Parameters for Flow Control*

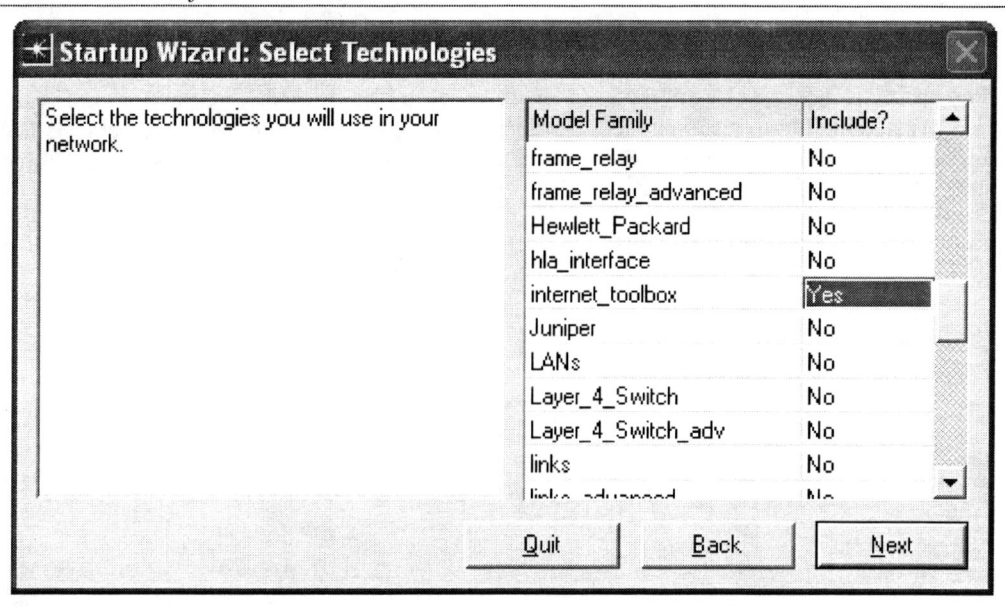

6. In the **Start up Wizard Review** dialog box, verify that the **internet_toolbox** technologies are chosen, that the scale is World, and the map is World. Click **OK**. You should see the **Object Palette** window.

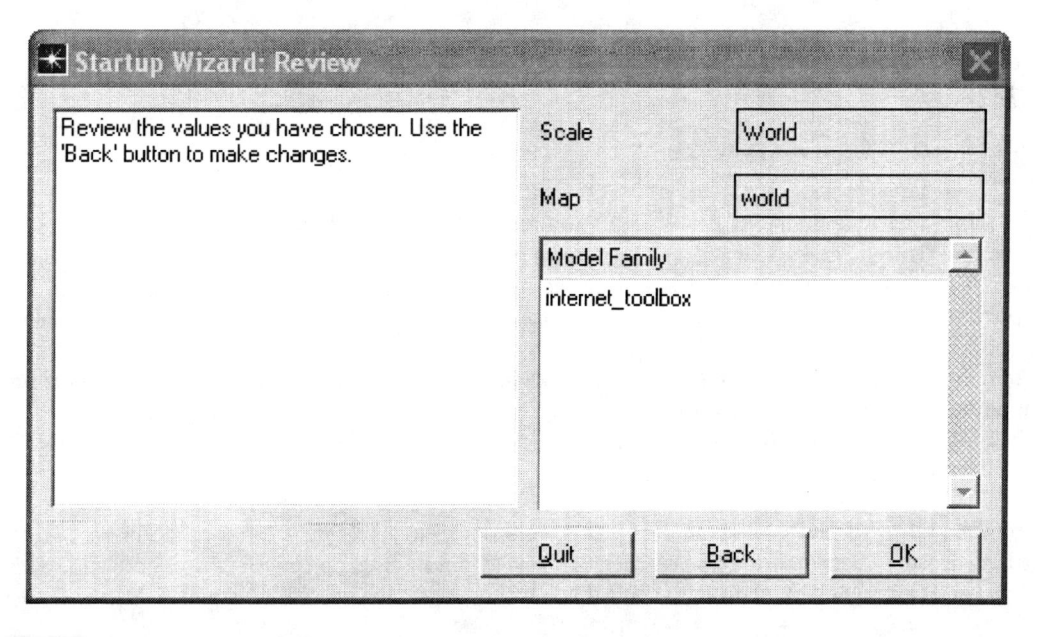

7. We would like to create a network where the server is a Point-to-Point (PPP) server and the client is a PPP advanced workstation. To include these objects, we need to configure the object palette. In the **Object Palette** window, click on the **Configure Palette** button. A new window will appear.

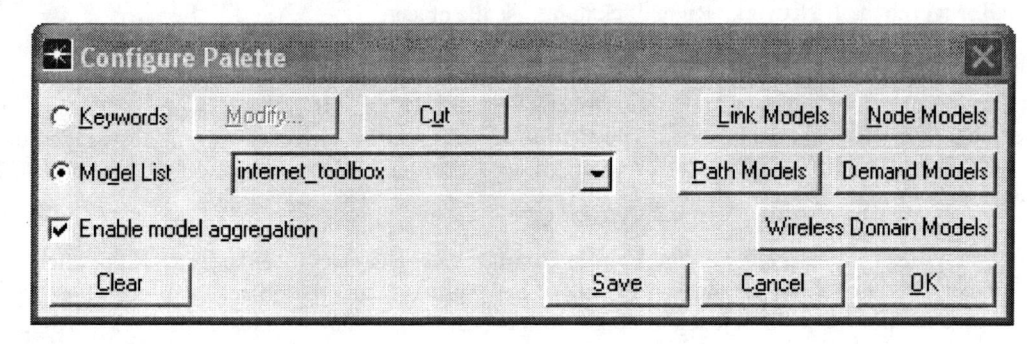

8. Click on the **Node Models** button. This will bring up a window with an alphabetical list of all the possible object nodes that you can use in your projects. Scroll down until you find the PPP objects. Click on the value next to **ppp_server_adv** to include it in the project. Click on the value next to **ppp_workstation_adv** to also include it in the project. We will use the advanced objects because these allow us to select from a variety of transport layer protocols. Click **OK** to save the **object Toolbox Configuration**.

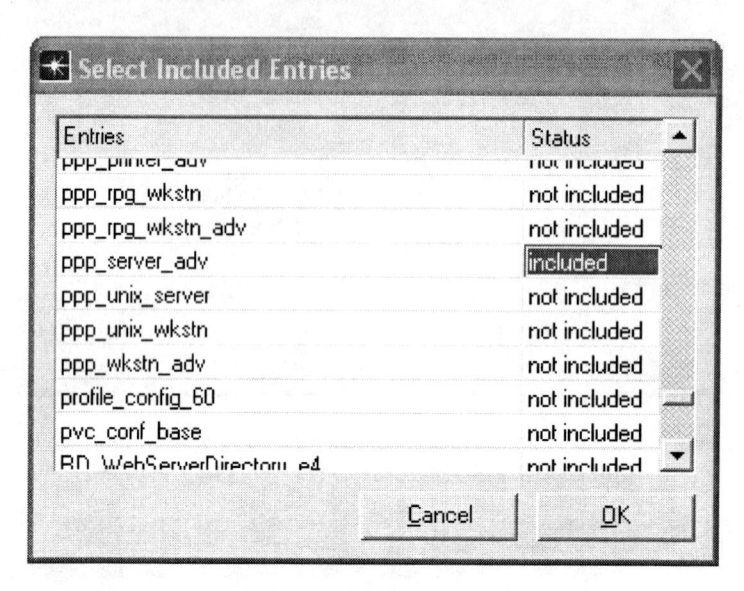

Building the Client/Server Network

1. We will now create the network. We will put our email server in India and our email client in California. From the **Object Palette**, click on the **ppp_server_adv** object to select it. Drag it to the project grid in the area of India. Click to paste a copy of it on the map. Right click to release the object. Right click on the **ppp_server_adv** object and choose **Set Name** from the pull-down menu. Name the object **EmailServer**.

2. From the **Object Palette**, click on the **ppp_workstation_adv** object to select it. Drag it to the project grid in the area of California. Click to paste a copy of it on the map. Right click to release the object. Right click on the **ppp_workstation_adv** object and choose **Set Name** from the pull-down menu. Name the object **EmailClient**.

3. From the **Object Palette**, click on the **IP32_Cloud** object to select it. Drag it to the project grid and paste it between the client and the server. Right click to release the object. Right click on the **IP32_Cloud** object and choose **Set Name** from the pull-down menu. Name the object **IPNetwork.**

4. We will use a DS1 link to connect our client and server to the IP Network. DS1 links are made up of 24 DS0s (64Kbps); therefore, they have a speed of 1.544 Mbps. From the **Object Palette**, click on the **PPP_DS1 link** object to select it. Connect the **EmailClient** to the IPNetwork object. Next, connect the **EmailServer** to the IPNetwork. Right click to release the object.

5. Now we will add an **Application Configuration** and **Profile Configuration** object to the workspace. The application object will allow us to choose the applications we want to include in our simulation. We will use the **Email (Heavy)** application. The **Profile Configuration** object allows us to use default traffic profiles for different types of users or to create new profiles. We will create our own profile for this project.

6. From the **Object Palette**, click on the **Profile Config** object and drag it to the grid. Right click to release the object. Right click on the object and choose **Set Name**. Name the object **ProfileConfig**. Next, click on the **Application Config** object and drag it to the grid. Right click to release the object. Right click on

Lab 4   TCP Parameters for Flow Control

the object and choose **Set Name**. Name the object **ApplicationConfig**. Verify that your project looks like the following figure:

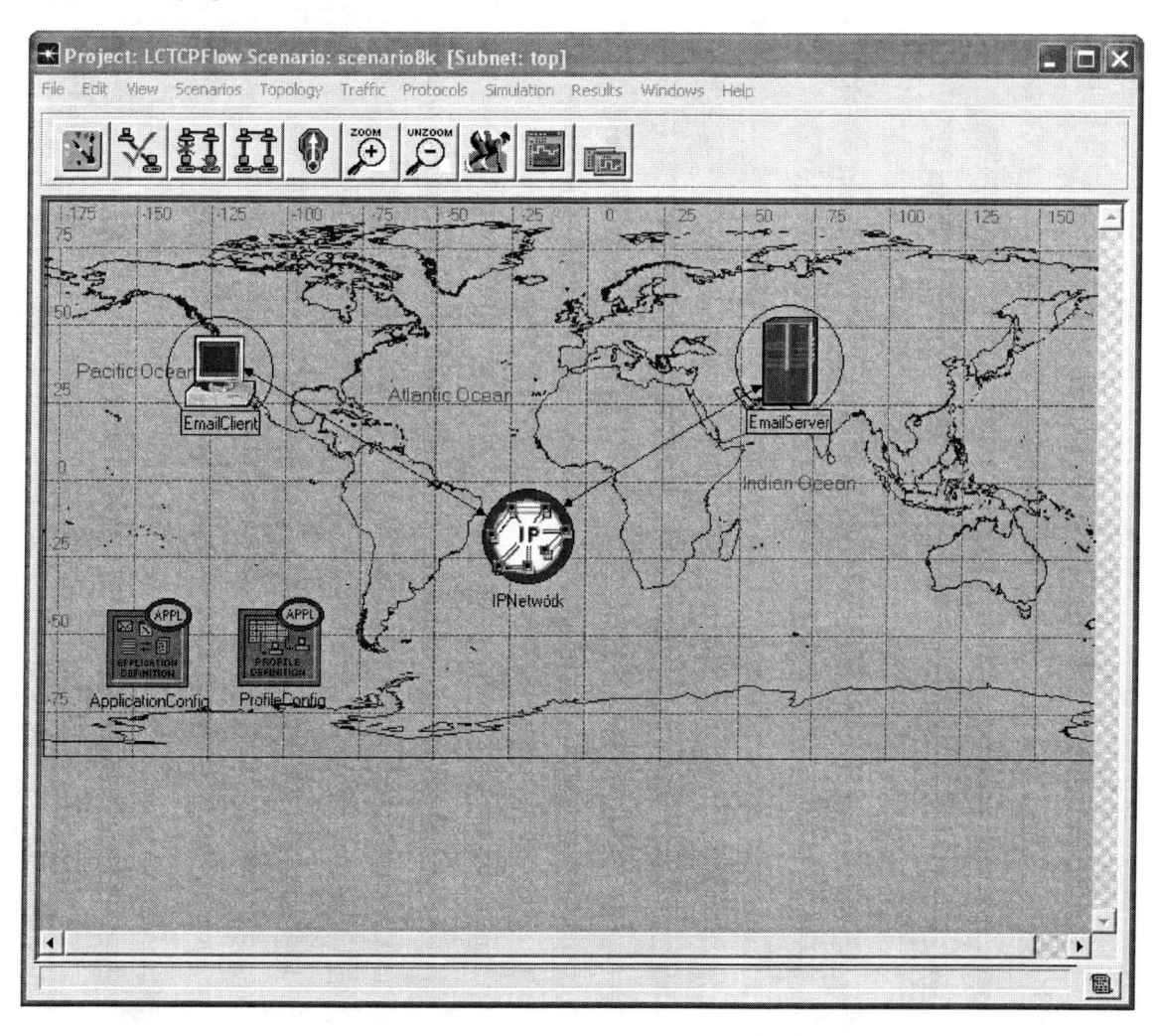

7. From the main menu, choose **File/Save** to save your project.

Configuring the Network

1. We will start by configuring the application object to provide the application we are interested in using in our simulation. OPNET IT Guru allows you to define new applications or use default descriptions for standard applications. We are interested in the **Email (Heavy)** application, that is predefined; however, we will modify the size of the email message that is transferred, as well as how often email messages are sent. To simplify our study, we will also modify the application so that email is sent from the client to the server, but not from the server to the client. When evaluating our results, we will, therefore, look at Email Upload response time.

   Right click on the **ApplicationConfig** object and choose **Edit Attributes**. Click on the value next to **Application Definition**. Choose **Default** from the pull-down menu.

2. Click on the + next to **Application Definitions** to expand the parameters. Click on the + next to **Email (Heavy)**. Click on the field next to **Description** and choose **Edit** from the pull-down menu.

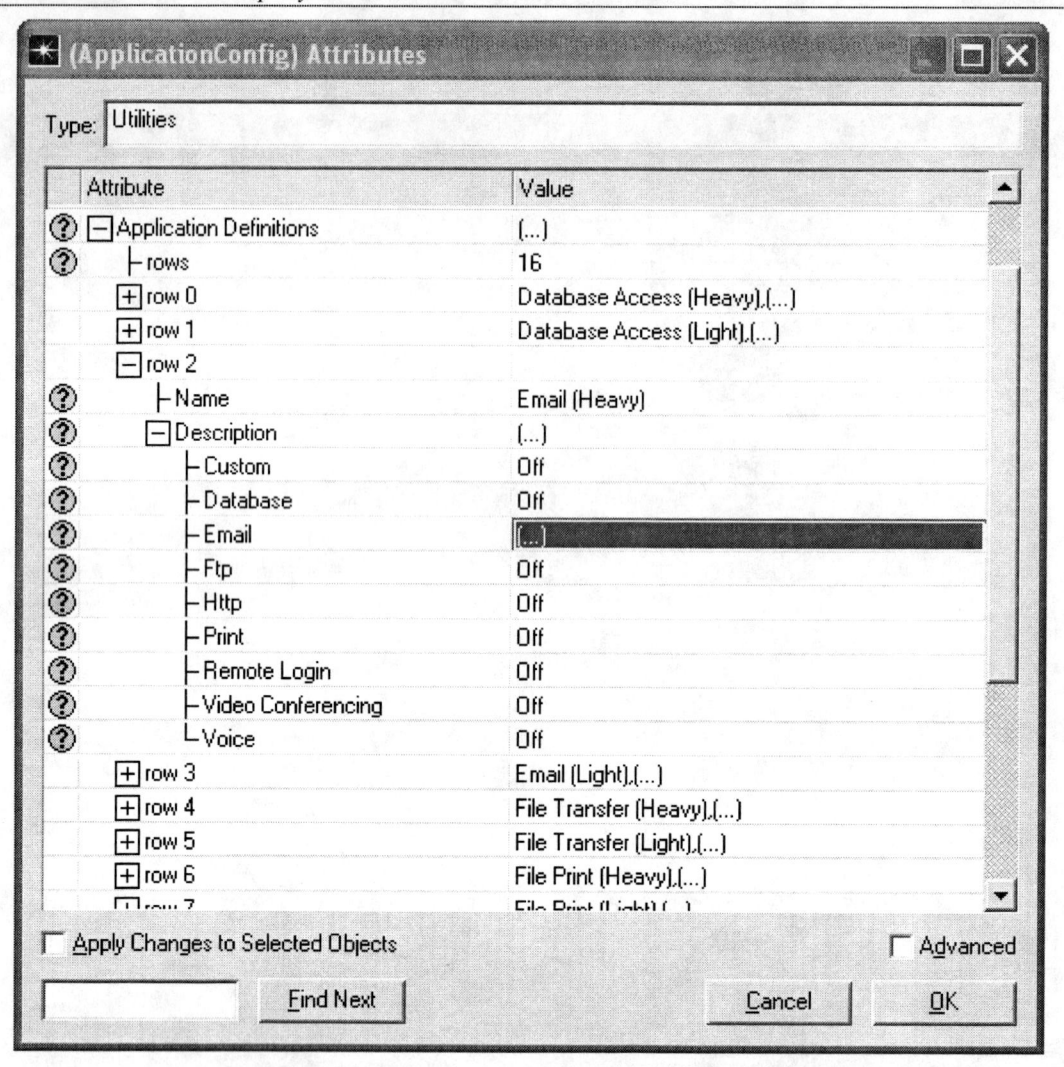

3. First, we will modify the send arrival time. Double click on the value next to the **Email** attribute. Click on the value next to **Send Arrival Time (Seconds)**. A new window will appear. Use the pull-down menu to change the **Special Value** field to **Not Used**. Set the **Distribution Name** to **exponential** and the **Mean Outcome** value to **1000**. This means that the amount of time between each email message will follow an exponential distribution with the mean at 1000 seconds. Click **OK**.

Lab 4   TCP Parameters for Flow Control

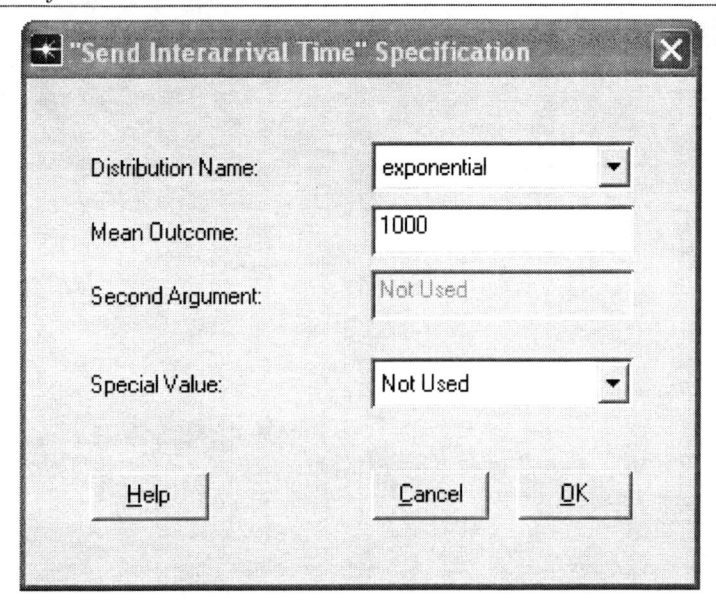

4. We do not care about the Receive Interarrival Time because we will not be receiving messages from the server. Our client will only upload messages. To enforce this, click on the value next to **Receive Group Size**. A new window will appear. Use the pull-down menu to change the **Special Value** field to **Not Used**. Set the **Distribution Name** to **constant** and the **Mean Outcome** value to **0** and click **OK**.

5. Next, we will change the size of the email message to 1 MB. We want a fairly large email message that will create many TCP segments. This way we will be able to observe the effect of window and segment size on our transfer. Click on the value next to **E-mail Size (bytes)**. A new window will appear. Use the pull-down menu to change the **Special Value** field to **Not Used**. Set the **Distribution Name** to **constant** and the **Mean Outcome** value to **1000000** and click **OK**. You should now have the values listed on the following screen.

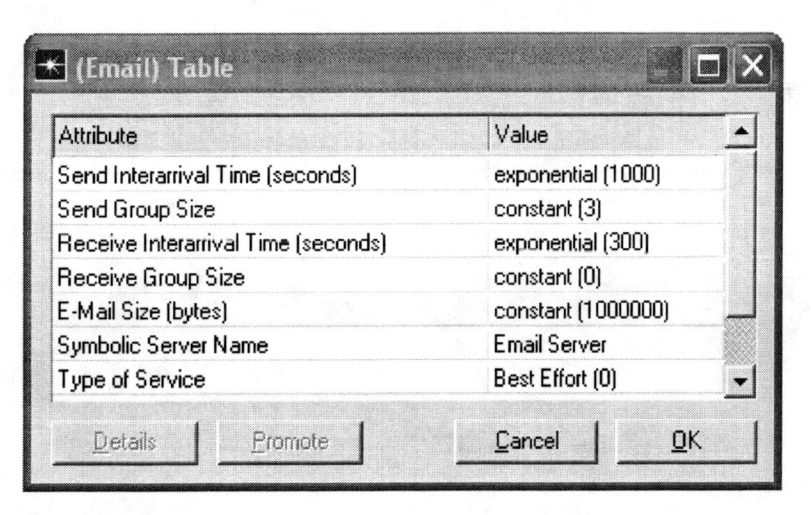

6. Click **OK** to close the **Application Configure** window.

7. Now we will create a new traffic profile. We will call the profile **Email.** The profile will include the **Email Transfer (Heavy)** application. Right click on the **Profile Definition** object and choose **Edit Attributes**. Click on the + next to **Profile Configuration**. Click on the value next to **rows** and choose **Edit**. Change the number of rows to **1**. Click on the value next to the newly added row and choose **Expand Row**. Next to **Profile Name** type **Email**. Choose **Simultaneous** under **Operation Mode**.

67

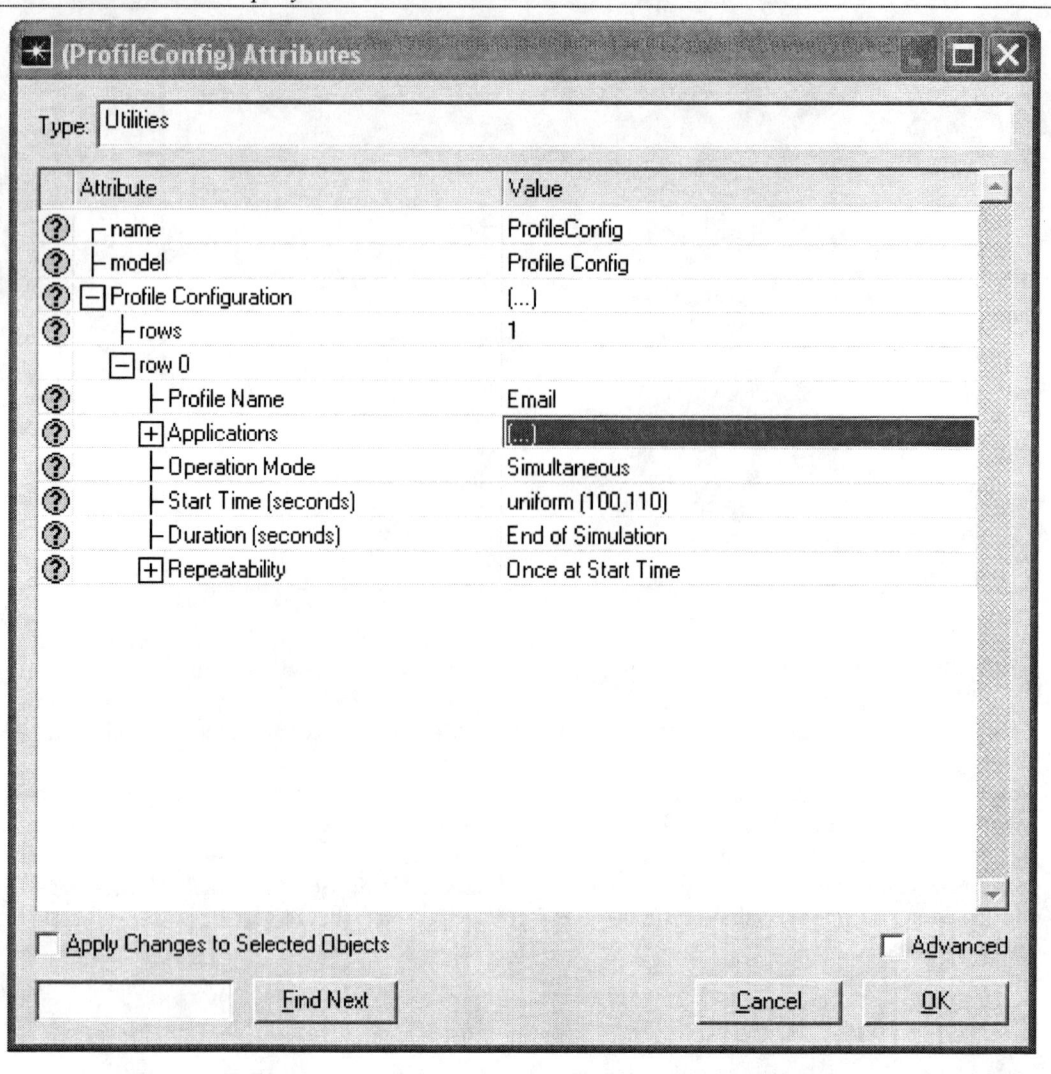

8. Click on the value next to **Applications** and choose **Edit** from the pull-down menu. A new window will appear. Next to **Rows** at the bottom of the window, choose **1** from the pull-down menu. A new row will appear. Click on the value under the **Profile Name** column and choose **Email** from the pull-down menu and click **OK.** Click **OK** again to close the **Profile Configuration** window.

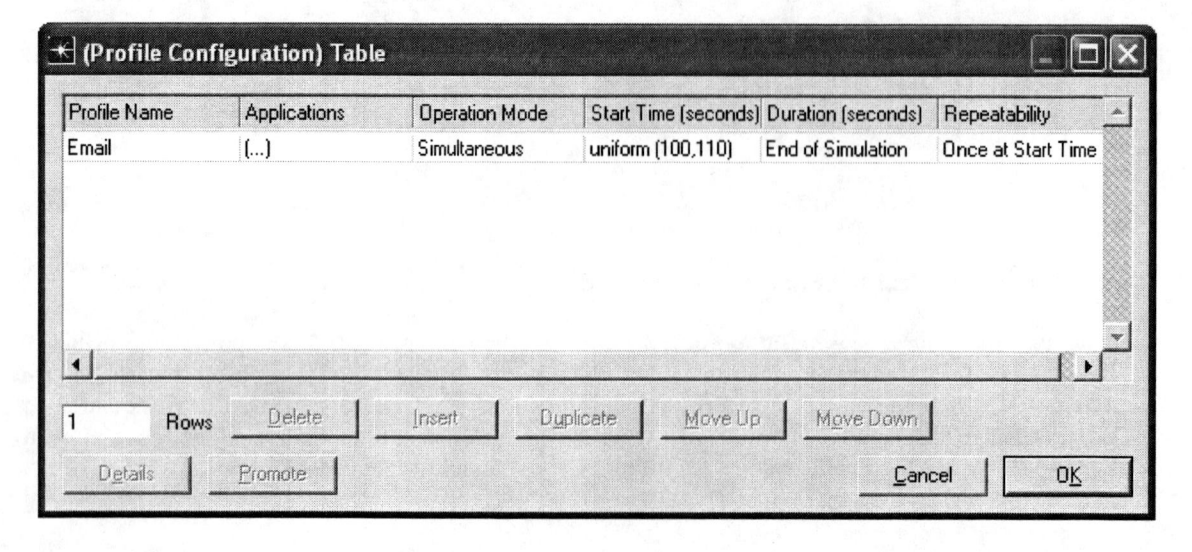

Lab 4 TCP Parameters for Flow Control

9. Now we will give the California **EmailClient** the Email profile. Right click on the **EmailClient** object and choose **Edit Attributes**. Next, click on the **+** next to **Application Supported Profiles** to expand the parameters. Click on the value next to **Application Supported Profiles** and choose **Edit**. A new window will appear.

10. Click on the value listed next to **Rows** and choose **1** from the pull-down menu. You will see a row added to the table. Now click on the value under **Profile Name**. Choose **Email** from the pull-down menu. This configures the client to generate traffic with the Email profile. Click **OK** to close the window.

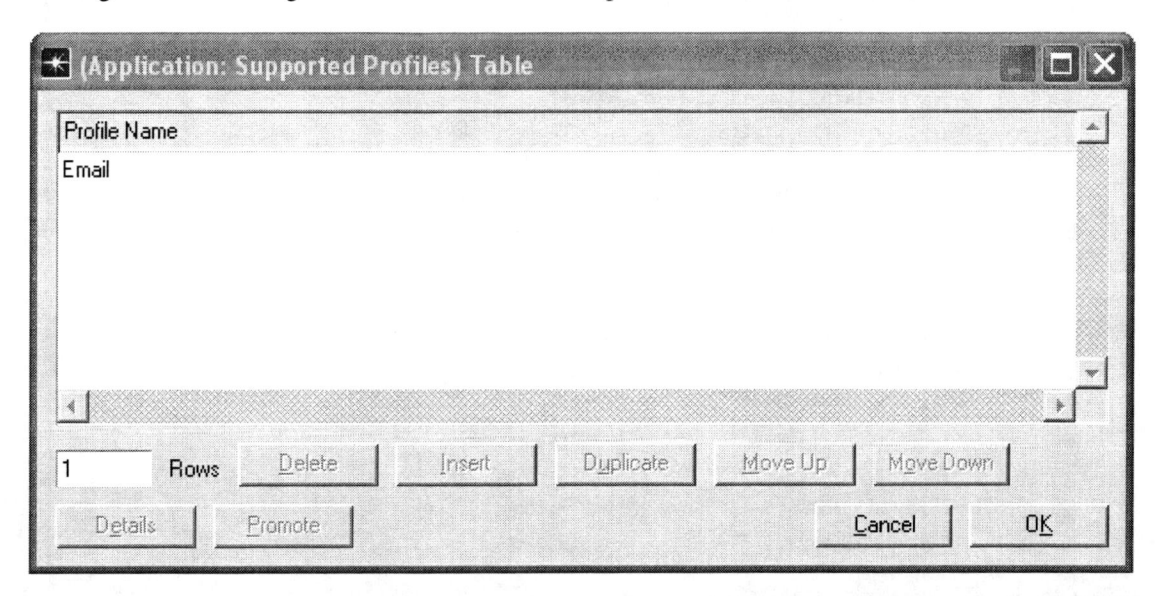

11. Next, we will configure the **Email Client's** TCP parameters. We will use a small segment size of 512 bytes. We'll set the receive buffer to the smallest size it can be (8KB), even though we don't expect to receive anything from the Email server. In the **Email Client Attribute** window, scroll down until you see **TCP Parameters**. Click on the + to expand the parameter list. Click on the value next to **Maximum Segment Size (bytes)** and choose **Edit** from the pull-down menu. Change the value to **512**. Click on the value next to **Receive Buffer (bytes)**. Choose **8760** from the pull-down menu.

69

OPNET Lab Manual to accompany Business Data and Communications

(EmailClient) Attributes

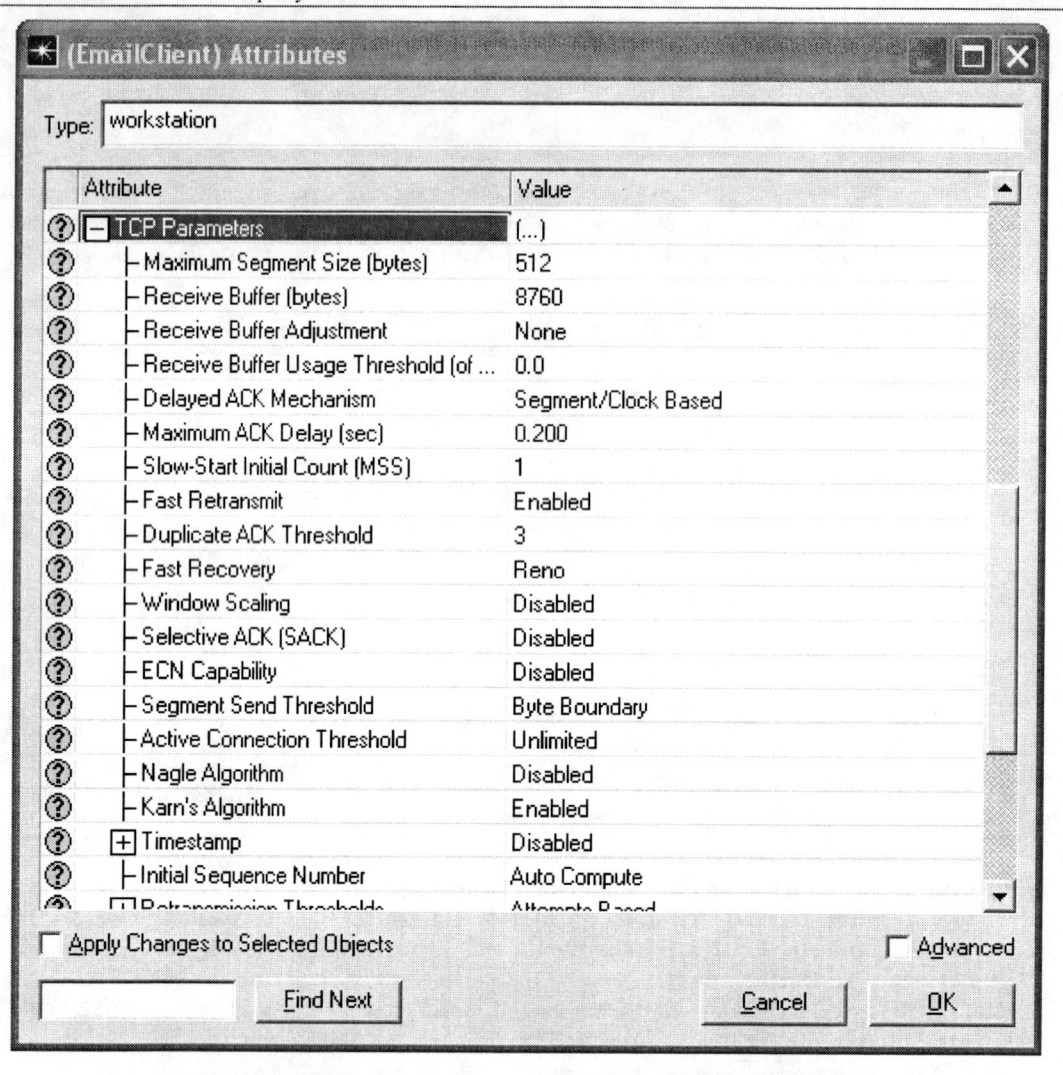

Type: workstation

Attribute	Value
⊟ TCP Parameters	(...)
├ Maximum Segment Size (bytes)	512
├ Receive Buffer (bytes)	8760
├ Receive Buffer Adjustment	None
├ Receive Buffer Usage Threshold (of ...	0.0
├ Delayed ACK Mechanism	Segment/Clock Based
├ Maximum ACK Delay (sec)	0.200
├ Slow-Start Initial Count (MSS)	1
├ Fast Retransmit	Enabled
├ Duplicate ACK Threshold	3
├ Fast Recovery	Reno
├ Window Scaling	Disabled
├ Selective ACK (SACK)	Disabled
├ ECN Capability	Disabled
├ Segment Send Threshold	Byte Boundary
├ Active Connection Threshold	Unlimited
├ Nagle Algorithm	Disabled
├ Karn's Algorithm	Enabled
⊞ Timestamp	Disabled
├ Initial Sequence Number	Auto Compute
⊞ Retransmission Thresholds	Attempts Based

☐ Apply Changes to Selected Objects   ☐ Advanced

[Find Next]   [Cancel]   [OK]

12. Now we will configure the India **EmailServer** to provide **Email** service. Right click on the **EmailServer** object and choose **Edit Attributes.** Next, click on the value next to **Application Supported Services** and choose **Edit.** A new window will appear.

13. Click on the value listed next to **Rows** and choose **1** from the pull-down menu. You will see a row added to the table. Click on the value under **Name.** Choose **Email (Heavy)** from the pull-down menu. Now the EmailServer will provide access to this application. Click **OK** to close the window.

Lab 4   *TCP Parameters for Flow Control*

(Application: Supported Services) Table

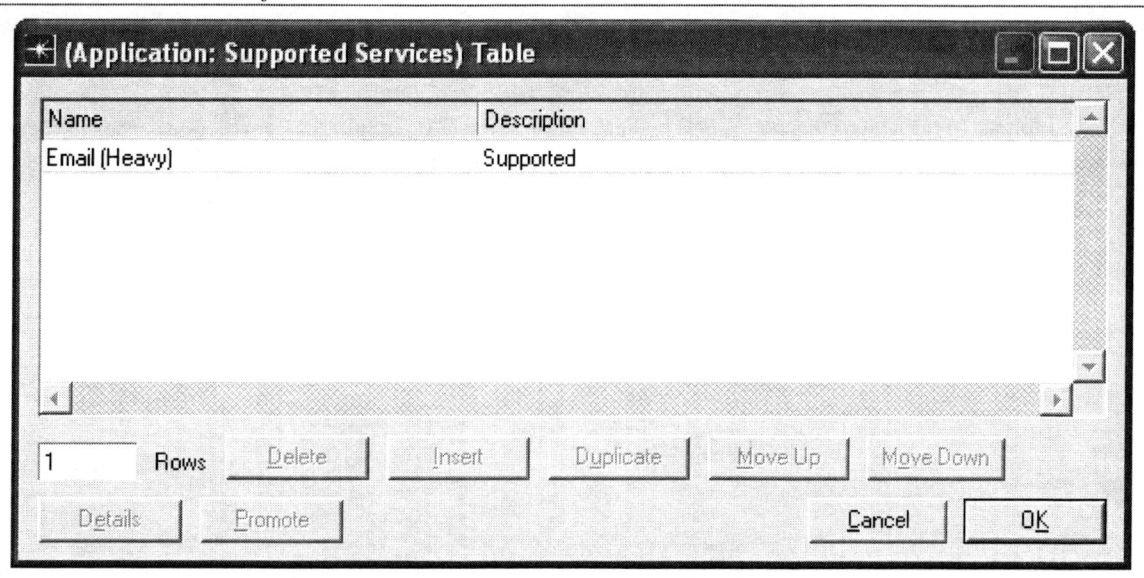

Name	Description
Email (Heavy)	Supported

14. Next, we will configure the **EmailServer's** TCP parameters to be the same as those used by the **EmailClient**. In the **EmailServer Attribute** window, scroll down until you see **TCP Parameters**. Click on the + to expand the parameter list. Click on the value next to **Maximum Segment Size (bytes)** and choose **Edit** from the pull-down menu. Change the value to **512**. Click on the value next to **Receive Buffer (bytes)**. Choose **8760** from the pull-down menu. Click **OK** again to close the **Attribute** window.

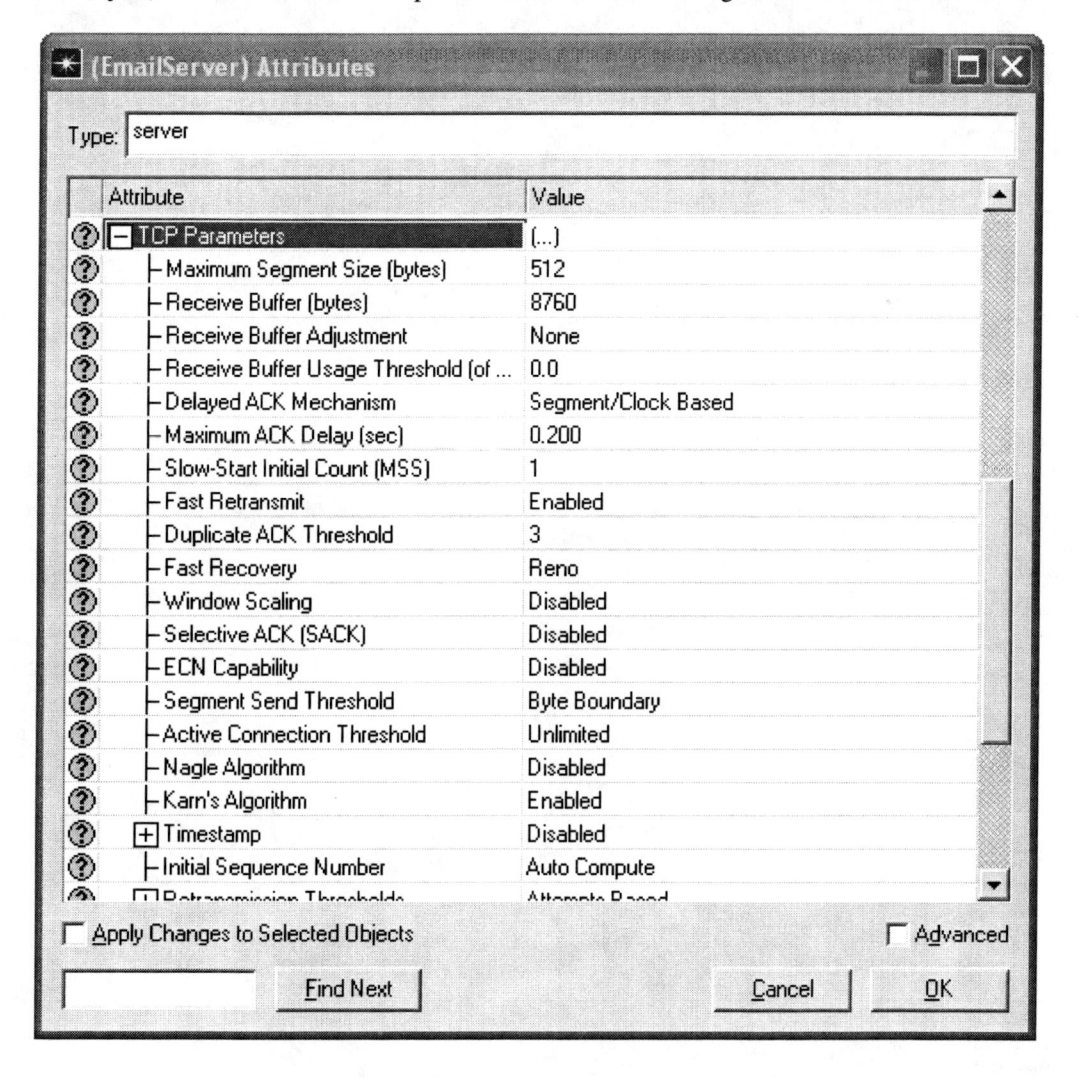

(EmailServer) Attributes

Type: server

Attribute	Value
⊟ TCP Parameters	(...)
├ Maximum Segment Size (bytes)	512
├ Receive Buffer (bytes)	8760
├ Receive Buffer Adjustment	None
├ Receive Buffer Usage Threshold (of ...	0.0
├ Delayed ACK Mechanism	Segment/Clock Based
├ Maximum ACK Delay (sec)	0.200
├ Slow-Start Initial Count (MSS)	1
├ Fast Retransmit	Enabled
├ Duplicate ACK Threshold	3
├ Fast Recovery	Reno
├ Window Scaling	Disabled
├ Selective ACK (SACK)	Disabled
├ ECN Capability	Disabled
├ Segment Send Threshold	Byte Boundary
├ Active Connection Threshold	Unlimited
├ Nagle Algorithm	Disabled
├ Karn's Algorithm	Enabled
├ ⊞ Timestamp	Disabled
├ Initial Sequence Number	Auto Compute
⊞ Retransmission Thresholds	Attempts Based

☐ Apply Changes to Selected Objects   ☐ Advanced

15. Finally, we will configure the **IPNetwork**. Right click on the **IPNetwork** object and choose **Edit Attributes.** Click on the value next to **Packet Latency**. A new window will appear. Choose **Not Used** from the pull-down menu next to the **Special Value** field. For the **Distribution Name** choose **exponential**, and for the **Mean Outcome** enter **.25**. The latency for our packets in this project will always be .25 second.

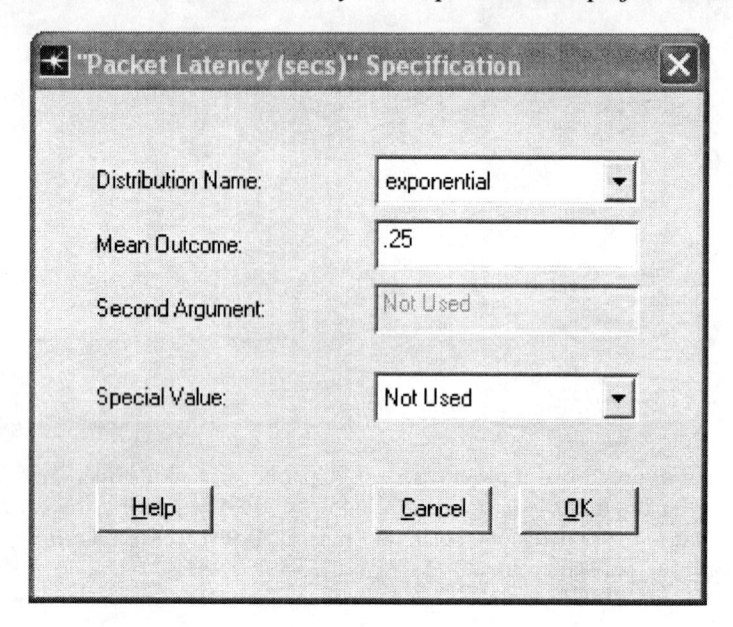

Lab 4 *TCP Parameters for Flow Control*

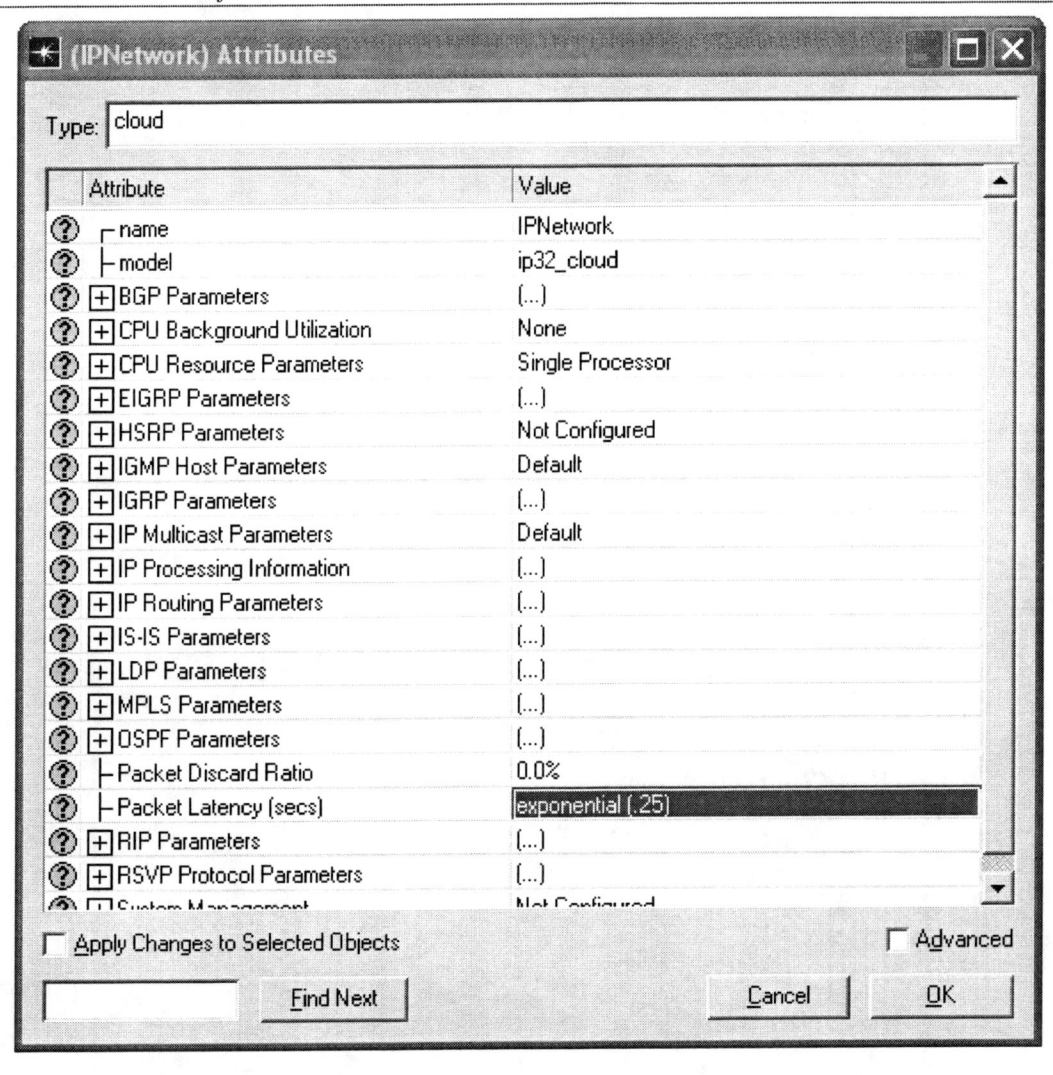

[Screenshot of (IPNetwork) Attributes dialog box]

Type: cloud

Attribute	Value
⌐name	IPNetwork
├model	ip32_cloud
⊞ BGP Parameters	(...)
⊞ CPU Background Utilization	None
⊞ CPU Resource Parameters	Single Processor
⊞ EIGRP Parameters	(...)
⊞ HSRP Parameters	Not Configured
⊞ IGMP Host Parameters	Default
⊞ IGRP Parameters	(...)
⊞ IP Multicast Parameters	Default
⊞ IP Processing Information	(...)
⊞ IP Routing Parameters	(...)
⊞ IS-IS Parameters	(...)
⊞ LDP Parameters	(...)
⊞ MPLS Parameters	(...)
⊞ OSPF Parameters	(...)
├Packet Discard Ratio	0.0%
├Packet Latency (secs)	exponential (.25)
⊞ RIP Parameters	(...)
⊞ RSVP Protocol Parameters	(...)
⊞ System Management	Not Configured

☐ Apply Changes to Selected Objects   ☐ Advanced

[Find Next]   [Cancel]   [OK]

Configuring the Simulation and Choosing Statistics

Now we will choose the statistics that will let us evaluate the performance of our email application given our chosen segment and receive window sizes.

1. Right click somewhere on the project workspace and select **Choose Individual Statistics** from the pull-down menu. Click on the **+** next to **Global** to expand the parameters. Click on the **+** next to **Email**. Click on the box next to **Download Response Time (sec)**, **Upload Response Time (sec)**, **Traffic Received (bytes/sec)**, and **Traffic Sent (bytes/sec)**.

2. Click on the **+** next to **TCP**. Click on the box next to **Delay (sec)**, **Retransmission Count**, and **Segment Delay (sec)**. Click **OK** to close the window.

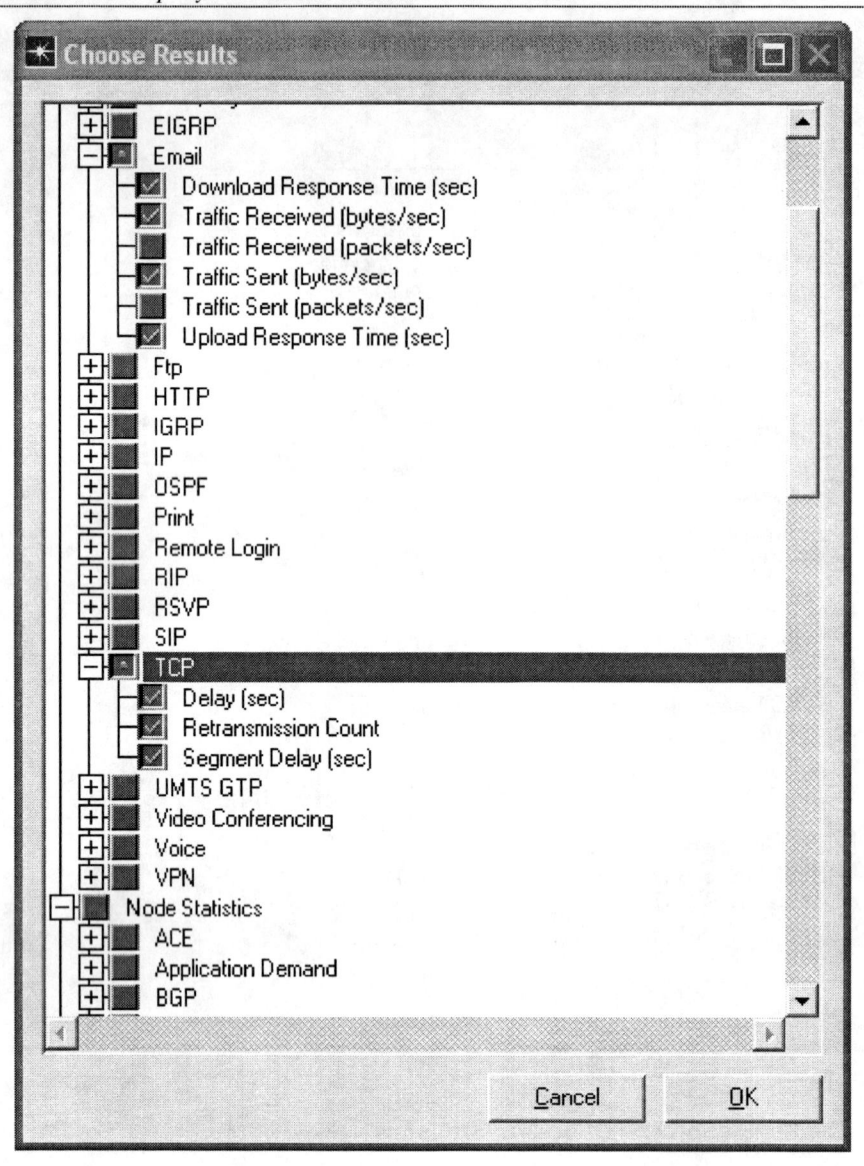

OPNET Lab Manual to accompany Business Data and Communications

[Choose Results dialog showing checked items: Email (Download Response Time (sec), Traffic Received (bytes/sec), Traffic Sent (bytes/sec), Upload Response Time (sec)) and TCP (Delay (sec), Retransmission Count, Segment Delay (sec))]

Running the Simulation

1. Click on the **Configure and Run** Button.  Set the **Duration** value to **10** and choose **minute(s)** for the time unit. Click the **Run** button to run the simulation. Depending on your processor, the simulation should take several minutes to run.

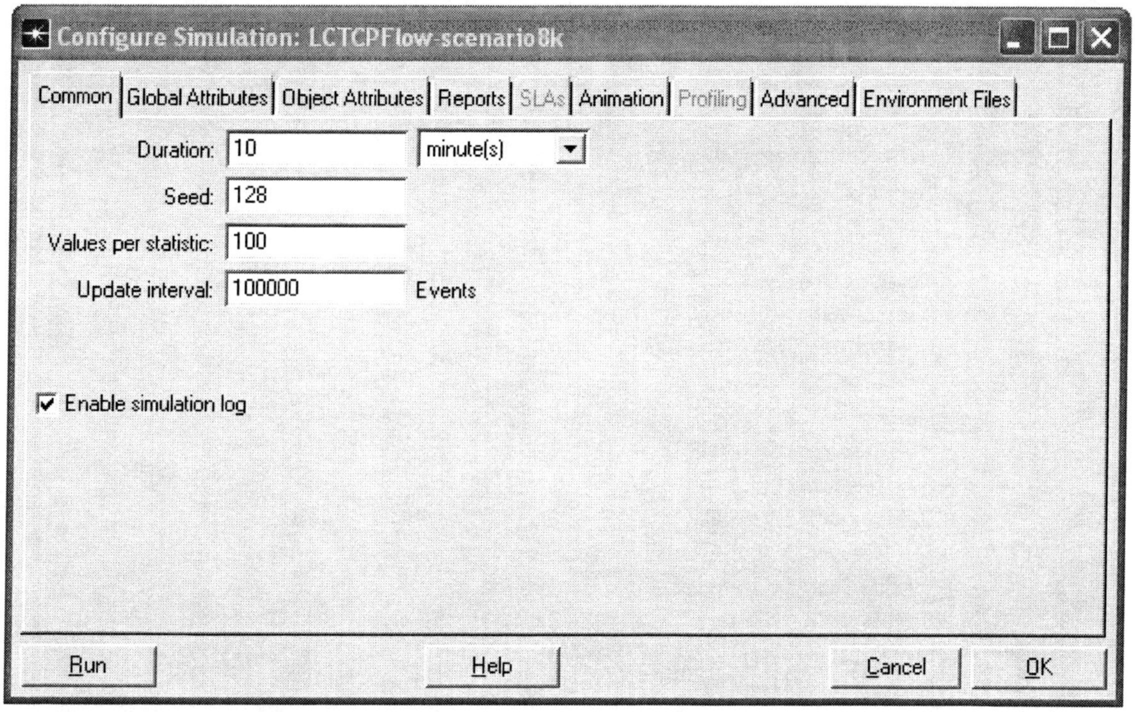
Configure Simulation: LCTCPFlow-scenario8k
Common
Global Attributes
Object Attributes
Reports
SLAs
Animation
Profiling
Advanced
Environment Files
Duration: 10
minute(s)
Seed: 128
Values per statistic: 100
Update interval: 100000
Events
Enable simulation log
Run
Help
Cancel
OK

2. Choose **File/Save** from the Main Menu to save your project.

Creating the Scenarios for 32K Receive Windows

1. Now we will duplicate the scenario and change the TCP Parameters to use a larger receive window. From the Main Menu, choose **Scenarios/Duplicate Scenario**. Give the new scenario a name such as **scenario2_32K.**

2. Right click on the **EmailServer** object and choose **Edit Attributes.** Scroll down until you see **TCP Parameters**. Click on the + to expand the parameter list. Click on the value next to **Receive Buffer (bytes)**. Choose **32768** from the pull-down menu. Click **OK** again to close the **Attribute Window.** Repeat this step to change the **EmailClient's** Receive Buffer to **32768** as well.

OPNET Lab Manual to accompany Business Data and Communications

(EmailServer) Attributes

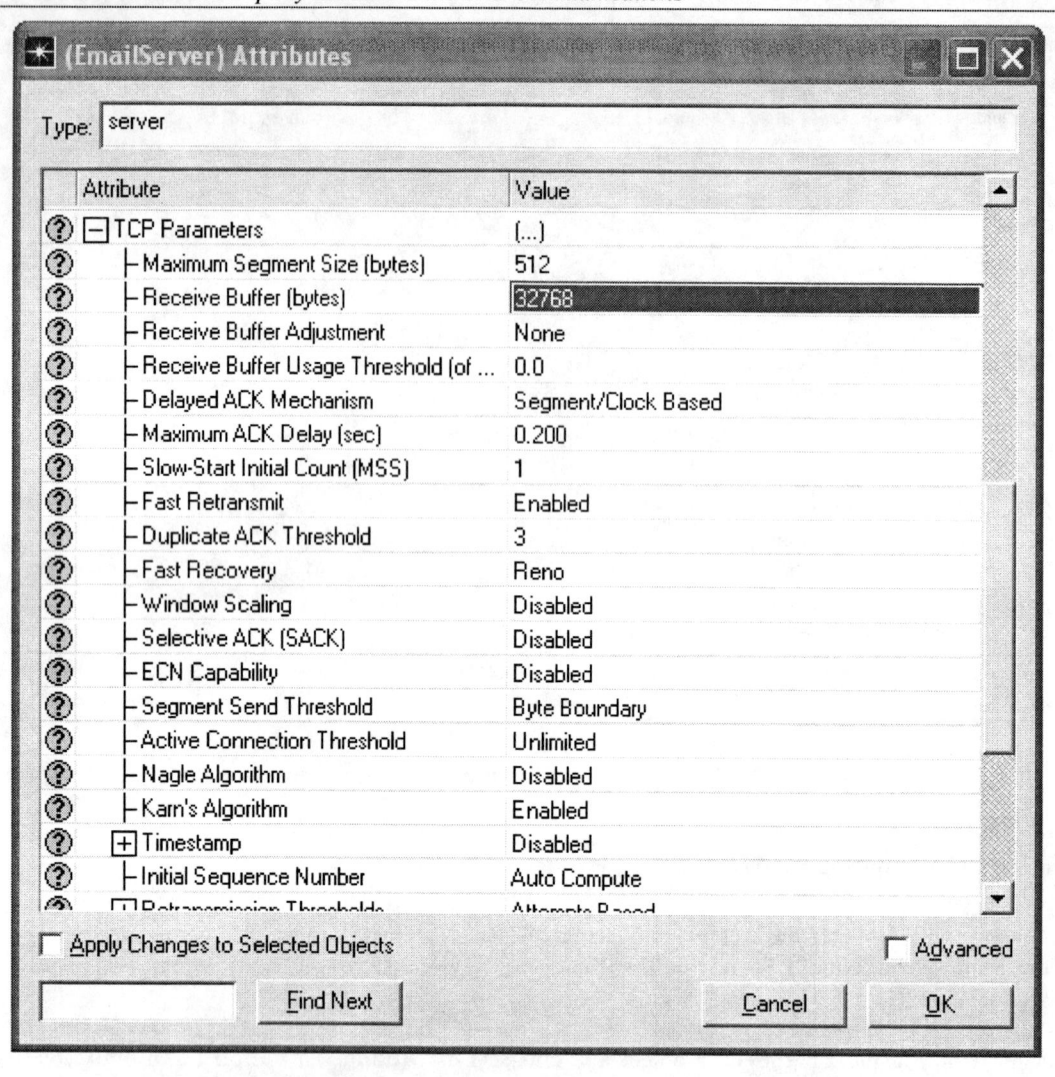

Type: server

Attribute	Value
⊟ TCP Parameters	(...)
— Maximum Segment Size (bytes)	512
— Receive Buffer (bytes)	32768
— Receive Buffer Adjustment	None
— Receive Buffer Usage Threshold (of ...)	0.0
— Delayed ACK Mechanism	Segment/Clock Based
— Maximum ACK Delay (sec)	0.200
— Slow-Start Initial Count (MSS)	1
— Fast Retransmit	Enabled
— Duplicate ACK Threshold	3
— Fast Recovery	Reno
— Window Scaling	Disabled
— Selective ACK (SACK)	Disabled
— ECN Capability	Disabled
— Segment Send Threshold	Byte Boundary
— Active Connection Threshold	Unlimited
— Nagle Algorithm	Disabled
— Karn's Algorithm	Enabled
⊞ Timestamp	Disabled
— Initial Sequence Number	Auto Compute

☐ Apply Changes to Selected Objects   ☐ Advanced

[Find Next]   [Cancel] [OK]

3. Choose **File/Save** from the Main Menu to save your project.

4. Again from the Main Menu, choose **Simulation/Run Discrete Event Simulation**.

Comparing Results

To view results, choose **Results/Compare Results** from the Main Menu. To view results for **Email upload** response time, click on the + next to **Global** to expand the parameters. Click on the + next to **Email** and click next to **Upload Response Time (Seconds)**. You can view the results **As Is** or select **time_Average** from the pull-down menu. Choose **Time Average**. Click on the **Show** button to display the graph in a new window. Your graph should resemble the following figure:

76

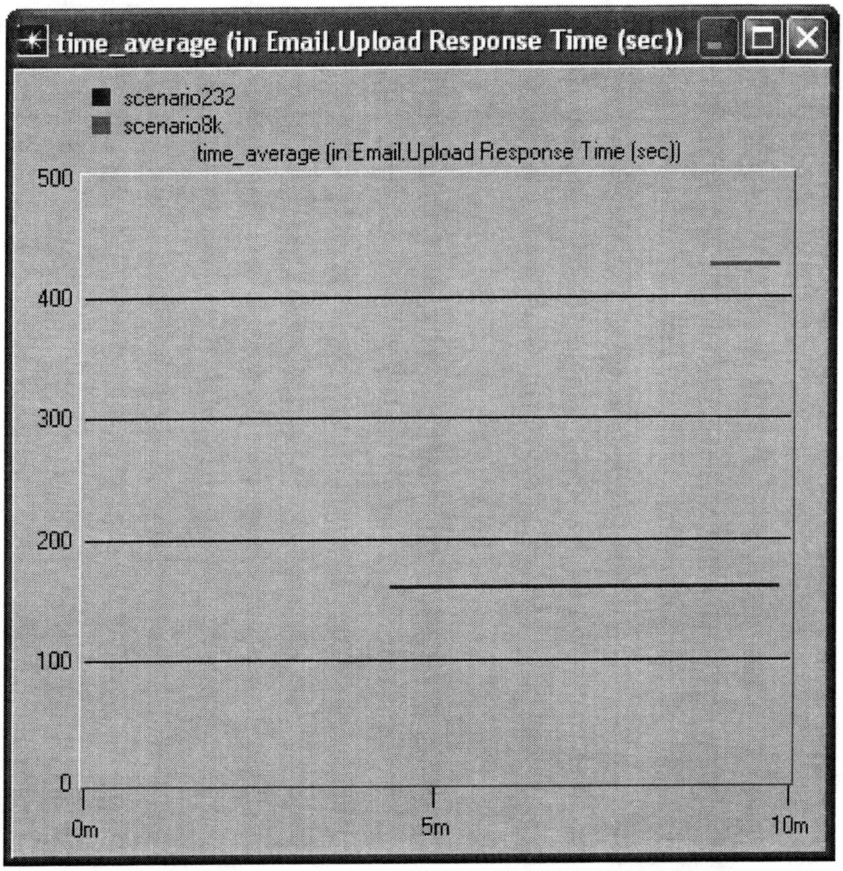
time_average (in Email.Upload Response Time (sec))
scenario232
scenario8k
time_average (in Email.Upload Response Time (sec))
500
400
300
200
100
0
0m
5m
10m

Notice that the response time is much higher for the scenario with the 8K receive window.

Click on the **X** to close the window. The **Close Analysis** window will appear asking you if you want to hide the panel or permanently delete it. Click on the **Delete** button to delete the window.

You may discard the results of your simulations, rerun them, and manage your scenarios by doing the following:

Choose **Scenarios/Manage Scenarios** from the Main Menu. A window will appear with the names of your scenarios, the length of time that the simulation ran, and the status of the data.

By clicking on a scenario name, you may choose to **Discard Results** or **Collect Results**. By clicking on the values under the **Results** column, you may choose **recollect** or **collect** from the pull-down menu. Click **OK** to close the window.

Use the results you obtained in your simulations to answer the following questions.

Questions

1. Based on your observations of email upload response time for window sizes of 8KB and 32KB, are larger or smaller windows better? Create a new scenario and change the receive window to its maximum size of 65535. What is the new upload response time? How does it compare to your previous values?
2. Look at the upload response time for the scenario where the receive window was set to 8KB. Calculate the TCP throughput for this scenario by dividing the email size (1MB) by this value. In the same manner, calculate the TCP throughput for the scenario when the receive window was set to 32KB. Which window size provides the largest throughput?
3. For the 8KB and 32KB receive window scenarios, change your simulation to capture node statistics. Choose Email Client statistics. Run your simulation again and look at the **Traffic Sent** (bytes/sec). Does this value match the throughput you calculated in the preceding problem?

4. For the 8KB and 32KB receive window scenarios, look at the statistics for TCP Delay and TCP Segment Delay. Explain what you observe.
5. Duplicate the scenario that uses receive windows of 32KB. In the new scenario, increase the TCP segment size to 1024 bytes. Does increasing the segment size increase or decrease response time?
6. Would the simulation results be different if the link speed was increased from DS1 to DS3 (45Mbps)? Change the link speed and verify your answer.

Lab 5   ATM and Quality of Service

Objective
1. To compare voice and video application performance under various Quality of Service (QoS) parameters in an ATM Wide Area Network (WAN).

Motivation
Telemedicine involves medical experts collaborating on medical research and remote diagnosis. The Cleveland Clinic located in Cleveland, Ohio and the Moscow Clinic located in Moscow, Russia have recently joined together in diagnosing and fighting acute heart disease. Video Conferencing and Voice over IP (VoIP) services will allow the two clinics to communicate effectively and efficiently. The clinics have contracted with a network service provider that will carry their voice and video conferencing traffic over an Asynchronous Transfer Mode (ATM) WAN. ATM offers several quality of service levels to support real-time traffic. The clinics would like to run a test to discover which quality of service profile is best for the applications they will be using.

Description
Asynchronous Transfer Mode (ATM) was designed to offer various classes of Quality of Service (QoS) for real-time data. ATM differs from IP networks in that it is connection-oriented and cell-switched. ATM cells are small and have a fixed length of 53 bytes. Fixed length cells are easy to process and allow for continuous delivery, which is specifically important for real-time data. When IP variable length packets are sent over ATM networks, the ATM Adaptation Layer (AAL) must segment the packets into cells and reassemble them at the destination. There are several AAL protocols that have been designed for different types of data. AAL3/4 and AAL5 were designed for IP packet data over ATM. AAL1 and AAL2 were designed for fixed bit rate applications, such as voice traffic. There are five ATM QoS classes: Constant Bit Rate (CBR), Unspecified Bit Rate (UBR), Available Bit Rate (ABR), Variable Bit Rate Real-Time (VBR-rt), and Variable Bit Rate Non-Real-Time (VBR-nrt). CBR was designed for voice traffic and is the most commonly used class. UBR is used for packet data and is ATM's best effort service class.

In this lab, we will create a Wide Area Network (WAN) that spans from a subnet in Cleveland, Ohio to a subnet in Moscow, Russia. The two subnets will be configured to send video conferencing and voice traffic. Traffic sent and delay for these applications will be observed for both CBR and VBR service classes. We will also observe how performance varies when AAL2 and AAL5 protocols are used.

Follow the next set of directions to create an OPNET model, run simulations, and view results. Questions regarding the lab are listed at the end.

Create a New Project

1. Start the **OPNET IT Guru Academic Edition Application.** Create a **New Project** by selecting **File/New/Project** from the Main Menu. Click **OK**.

2. Give the project a name such as **your_initials_atm.** Give the scenario a name such as **CBR**. Click **OK**. You should see the **Startup Wizard Initial Topology** dialog box.

3. Verify that **Create Empty Scenario** is selected in the **Initial Topology Dialog Box**. Click **Next**. You will now see the **Network Scale** dialog box.

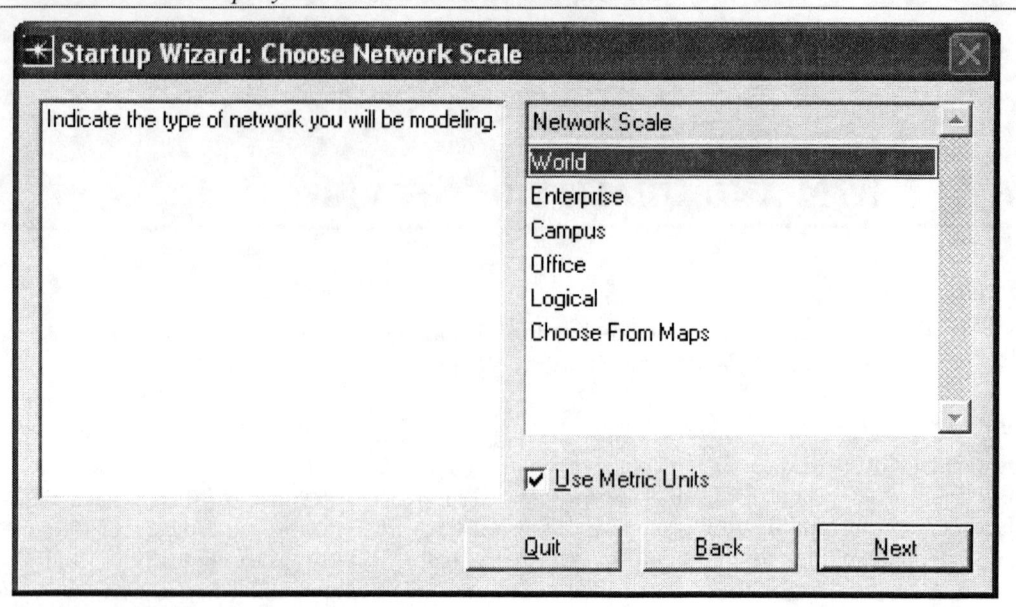

4. Choose **World** from the **Network Scale** list. Click **Next**. Choose **world** as your map and click **Next**. You should see the **Select Technologies** dialog box.

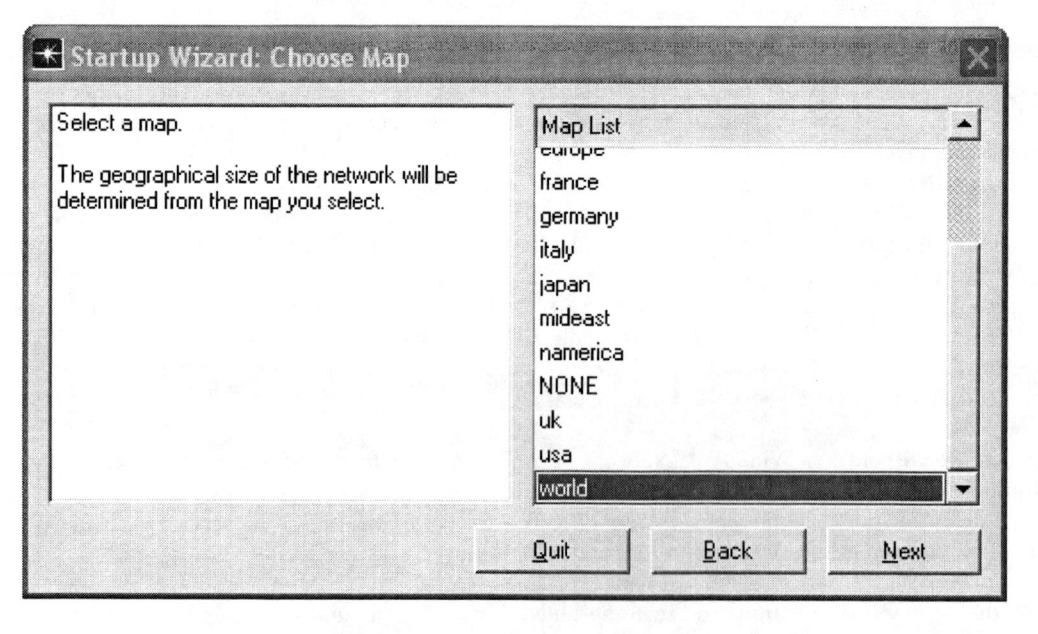

5. Scroll down until you see **atm_advanced** under the **Model Family** column. Under the **Include column**, click on the box next to **atm_advanced**. This will change the "no" to "yes" indicating that **atm_advanced** technologies will be included. Click **Next**.

Lab 5   ATM and Quality of Service

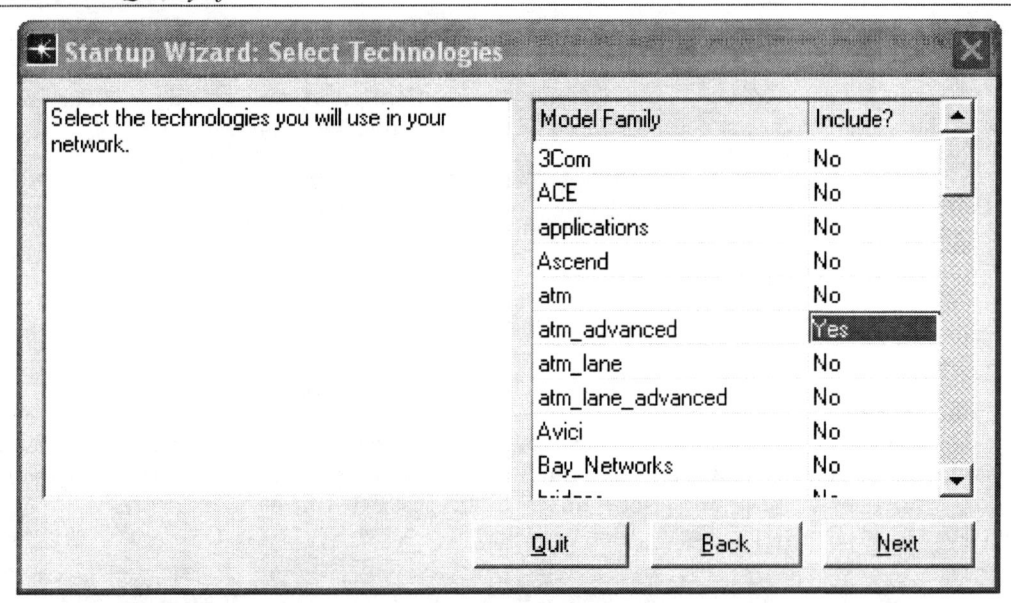

6. In the **Start up Wizard Review** dialog box, verify that the **atm_advanced** technologies are chosen and that the scale is **World** and the map is world. Click **OK**. You should see the **Object Palette** window.

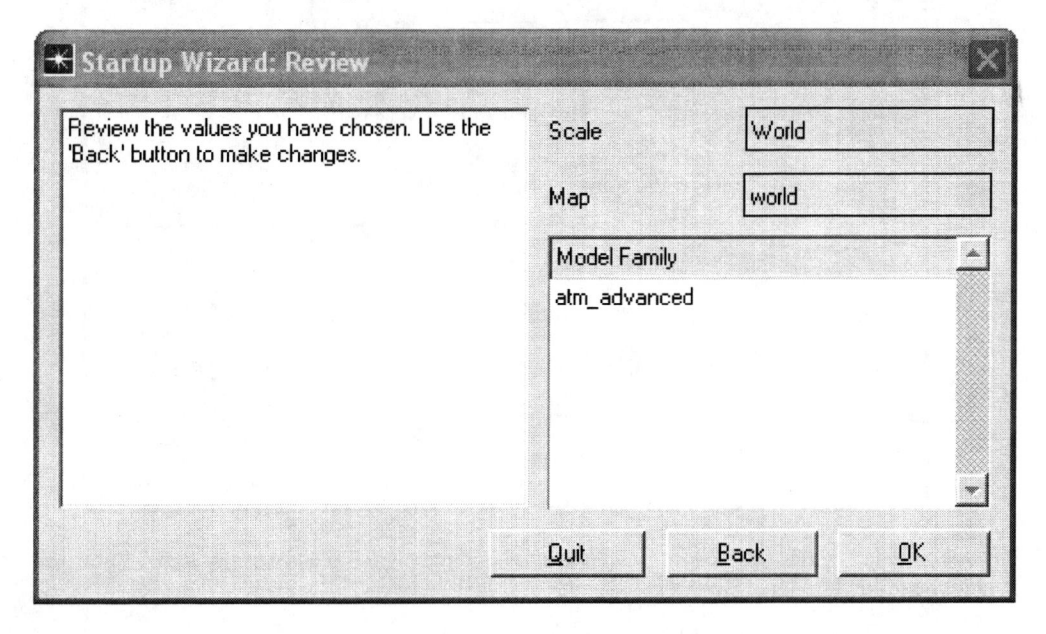

Building the ATM Network

1. We will now create the ATM network. We will start by placing one subnet object on the workspace in the area of Cleveland, Ohio. From the **Object Palette**, click on the **subnet** object to select it. Drag it to the project grid in the area of Ohio and click to paste a copy of it on the map. Right click to release the subnet object when you are through.

2. Right click on the **subnet** and choose **Set Name** from the pull-down menu. Name the subnet **ClevelandClinic**.

3. We will begin by creating the Cleveland Clinic's subnet, which will consist of ATM workstations and an ATM switch. The switch will be connected to the ATM WAN. We will later duplicate this subnet to create the MoscowClinic subnet. Double click on the **ClevelandClinic subnet** object to descend into the subnet.

4. We will create the network using **Rapid Configuration**. We will create a star network with eight ATM clients connected to an ATM switch. We will choose a type of ATM link that allows us to configure its speed. From the Main Menu, choose **Topology/Rapid Configuration**. A window will appear prompting you to choose the topology or network shape. Choose **Star** from the pull-down menu.

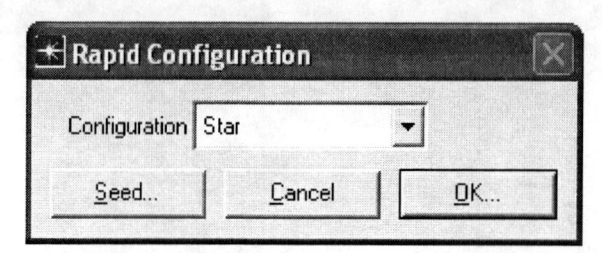

5. A new window will appear which will allow you to choose the nodes in the network. For the **Center Node Model** choose **atm8_crossconnect_adv** from the pull-down menu. This is our switch. For the **Periphery Node Model** choose **atm_uni_client_adv** and change the **Number** to **8**. This will create eight workstations on the network. For the **Link Model**, choose **ATM_adv**. Click **OK**.

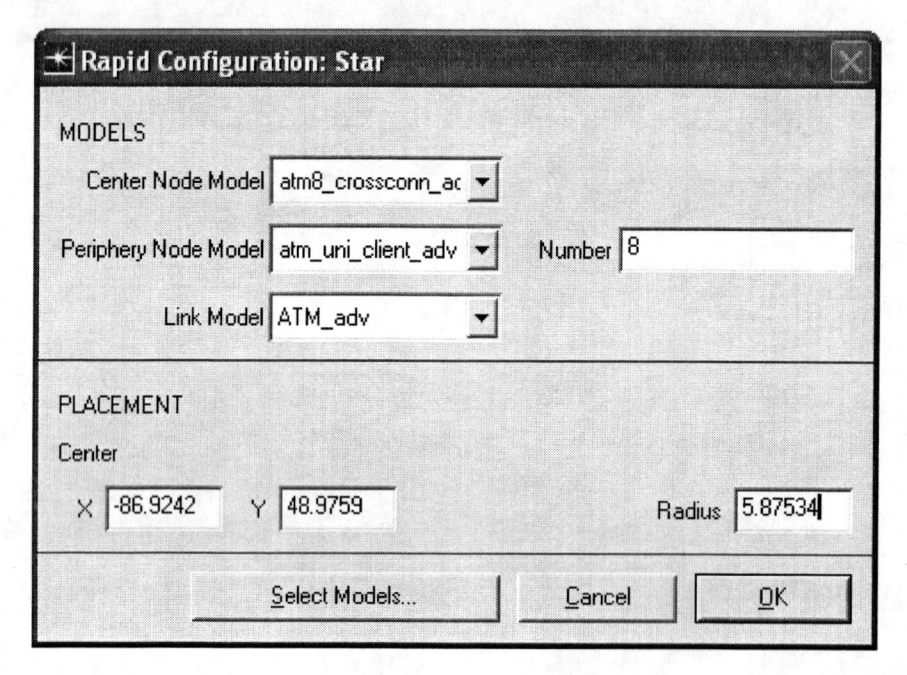

6. Right click on the switch and choose **Set Name**. Name the switch **ClevelandSwitch**. Your project should resemble the following figure:

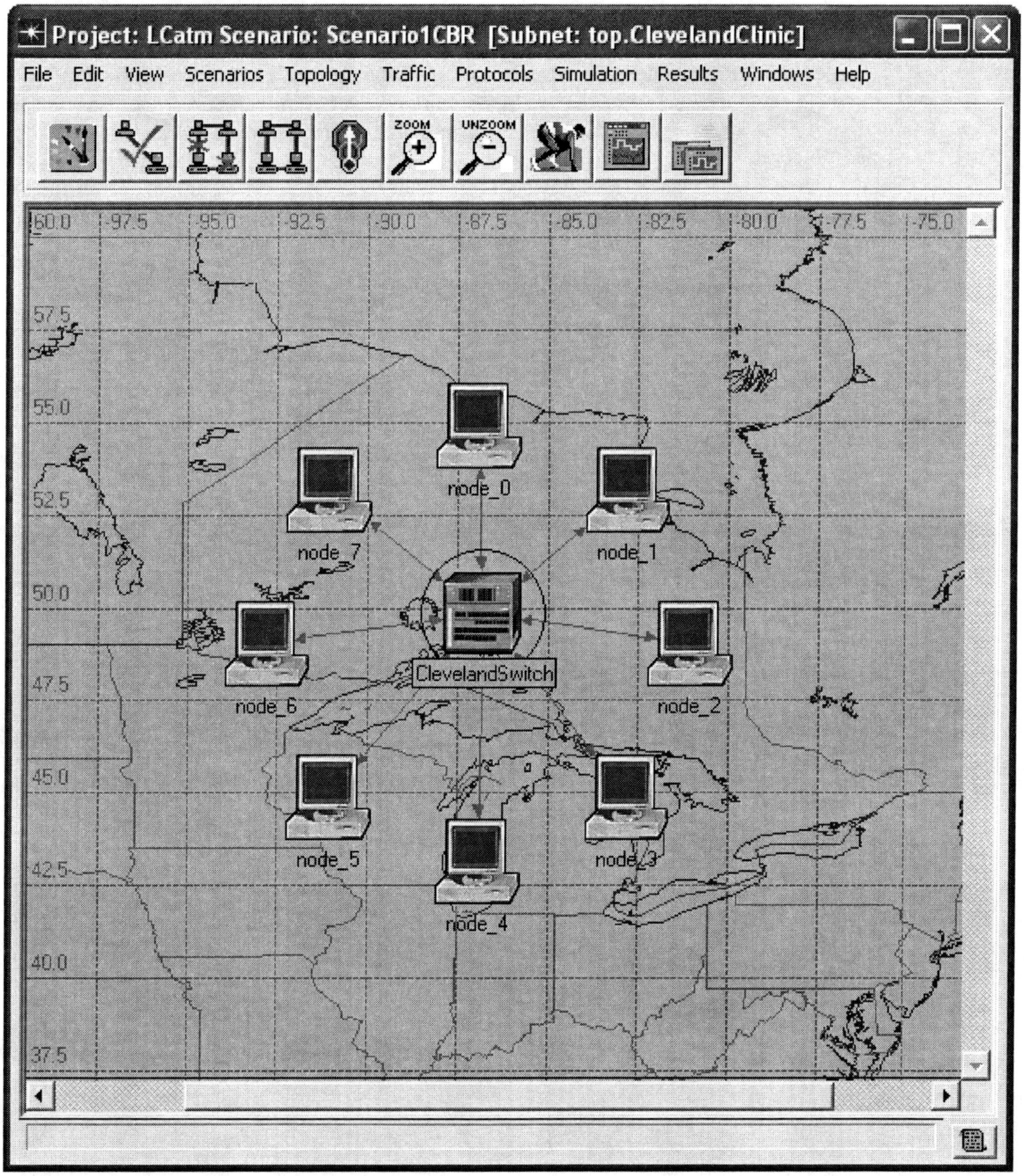
Project: LCatm Scenario: Scenario1CBR [Subnet: top.ClevelandClinic]
File Edit View Scenarios Topology Traffic Protocols Simulation Results Windows Help
ZOOM
UNZOOM
node_0
node_1
node_2
node_3
node_4
node_5
node_6
node_7
ClevelandSwitch

7. Click on the **Arrow** button  to go up to the next level of the project hierarchy.

8. Now we will create the Moscow subnet by copying the Cleveland subnet and pasting it in the appropriate area on the map. Click on the Cleveland subnet object and choose **Edit/Copy** from the Main Menu. Choose **Edit/Past** and paste a copy of the subnet in the area of Moscow, Russia. Right click to release the object. Right click on the new **subnet** object and choose **Set Name** from the pull-down menu. Name the subnet **MoscowClinic**. Double click on the subnet object to descend into the subnet. Right click on the switch object and choose **Set Name**. Name the switch **MoscowSwitch**.

9. Click on the **Arrow** button  to go up to the next level of hierarchy in the project.

10. From the **Object Palette**, click on the **atm32_cloud_adv** object to select it. Drag it to a spot between the two subnets and click to paste it. Right click to release the object. Right click on the **atm32_cloud_adv** object and choose **Set Name** from the pull-down menu. Name the object **ATM Network**.
11. We will now connect our subnets to the ATM network via **ATM_adv** links. This type of link is configurable, so we can select a particular bit rate for our simulation. From the **Object Palette**, click on the **ATM_adv** object to select it. Connect the ATM network object to the **ClevelandClinic** subnet. The **Select Node** window will appear. This allows you to choose which object to link to within the subnet. Choose the **ClevelandClinic.ClevelandSwitch** as shown next. Repeat the step to connect the ATM network object to the **MoscowClinic.MoscowSwitch**. Right click to release the object.

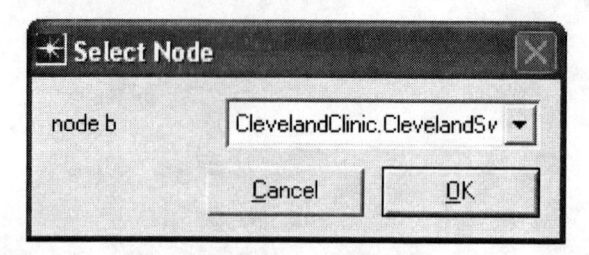

12. We will add an **Application Config** and a **Profile Config** object to the workspace. The application object will allow us to choose the applications we want to include in our simulation. We will use the voice and video conferencing applications. The Profile Configuration object allows us to use default traffic profiles for different types of users or to create new profiles. We will use the Multimedia User profile.

13. From the **Object Palette**, click on the **Profile Config** object, and click to paste it on the project grid. Right click to release the object. Right click on the **Profile Config** object and choose **Set Name**. Name the object **Profile**. Again from the **Object Palette**, click on the **Application Config** object and click to paste it on the project grid. Right click to release the object. Right click on the **Application Config** object and choose **Set Name**. Name the object **Application**. Verify that your project looks like the following figure:

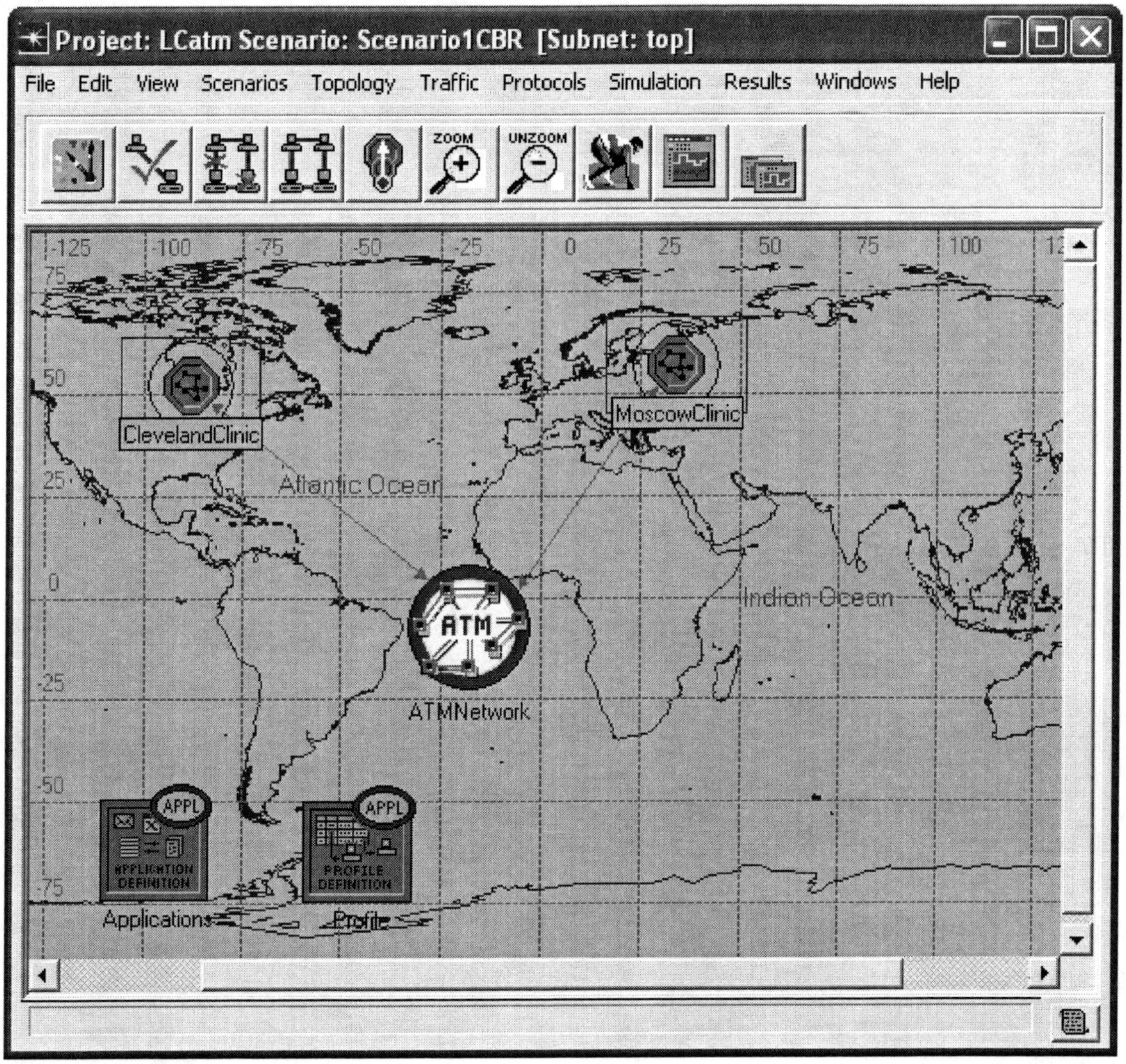
Project: LCatm Scenario: Scenario1CBR [Subnet: top]
File Edit View Scenarios Topology Traffic Protocols Simulation Results Windows Help
ZOOM
UNZOOM
ClevelandClinic
MoscowClinic
Atlantic Ocean
Indian Ocean
ATM
ATMNetwork
APPL
APPLICATION DEFINITION
PROFILE DEFINITION
Applications
Profile

14. From the main menu, choose **File/Save** to save your project.

Configuring the Network

1. We will start by configuring the application object to provide the applications we are interested in using in our simulation. OPNET IT Guru allows you to define new applications or use default descriptions for standard applications. We are interested in videoconferencing and voice applications, which are predefined. Right click on the **Application** object and choose **Edit Attributes.** Click on the value next to **Application Configuration.** Choose **Default** from the pull-down menu. This will allow us to use default applications during our simulation. Click **OK**.

2. Now, we will configure the **Profile** object to use the default profiles. Right click on the **Profile** object and choose **Edit Attributes.** Click on the value next to **Profile Configuration.** Choose **Sample Profiles** from the pull-down menu. This will include a **Multimedia User** profile, which will include **Video Conferencing (Light)** and **Voice over IP Call (PCM Quality Heavy)** applications.

(Profile) Attributes

Type: Utilities

Attribute	Value
⌐ name	Profile
├ model	Profile Config
─ Profile Configuration	(...)
├ rows	5
⊞ row 0	Engineer,(...),Simultaneous,uniform (100,110),En...
⊞ row 1	Researcher,(...),Simultaneous,uniform (100,110),...
⊞ row 2	E-commerce Customer,(...),Simultaneous,uniform ...
⊞ row 3	Sales Person,(...),Simultaneous,uniform (100,110...
⊟ row 4	
├ Profile Name	Multimedia User
⊞ Applications	(...)
├ Operation Mode	Simultaneous
├ Start Time (seconds)	uniform (100, 110)
├ Duration (seconds)	End of Simulation
⊞ Repeatability	Once at Start Time

☐ Apply Changes to Selected Objects   ☐ Advanced

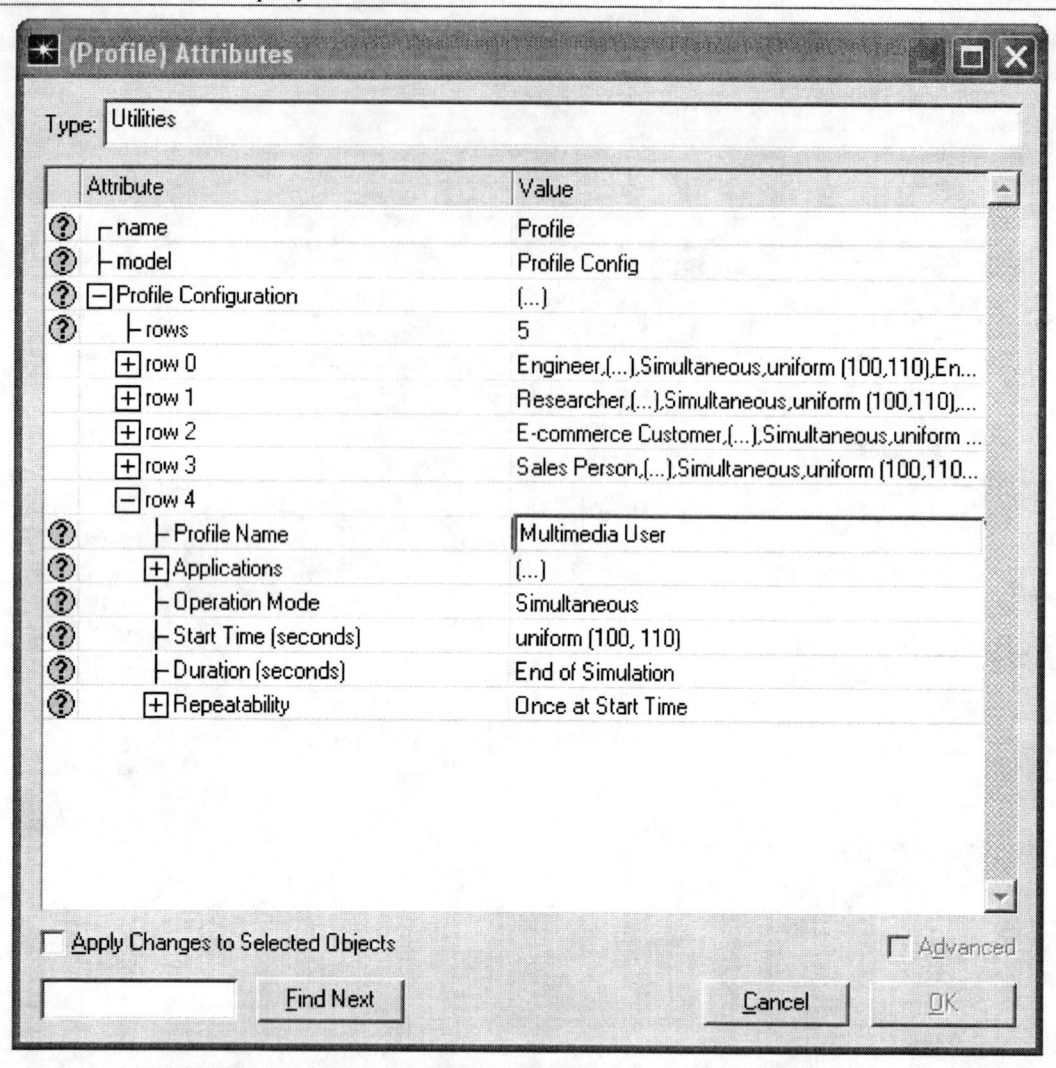

[Find Next]   [Cancel] [OK]

3. Click on the + to expand the **Multimedia User** Parameters. We will change the start time so that our simulation starts up more quickly. Click on **Start Time**. A new window will appear. Change the **Distribution Name** to **uniform**, the **Minimum Outcome** to **5**, and the **Maximum Outcome** to **10**. Now the simulation will start after 5 to 10 seconds. Click **OK** to close the window. Click **OK** to close the **Profile Attribute** window.

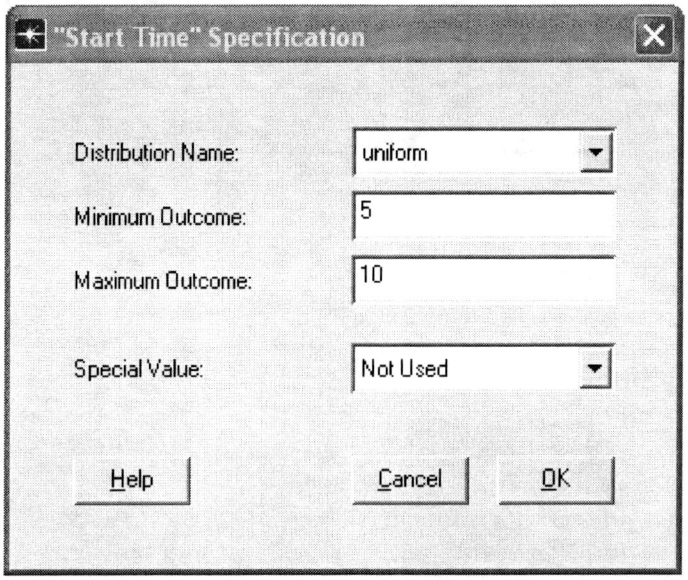
"Start Time" Specification
Distribution Name:
uniform
Minimum Outcome:
5
Maximum Outcome:
10
Special Value:
Not Used
Help
Cancel
OK

4. Now we will give the nodes in the Moscow Clinic the Multimedia User Profile. Double click on the Moscow Clinic subnet to descend into the subnet. Click on one of the **workstation** objects and choose **Select Similar Nodes**. Right click on one of the selected workstation objects and choose **Edit Attributes.** Click on the box next to **Apply Changes to Selected Objects.** Next, click on the + next to **Application Supported Profiles** to expand the parameters. Click on the value next to **Application Supported Profiles** and choose **Edit.** A new window will appear.

5. Click on the value listed next to **Number of Rows** (displayed in the bottom left hand corner), and choose **1** from the pull-down menu. You will see a row added to the table. Now click on the value under **Profile Name.** Choose **Multimedia User** from the pull-down menu. This configures the workstations to create traffic with the Multimedia User traffic profile. Click **OK.**

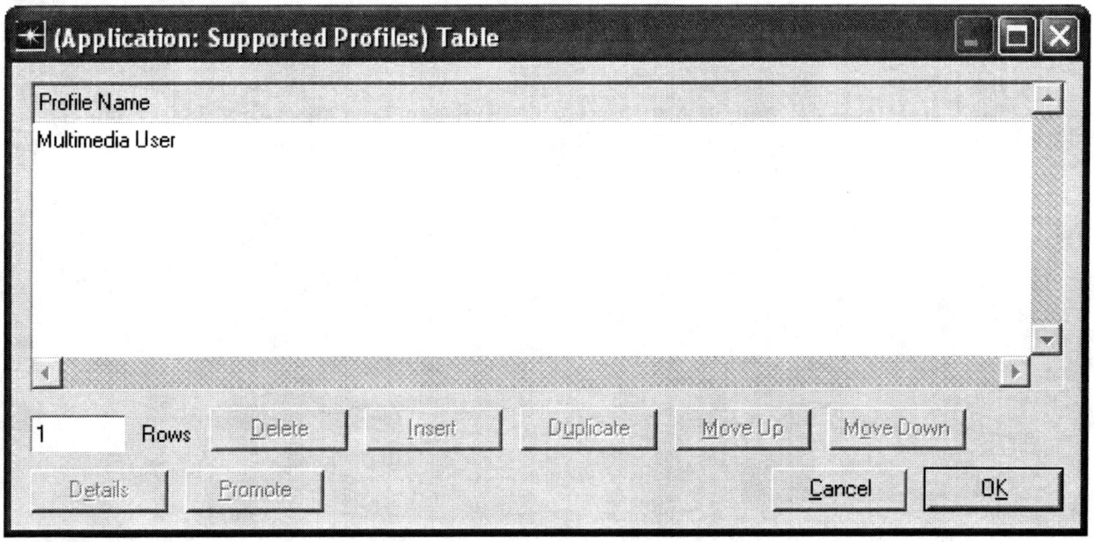
(Application: Supported Profiles) Table
Profile Name
Multimedia User
1
Rows
Delete
Insert
Duplicate
Move Up
Move Down
Details
Promote
Cancel
OK

6. Next, we must choose the ATM Quality of Service that we would like for this simulation. We will start with **Constant Bit Rate (CBR)**. Click on the value next to **ATM Application Parameters,** and choose **CBR Only**. Click on the + next to **ATM Application Parameters,** and verify that **CBR** is listed next to **Category** and that **AAL5** is listed next to **Application: Transport Protocol.** Click on **OK** to close the window.

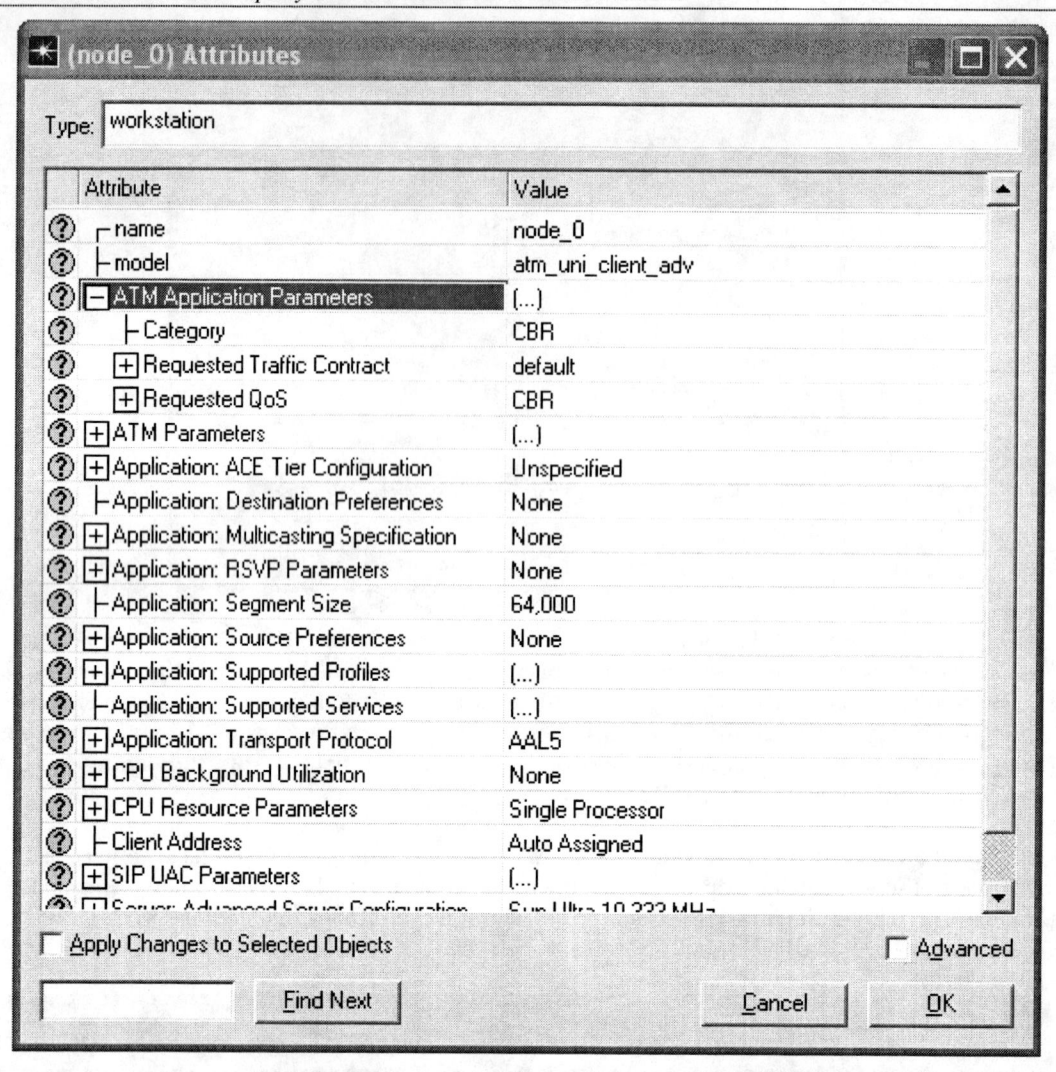

Attribute	Value
ⓘ ─name	node_0
ⓘ ─model	atm_uni_client_adv
ⓘ ⊟ ATM Application Parameters	(...)
ⓘ ─Category	CBR
ⓘ ⊞ Requested Traffic Contract	default
ⓘ ⊞ Requested QoS	CBR
ⓘ ⊞ ATM Parameters	(...)
ⓘ ⊞ Application: ACE Tier Configuration	Unspecified
ⓘ ─Application: Destination Preferences	None
ⓘ ⊞ Application: Multicasting Specification	None
ⓘ ⊞ Application: RSVP Parameters	None
ⓘ ─Application: Segment Size	64,000
ⓘ ⊞ Application: Source Preferences	None
ⓘ ⊞ Application: Supported Profiles	(...)
ⓘ ─Application: Supported Services	(...)
ⓘ ⊞ Application: Transport Protocol	AAL5
ⓘ ⊞ CPU Background Utilization	None
ⓘ ⊞ CPU Resource Parameters	Single Processor
ⓘ ─Client Address	Auto Assigned
ⓘ ⊞ SIP UAC Parameters	(...)

7. The switch must also be configured to use **CBR**. Right click on the **switch** object and choose **Edit Attributes**. Click on the + next to **ATM parameters** to expand the values. Click on the value next to **Queue Configuration** and choose **CBR Only**. Click **OK** to close the window.

Lab 5   ATM and Quality of Service

(MoscowClinic) Attributes

Type: switch

Attribute	Value
─ name	MoscowClinic
├ model	atm8_crossconn_adv
─ ATM Parameters	(...)
├ Address	Auto Assigned
├ Queue Configuration	CBR Only
├ Per-Port Configuration	(...)
├ VC Routes Report	Do Not Export
├ SPVC Reroute Parameters	Default
├ Policing Parameters	Disabled
├ ABR Feedback Scheme	None
├ CAC Algorithm	Default
├ Connection Limit	Unlimited
├ Processing Parameters	Default
├ SSCOP Parameters	Default
─ ATM Routing Parameters	(...)

☐ Apply Changes to Selected Objects   ☐ Advanced

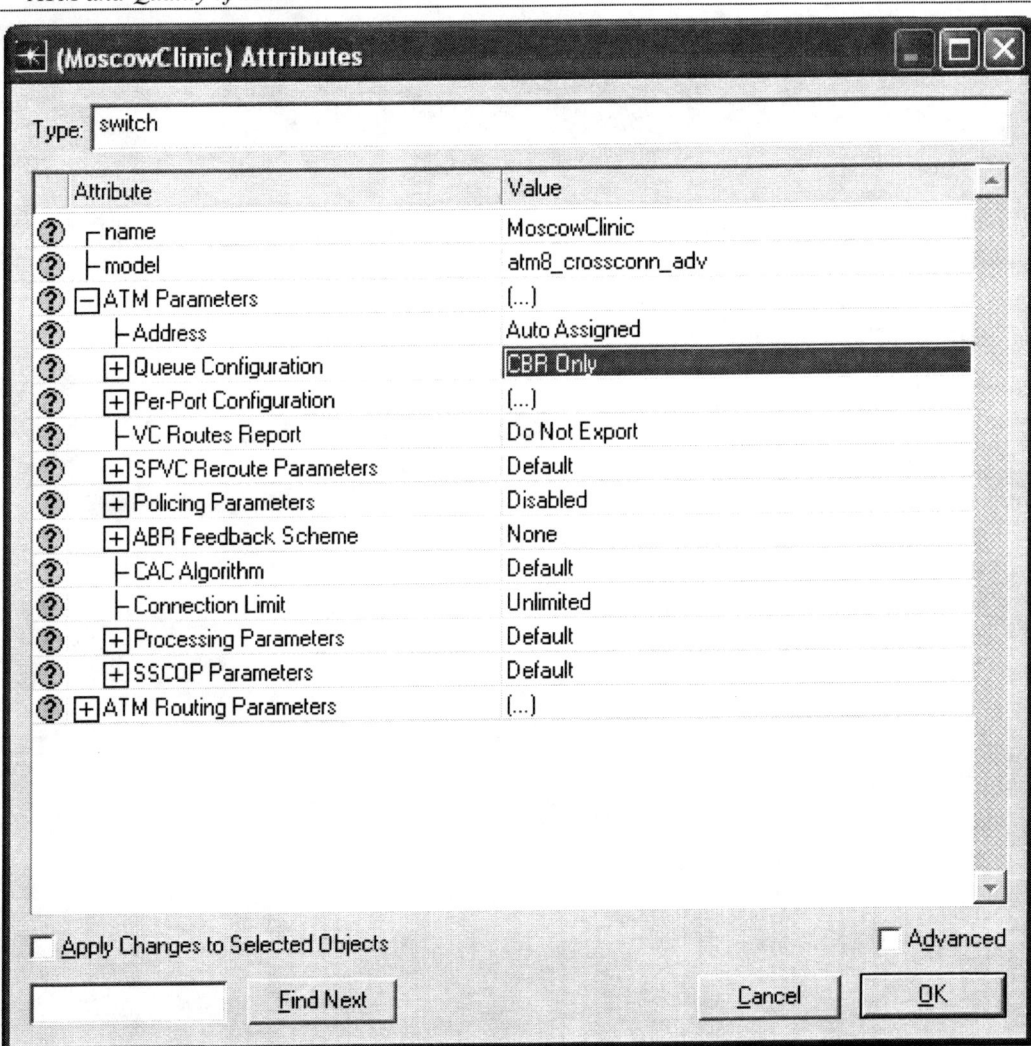

[Find Next]   [Cancel]   [OK]

8. Now click on one of the links in the network and choose **Select Similar Links**. Right click on the link, and choose **Edit Attributes**. Click on the box next to **Apply Changes to Selected Objects**. Click on the value next to **data rate** and choose **DS1**. This will configure the links to run at speeds of 1.5 Mbps. Click **OK** to close the window.

Attribute	Value
name	duplex_7
model	ATM_adv
port a	MoscowClinic.ATM (P7)
port b	node_7.ATM (P0)
ATM Link Failure Report Export	Do Not Export
ATM Trunk Report Export	Do Not Export
Aggregation Token	0
Background Utilization	None
Fiber Segments	None
Propagation Speed	Speed of Light
data rate	DS1

☑ Apply Changes to Selected Objects ☐ Advanced

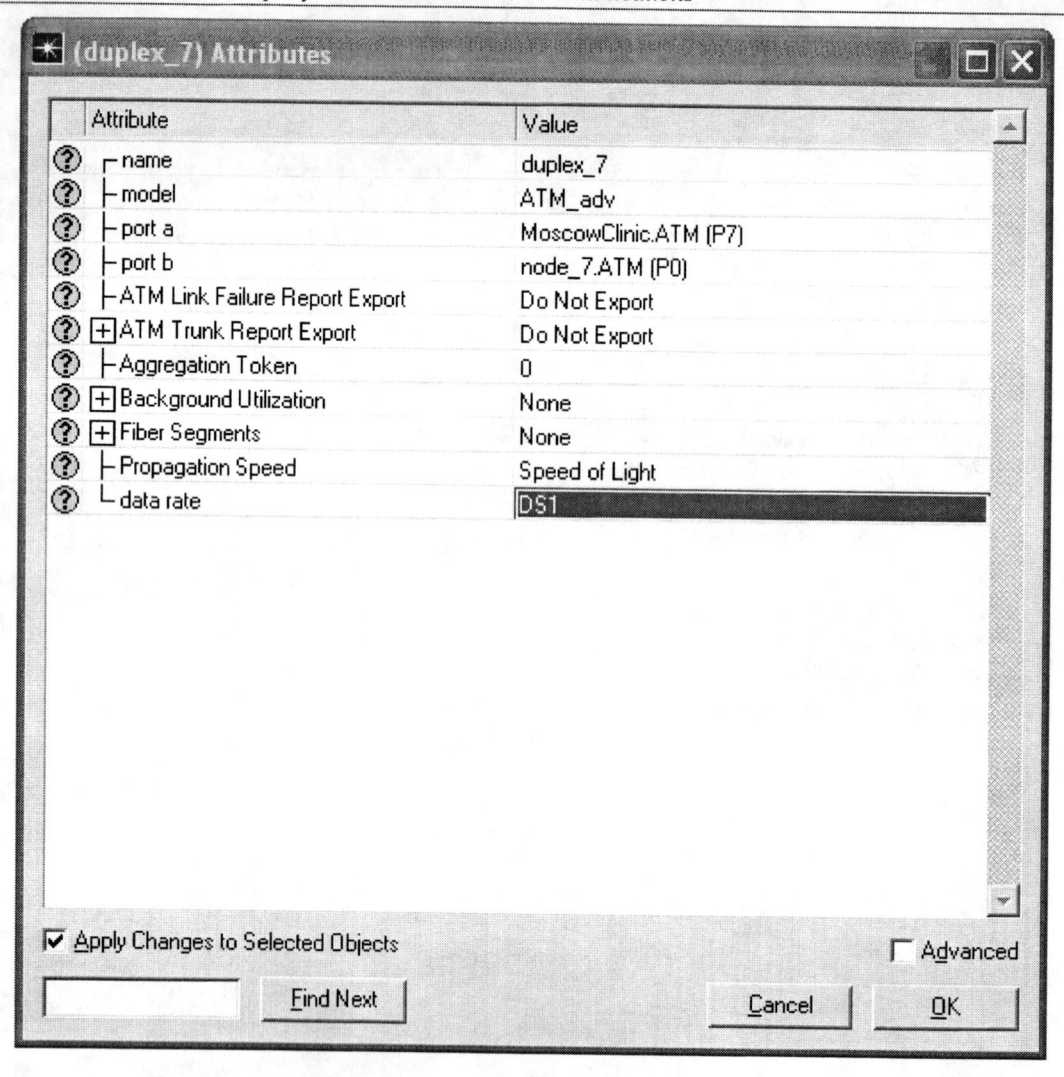

9. Click on the **Arrow** button to go up to the next level of hierarchy in the project. Choose **File/Save** from the Main Menu to save your project.

10. We now will configure the Cleveland Clinic subnet to provide Voice and Videoconferencing services. Double click on the Cleveland Clinic subnet to descend into the subnet. Click on one of the **workstation** objects and choose **Select Similar Nodes**. Right click on one of the selected workstation objects and choose **Edit Attributes**. Click on the box next to **Apply Changes to Selected Objects**. Next, click on the value next to **Application: Supported Services** and choose **Edit**.

Lab 5 ATM and Quality of Service

(node_1) Attributes

Type: workstation

Attribute	Value
─name	node_1
├─model	atm_uni_client_adv
⊞ ATM Application Parameters	(...)
⊞ ATM Parameters	(...)
⊞ Application: ACE Tier Configuration	Unspecified
├─Application: Destination Preferences	None
⊞ Application: Multicasting Specification	None
⊞ Application: RSVP Parameters	None
├─Application: Segment Size	64,000
⊞ Application: Source Preferences	None
⊞ Application: Supported Profiles	(...)
├─Application: Supported Services	(...)
⊞ Application: Transport Protocol	AAL5
⊞ CPU Background Utilization	None
⊞ CPU Resource Parameters	Single Processor
├─Client Address	Auto Assigned
⊞ SIP UAC Parameters	(...)
⊞ Server: Advanced Server Configuration	Sun Ultra 10 333 MHz
└─Server: Modeling Method	Simple CPU

☐ Apply Changes to Selected Objects   ☐ Advanced

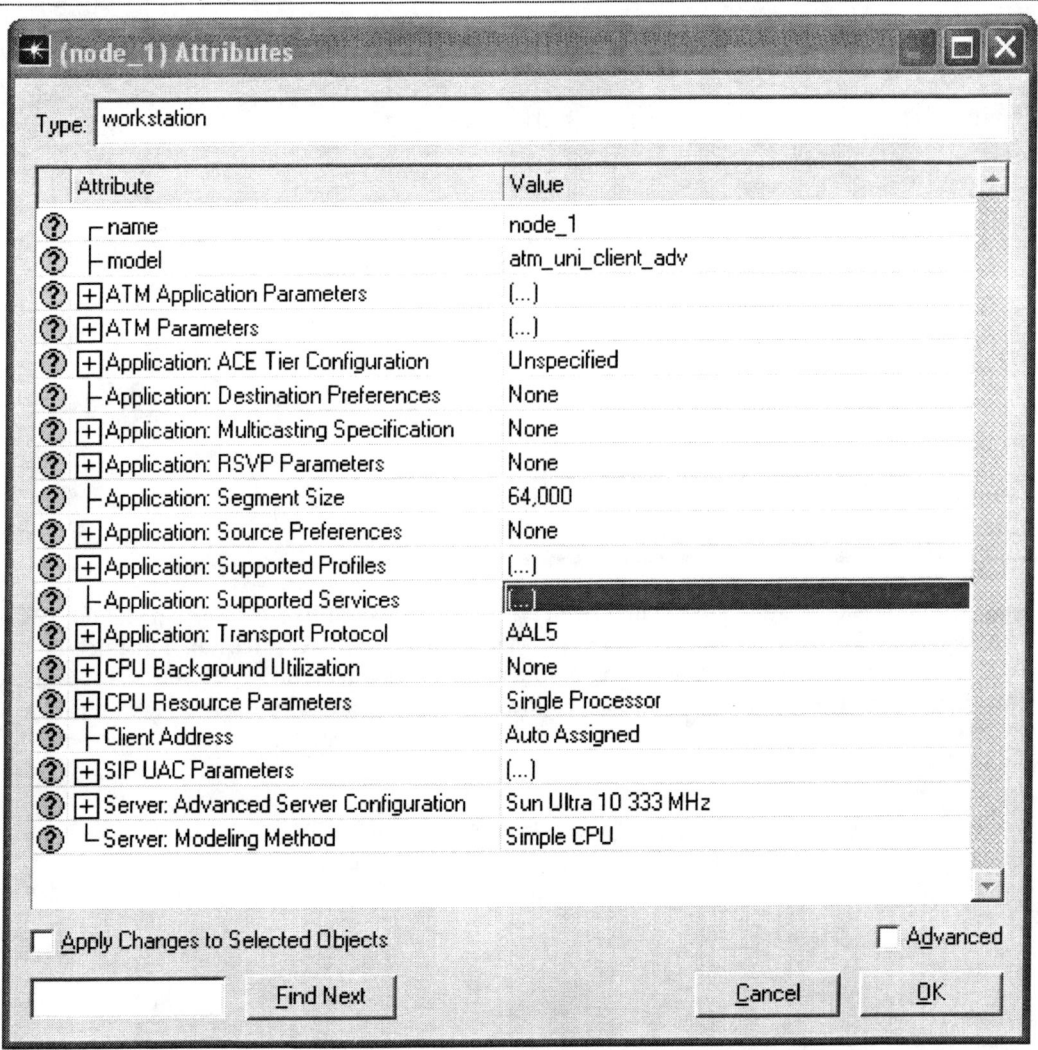

[Find Next]   [Cancel]   [OK]

11. Click on the value listed next to **Rows**, and choose **2** from the pull-down menu. You will see two rows added to the table. Now click on the value under **Application Name.** Choose **Video Conferencing (Light)** from the pull-down menu. In the second row under **Name**, choose **Voice over IP Call (PCM Quality)** from the pull-down menu. Now the Cleveland Clinic subnet will provide access to these applications. Click **OK.**

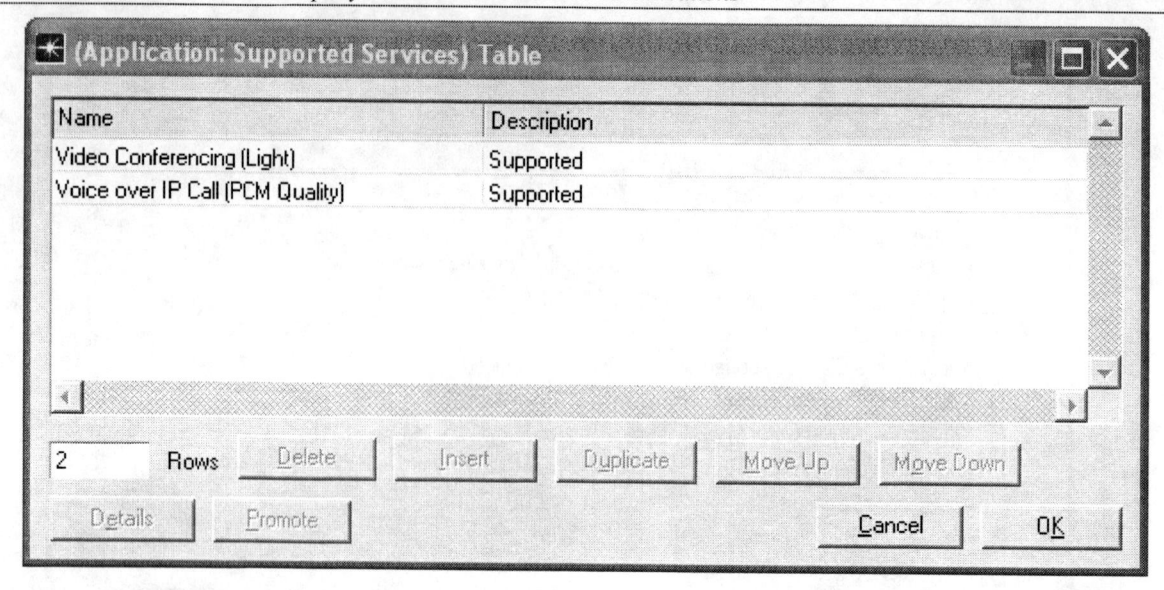

12. Next, we must choose the ATM Quality of Service that we would like for this simulation. We will start with **Constant Bit Rate (CBR)**. Click on the value next to **ATM Application Parameters,** and choose **CBR Only**. Click on the + next to **ATM Application Parameters,** and verify that **CBR** is listed next to **Category** and that **AAL5** is listed next to **Application: Transport Protocol.** Click on **OK** to close the window.

Lab 5 ATM and Quality of Service

Attribute	Value
⌐ name	node_0
├ model	atm_uni_client_adv
⊟ ATM Application Parameters	(...)
├ Category	CBR
⊞ Requested Traffic Contract	default
⊟ Requested QoS	(...)
⊞ ppCDV (millisec)	CBR
⊞ maxCTD (millisec)	CBR
⊞ CLR	CBR
⊞ ATM Parameters	(...)
⊞ Application: ACE Tier Configuration	Unspecified
├ Application: Destination Preferences	None
⊞ Application: Multicasting Specification	None
⊞ Application: RSVP Parameters	None
├ Application: Segment Size	64,000
⊞ Application: Source Preferences	None
⊞ Application: Supported Profiles	(...)
├ Application: Supported Services	(...)
⊞ Application: Transport Protocol	AAL5
⊞ CPU Background Utilization	None

(node_0) Attributes — Type: workstation

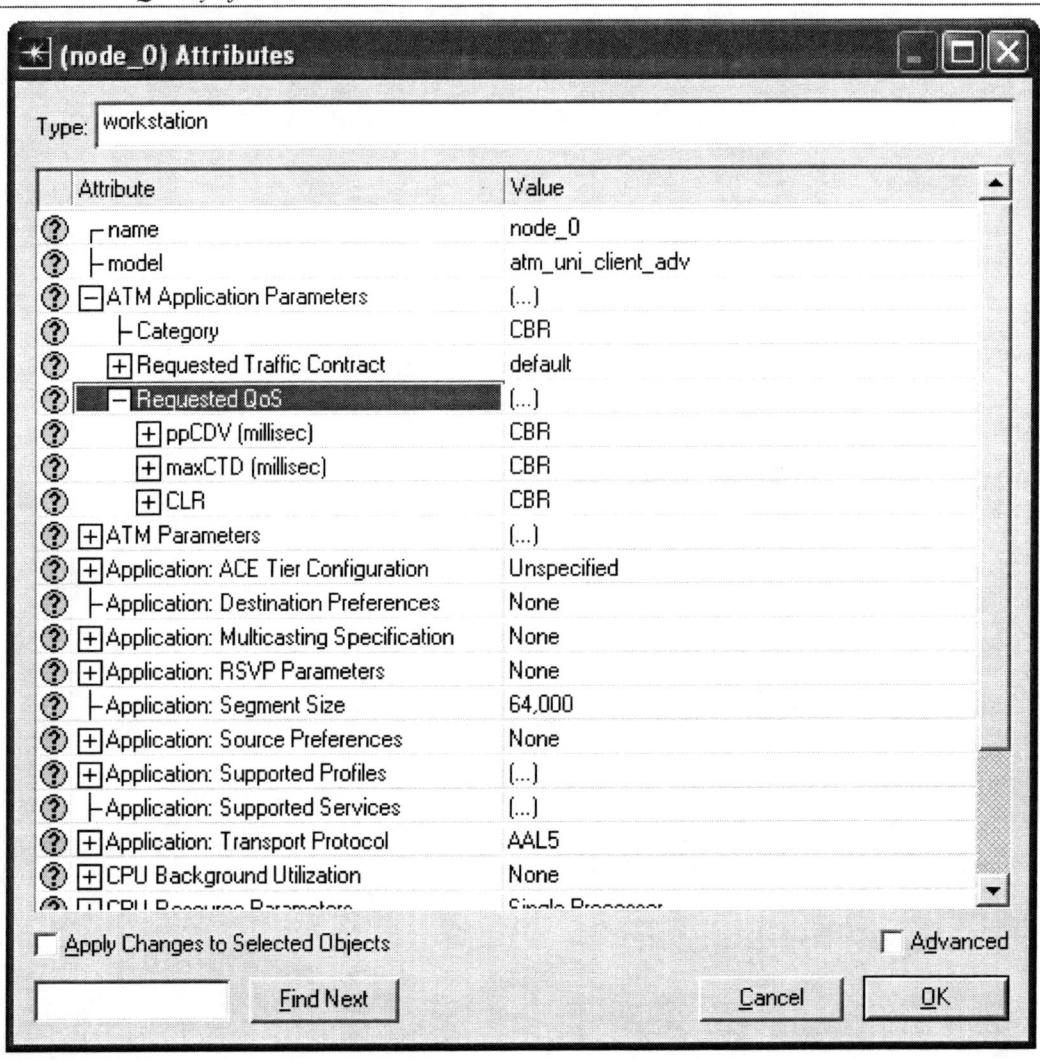

13. Again, we must configure the switch to use **CBR**. Right click on the **switch** object and choose **Edit Attributes**. Click on the + next to **ATM parameters** to expand the values. Click on the value next to **Queue Configuration** and choose **CBR Only**. Click **OK** to close the window.

93

OPNET Lab Manual to accompany Business Data and Communications

(ClevelandSwitch) Attributes

Type: switch

Attribute	Value
⌐ name	ClevelandSwitch
├ model	atm8_crossconn_adv
⊟ ATM Parameters	(...)
├ Address	Auto Assigned
⊞ Queue Configuration	CBR Only
⊞ Per-Port Configuration	(...)
├ VC Routes Report	Do Not Export
⊞ SPVC Reroute Parameters	Default
⊞ Policing Parameters	Disabled
⊞ ABR Feedback Scheme	None
├ CAC Algorithm	Default
├ Connection Limit	Unlimited
⊞ Processing Parameters	Default
⊞ SSCOP Parameters	Default
⊞ ATM Routing Parameters	Default

☐ Apply Changes to Selected Objects   ☐ Advanced

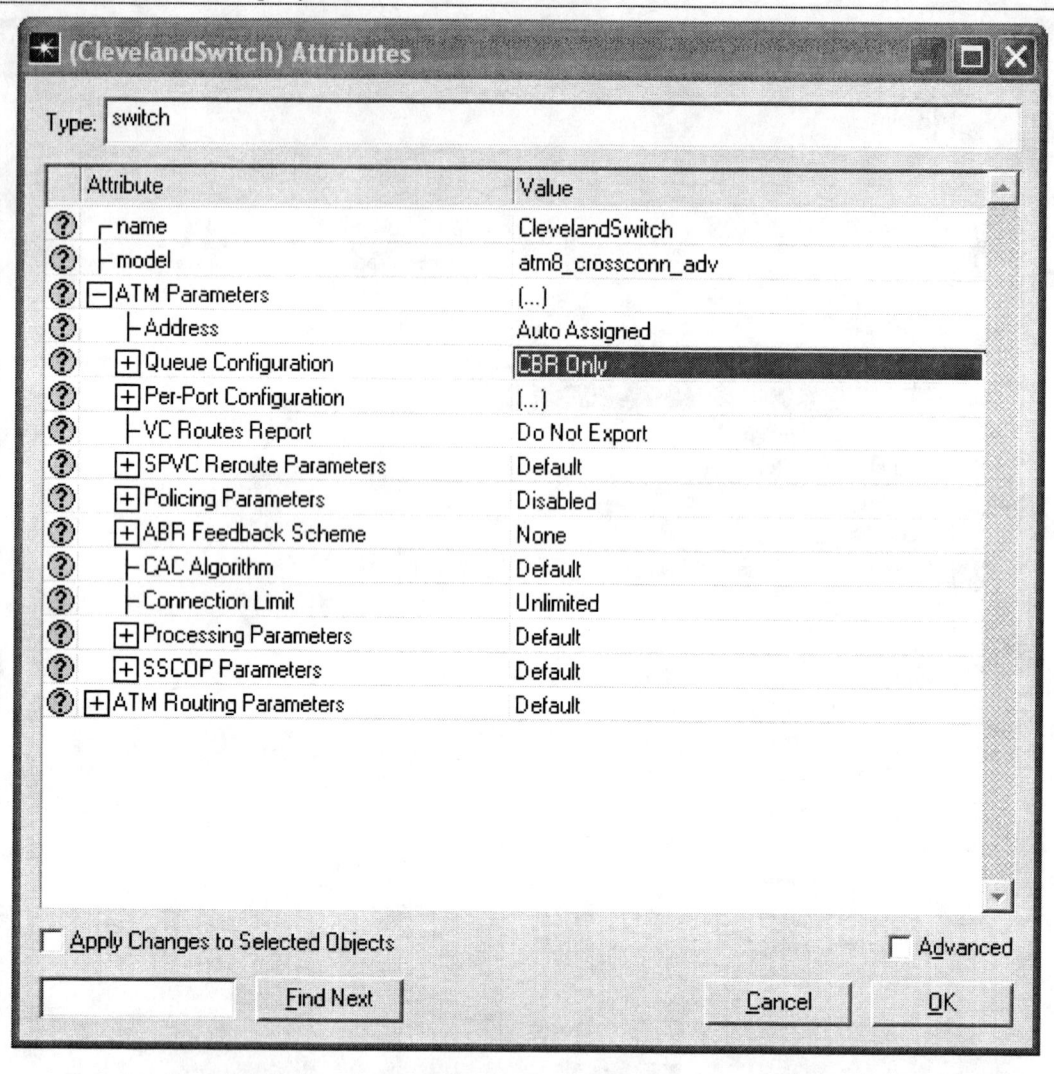

[Find Next]   [Cancel]   [OK]

14. Now we will configure the links to run at DS1 speeds as we did in the Moscow Clinic subnet. Click on one of the links in the network and choose **Select Similar Links**. Right click on the link and choose **Edit Attributes**. Click on the box next to **Apply Changes to Selected Objects**. Click on the value next to **data rate** and choose **DS1**. Click **OK** to close the window.

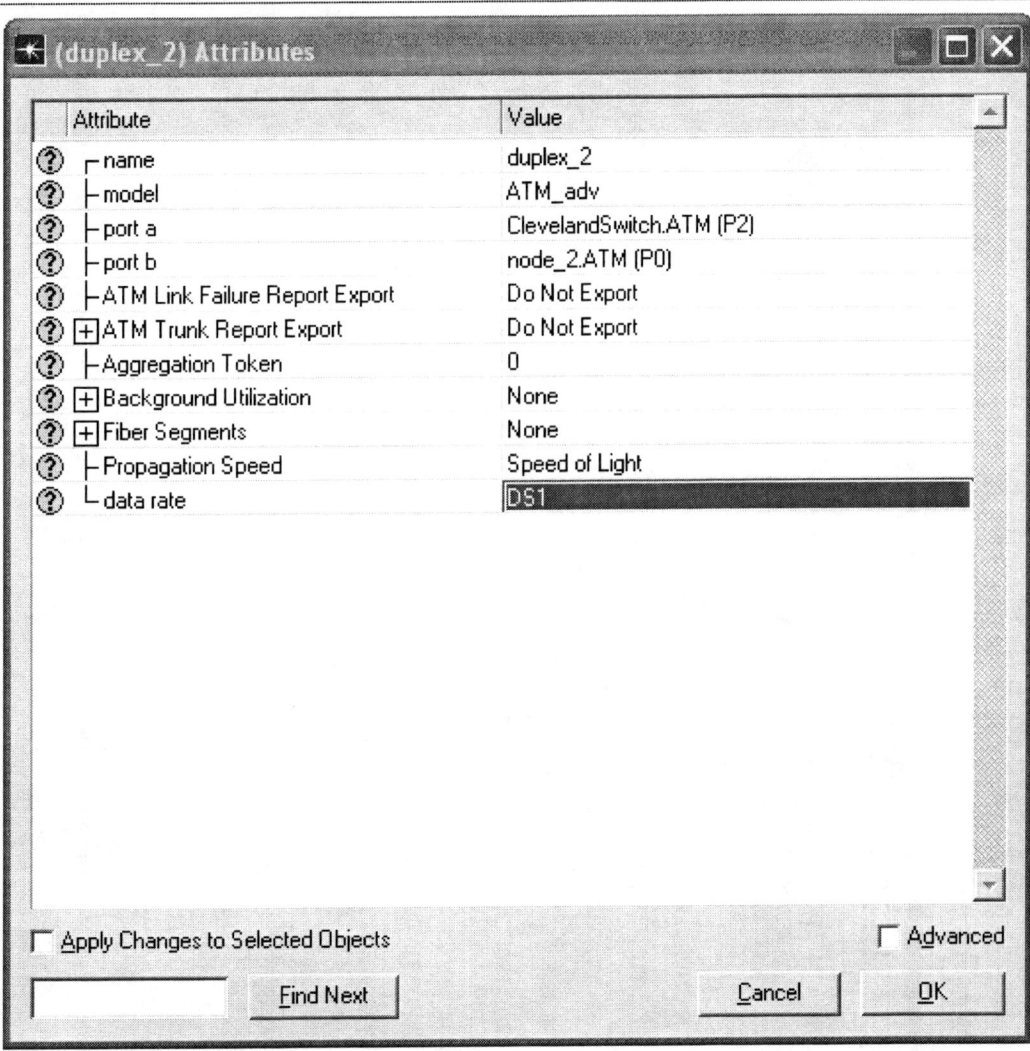

Lab 5   ATM and Quality of Service

15. Click on the **Arrow** button  to go up to the next level of hierarchy in the project.

16. Finally, we will configure the links between the subnets to run at OC3 speeds. This is a common ATM speed of 155 Mbps. Click on one of the links in the network and choose **Select Similar Links**. Right click on the link and choose **Edit Attributes**. Click on the box next to **Apply Changes to Selected Objects**. Click on the value next to **data rate** and choose **OC3**. Click **OK** to close the window.

17. Choose **File/Save** from the Main Menu to save your project.

Configuring the Simulation and Choosing Statistics

Now we would like to choose the statistics that will measure traffic sent and received and delay for our videoconferencing and voice applications. We will later evaluate the effect that CBR and UBR qualities of service have on these values.

1. Right click somewhere on the project workspace and select **Choose Individual Statistics** from the pull-down menu. Click on the + next to **Global** to expand the parameters. Click on the + next to **Video Conferencing**. Click on all the boxes listed under Video Conferencing to select all the statistics. Notice that the statistics reflect bytes sent and delay. Now click on the + next to **Voice** to expand the options. Again, choose all of the statistics.

Choose Results dialog

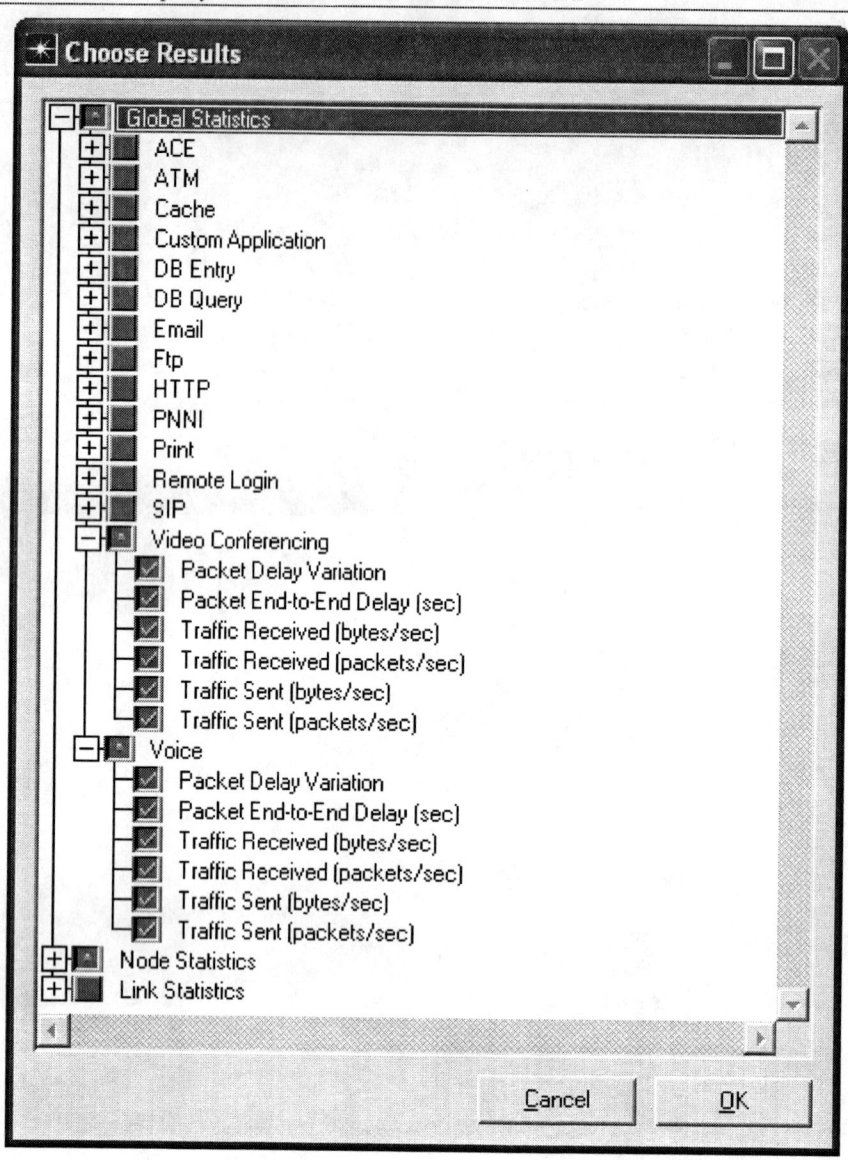

(Global Statistics tree showing: ACE, ATM, Cache, Custom Application, DB Entry, DB Query, Email, Ftp, HTTP, PNNI, Print, Remote Login, SIP, Video Conferencing [Packet Delay Variation, Packet End-to-End Delay (sec), Traffic Received (bytes/sec), Traffic Received (packets/sec), Traffic Sent (bytes/sec), Traffic Sent (packets/sec)], Voice [Packet Delay Variation, Packet End-to-End Delay (sec), Traffic Received (bytes/sec), Traffic Received (packets/sec), Traffic Sent (bytes/sec), Traffic Sent (packets/sec)], Node Statistics, Link Statistics)

Running the Simulation

1. Click on the **Configure and Run** Button. The applications that we chose generate quite a bit of traffic. This causes many events and can lead to long simulation times. To reduce the time it takes our simulation to run, we will reduce the traffic generated and run the simulation for a short period of time.

2. Click on the Tab labeled **Global Attributes**. Click on the value next to **Traffic Scaling Factor** and choose **Edit**. Change the value to **.2**. Click on the value next to **Traffic Scaling Mode** and choose **Edit**. Change the value to **All Traffic**. This forces our simulation to create only 20% of the events that would normally be generated.

Lab 5 ATM and Quality of Service

Configure Simulation: LCatm-Scenario1CBR

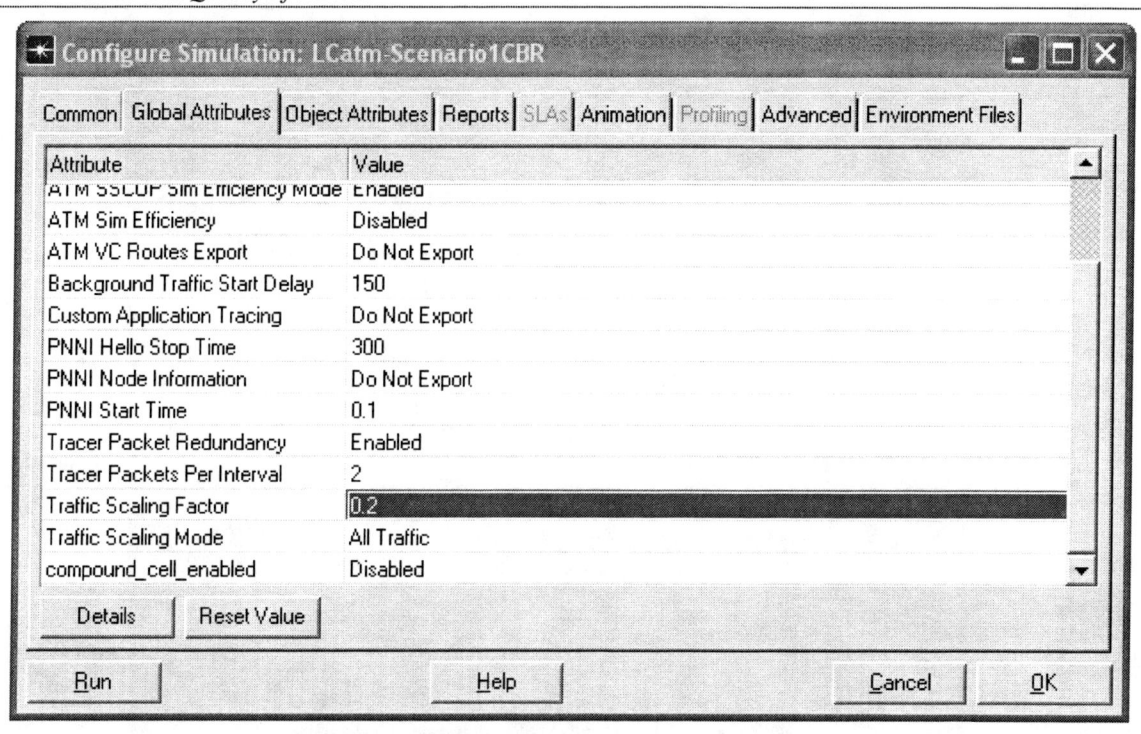

Common | Global Attributes | Object Attributes | Reports | SLAs | Animation | Profiling | Advanced | Environment Files

Attribute	Value
ATM SSCOP Sim Efficiency Mode	Enabled
ATM Sim Efficiency	Disabled
ATM VC Routes Export	Do Not Export
Background Traffic Start Delay	150
Custom Application Tracing	Do Not Export
PNNI Hello Stop Time	300
PNNI Node Information	Do Not Export
PNNI Start Time	0.1
Tracer Packet Redundancy	Enabled
Tracer Packets Per Interval	2
Traffic Scaling Factor	0.2
Traffic Scaling Mode	All Traffic
compound_cell_enabled	Disabled

Details | Reset Value

Run | Help | Cancel | OK

3. Change the **Duration** of the simulation to **1**. Choose **Minute(s)** from the pull-down menu. Click on the **Run** button to run the simulation. Depending on your processor it should take several minutes.

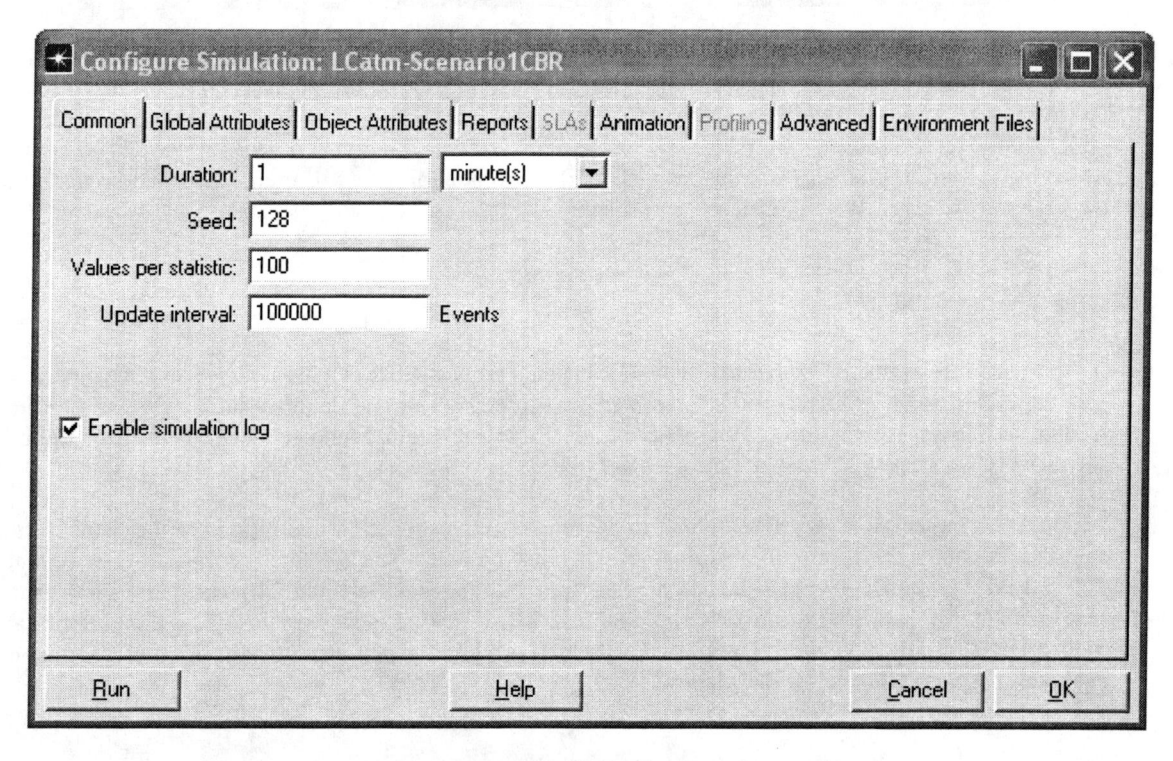

Configure Simulation: LCatm-Scenario1CBR

Common | Global Attributes | Object Attributes | Reports | SLAs | Animation | Profiling | Advanced | Environment Files

Duration: 1   minute(s)
Seed: 128
Values per statistic: 100
Update interval: 100000   Events

☑ Enable simulation log

Run | Help | Cancel | OK

4. To view results, choose **Results/View Results** from the Main Menu. To view results for Videoconferencing Traffic Sent (bytes), click on the + next to **Videoconferencing,** and click next to **Traffic Sent (bytes).** You can view the results **As Is.** Click on the **Show** button to display the graph in a new window. Your graph should resemble the one shown next.

97

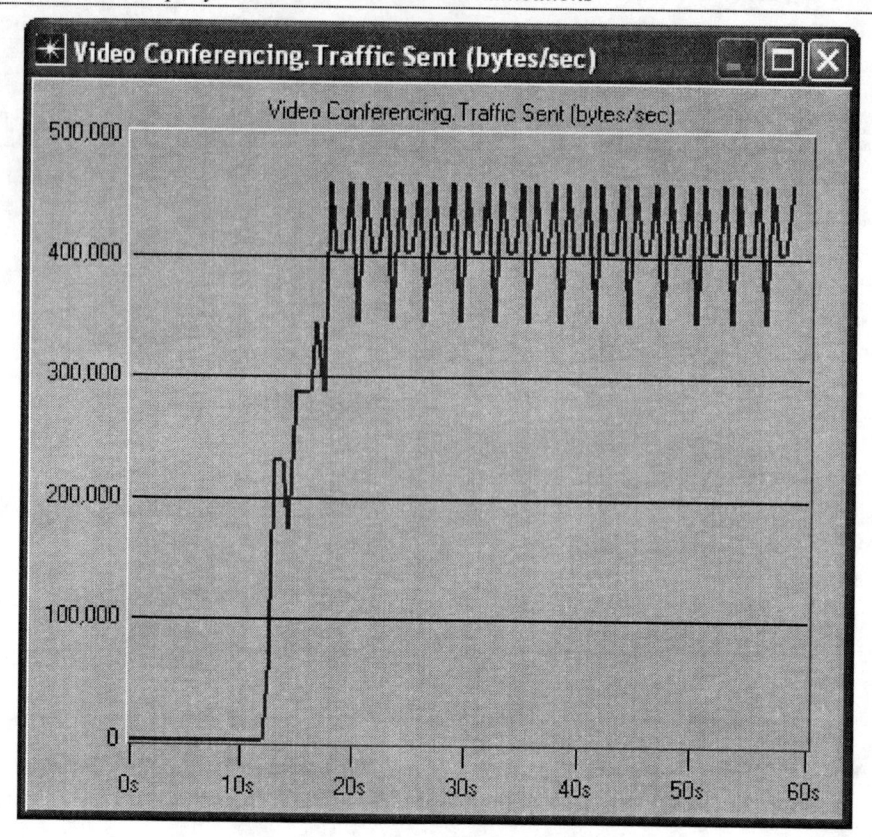

[Figure: Video Conferencing. Traffic Sent (bytes/sec) — graph showing oscillation between roughly 350,000 and 450,000 bytes/sec from about 20s to 60s]

Notice that traffic oscillates between around 350,000 and 450,000 bytes. Click on the **X** to close the window. The **Close Analysis** window will appear asking you if you want to hide the panel or permanently delete it. Click on the **Delete** button to delete the window.

Before we look at all the statistics for our original CBR Configuration, we will create a second scenario that uses the UBR service class. We will then compare results.

Creating a New Scenario

1. Our purpose is to evaluate whether the two clinics should request CBR or UBR classes of service based on how these classes affect traffic sent and received, end-to-end delay, and delay variation. We will create a second scenario that uses the UBR traffic class. From the Main Menu, choose **Scenarios/Duplicate Scenario**. Give the new scenario a name such as **UBR**.

2. Double click on the Moscow Clinic subnet to descend into the subnet. Click on one of the **workstation** objects and choose **Select Similar Nodes**. Right click on one of the selected workstation objects and choose **Edit Attributes**. Click on the box next to **Apply Changes to Selected Objects**. Click on the value next to **ATM Application Parameters** and choose **UBR**. Click on the + next to **ATM Parameters** and choose **UBR** for **Queue Configuration**. Verify that **AAL5** is listed next to **Application: Transport Protocol**. Click on **OK** to close the window.

Lab 5 ATM and Quality of Service

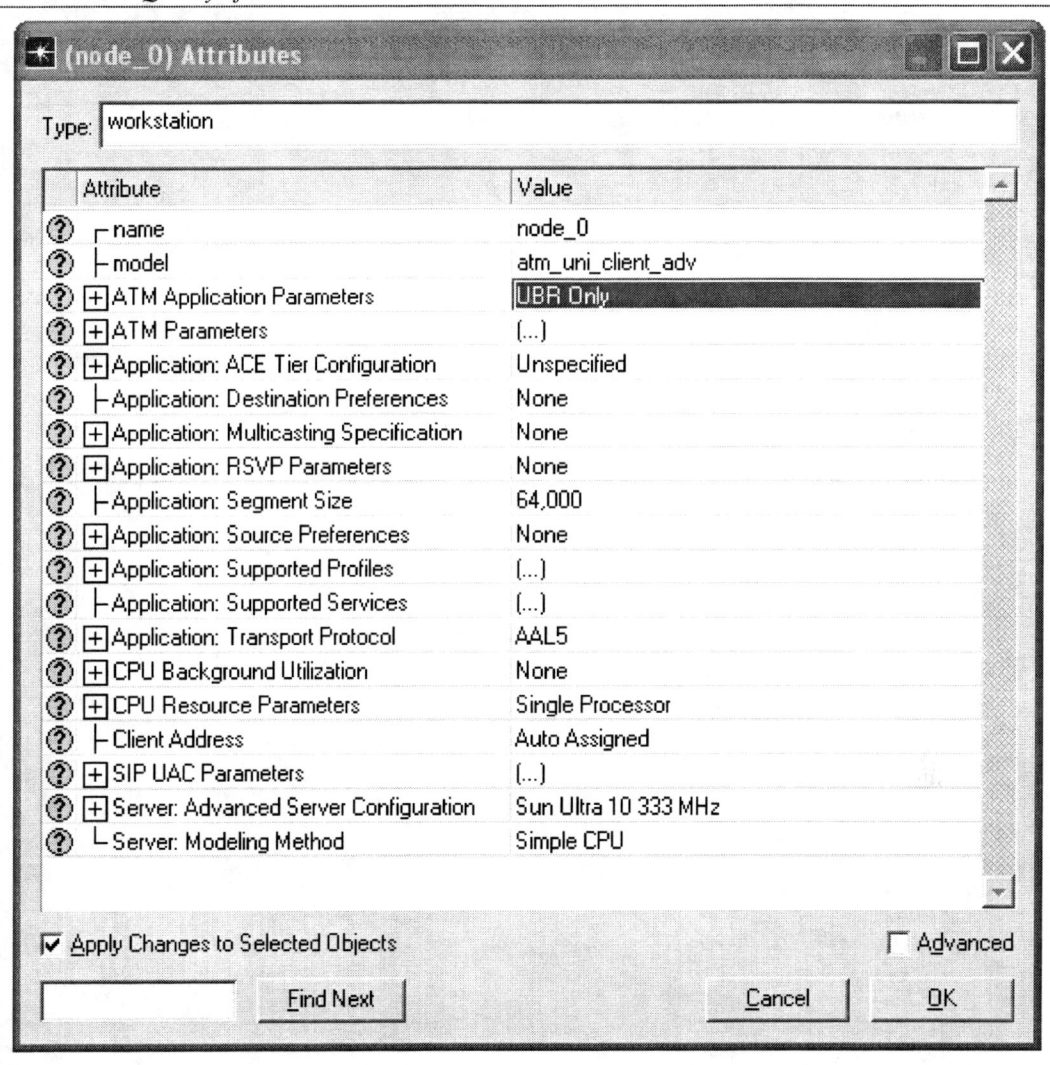

(node_0) Attributes

Type: workstation

Attribute	Value
name	node_0
model	atm_uni_client_adv
⊞ ATM Application Parameters	**UBR Only**
⊞ ATM Parameters	(...)
⊞ Application: ACE Tier Configuration	Unspecified
Application: Destination Preferences	None
⊞ Application: Multicasting Specification	None
⊞ Application: RSVP Parameters	None
Application: Segment Size	64,000
⊞ Application: Source Preferences	None
⊞ Application: Supported Profiles	(...)
Application: Supported Services	(...)
⊞ Application: Transport Protocol	AAL5
⊞ CPU Background Utilization	None
⊞ CPU Resource Parameters	Single Processor
Client Address	Auto Assigned
⊞ SIP UAC Parameters	(...)
⊞ Server: Advanced Server Configuration	Sun Ultra 10 333 MHz
Server: Modeling Method	Simple CPU

☑ Apply Changes to Selected Objects   ☐ Advanced

[Find Next]   [Cancel]   [OK]

3. Again, we must configure the switch to use UBR. Right click on the **switch** object and choose **Edit Attributes**. Click on the + next to **ATM Parameters** to expand the values. Click on the value next to **Queue Configuration** and choose **UBR.** Click **OK** to close the window.

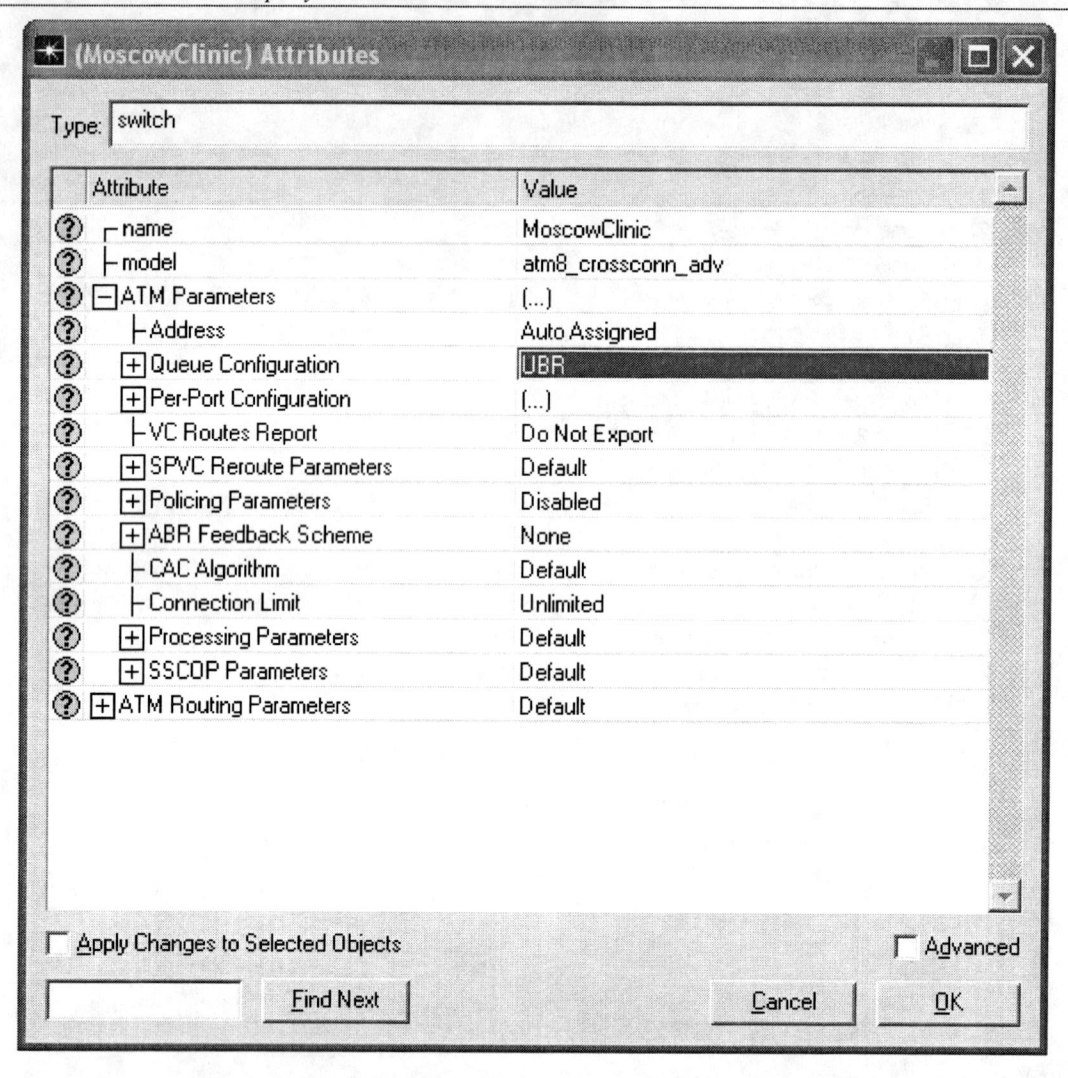

4. Repeat the same steps to reconfigure the switch and the workstations in the Cleveland Clinic to subnet to use UBR.

5. Choose **Simulation/Run Discrete Event Simulation** to gather results for the new scenario.

6. Choose **File/Save** from the Main Menu to save your project.

Comparing Results

We can now compare results for both scenarios. To do this, choose **Results/Compare Results** from the Main Menu. To compare results for **Voice Delay**, click on the **+** next to **Voice** and click on **Packet End-to-End Delay (Seconds)**. Choose **Time Average** from the pull-down menu and click **Show**. This will give you a graph with data from both scenarios as well as a legend for the graph. Your graph should resemble the following one:

Lab 5 ATM and Quality of Service

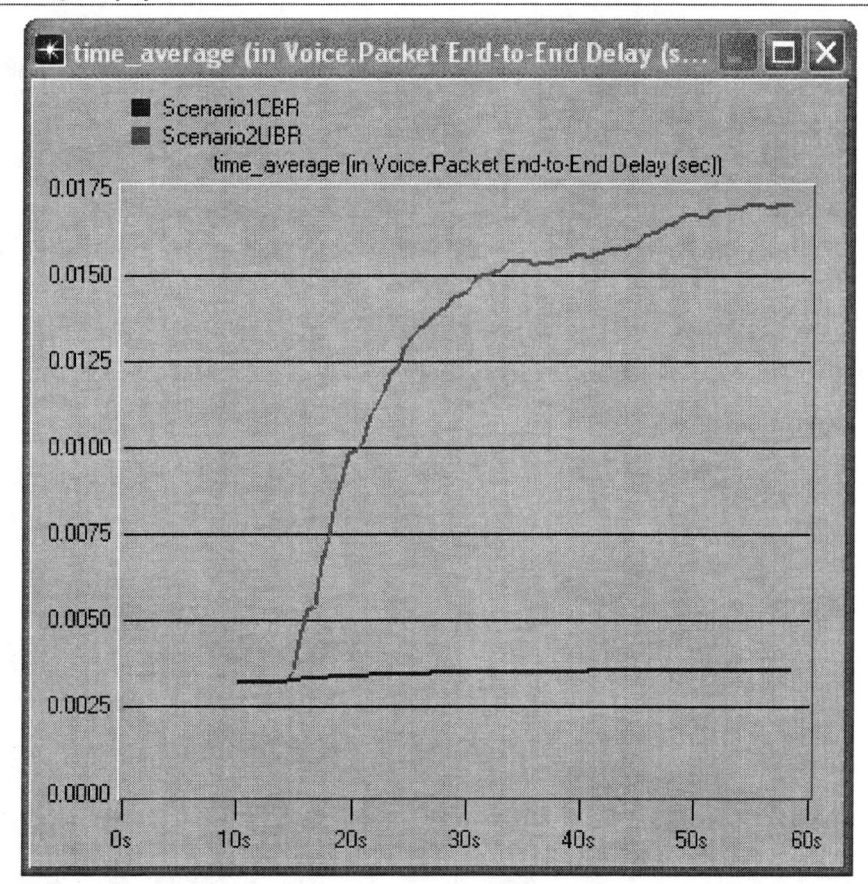

Notice that delay is much higher for the UBR scenario. Click on the **X** to close the window. The **Close Analysis** window will appear asking you if you want to hide the panel or permanently delete it. Click on the **Delete** button to delete the window.

You may discard the results of your simulations, rerun them, and manage your scenarios by doing the following:

Choose **Scenarios/Manage Scenarios** from the Main Menu. A window will appear with the names of your scenarios, the length of time that the simulation ran, and the status of the data.

By clicking on a scenario name, you may choose to **Discard Results** or **Collect Results**. By clicking on the values under the **Results** column, you may choose **recollect** or **collect** from the pull-down menu. Click **OK** to close the window.

Now use the results you obtained in your simulations to answer the questions below.

Questions

1. What is the average delay for the voice and videoconferencing applications in both the CBR and UBR scenarios? Based on what you see, can you make a recommendation for the service class that the clinics should use?
2. What was the average number of bytes sent for both the voice and videoconferencing applications in both the CBR and UBR scenarios. Based on this statistic, is your recommendation the same as your previous recommendation in Question 1?
3. Look at the videoconferencing End-to-End packet delay variation for the CBR and UBR scenarios. Which class creates more packet delay variation? Is this what you expected?

4. Calculate how many bytes per second the **Videoconferencing (Light)** application is sending. Look at the application description for **Videoconferencing (Light)**. It should state the number of frames being sent per second as well as the size in pixels of the screen. If each pixel is 16 bits, how many bytes will be sent per second? Remember that there will be two parties in a videoconference. Now, recalculate your answer based on the fact that we set the traffic to be .2 of what it normally would be. Does the value for Traffic Sent (bytes) match the number you calculated?
5. Duplicate the first CBR scenario and modify the workstations in each subnet to use the AAL2 transport protocol. This transport protocol was designed for voice. How do the traffic sent and delay values compare for the voice application in this scenario and the scenario that used AAL5 transport? Does the change in transport affect the videoconferencing application?
6. Duplicate the second UBR scenario and modify the workstations in each subnet to use AAL2 transport. How do the traffic sent and delay values compare for the videoconferencing application in this scenario and the scenario that used AAL5 transport? Which application is affected more by the change in transport protocol?
7. Create a third scenario that is a duplicate of the first CBR scenario. Change the links within the subnet to have an OC3 data rate. Does this affect the values that you observed previously?

LAB 6 FRAME RELAY

Objective
1. To evaluate the relationship between Frame Relay Permanent Virtual Circuit (PVC) speed and application response time in a Wide Area Network (WAN).

Motivation
CreditUSA processes credit card transactions for customers located across the United States. The primary processing center for CreditUSA is located in South Carolina. A secondary processing center, located in California, was established to handle customers on the West Coast. The two processing centers are linked via Permanent Virtual Circuits (PVCs) to a frame relay Wide Area Network (WAN). The California processing center uses the frame relay network to send database queries and updates, and to transfer customer information files to the primary processing center. CreditUSA initially contracted for a PVC Committed Information Rate (CIR) of 64 Kbps. Database and file transfer application response time has not been adequate; therefore, the company would like to evaluate two options for improvement. The first option is to increase the CIR and the second option is to keep the same CIR, but increase the bandwidth of the links from the processing center to the WAN.

Description
Frame Relay is a popular packet switching technique based on X.25, which was one of the first packet switching technologies offered by common carriers. Frame Relay utilizes virtual circuit technology. It is a stripped down version of X.25, providing a few basic quality of service and congestion avoidance services. In this lab, we will construct a network that consists of two subnets connected via Permanent Virtual Circuits (PVCs) to a frame relay Wide Area Network (WAN). We will configure the PVCs to provide a particular Committed Information Rate (CIR). This value is the amount of bandwidth that a customer needs at all times. Customers will contract and pay for this guaranteed bandwidth. PVCs can also be configured to allow for bursts of data above the CIR. Excess burst rates allow the customer to send extra data above his CIR rate when there is unused bandwidth available. In this lab, we will observe how link speeds, PVC CIRs, and burst rates affect database and file transfer application response times.

Follow the next set of directions below to create an OPNET model, run simulations, and view results. Questions regarding the lab are listed at the end.

Create a New Project

1. Start the **OPNET IT Guru Academic Edition Application.** Create a **New Project** by selecting **File/New/Project** from the Main Menu. Click **OK**.

2. Give the project a name such as **your_initials_FrameRelayPVC**. Give the scenario a name such as **PVCStartConfig**. Click **OK**. You should see the **Startup Wizard Initial Topology** dialog box.

3. Verify that **Create Empty Scenario** is selected in the **Initial Topology Dialog Box**. Click **Next**. You will now see the **Network Scale** dialog box.

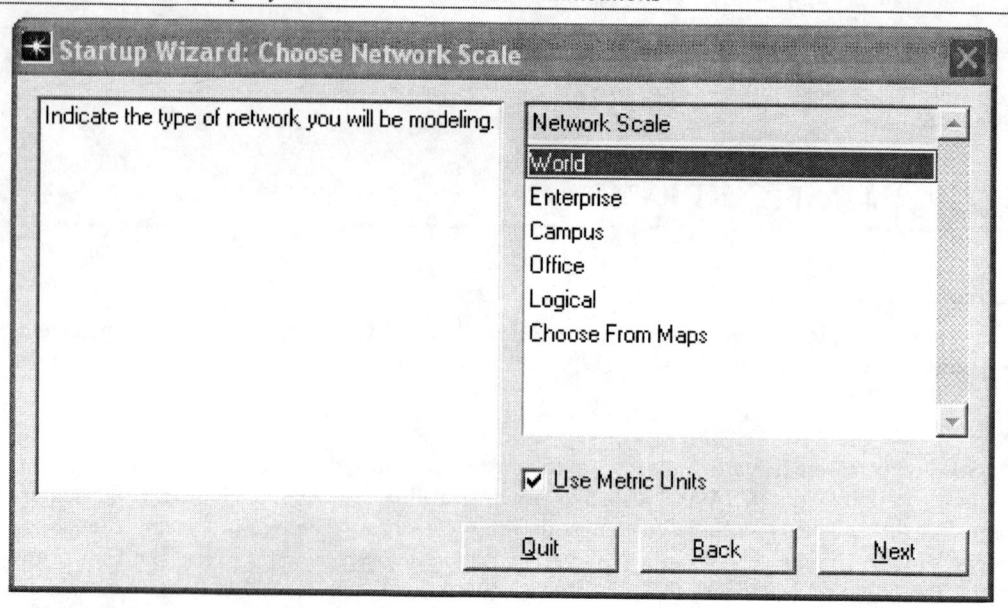

OPNET Lab Manual to accompany Business Data and Communications

[Screenshot: Startup Wizard: Choose Network Scale dialog, with Network Scale list showing World (selected), Enterprise, Campus, Office, Logical, Choose From Maps. "Use Metric Units" checkbox is checked. Buttons: Quit, Back, Next.]

4. Choose **World** from the **Network Scale** list. Click **Next**. Choose **usa** as your map and click **Next**. You should see the **Select Technologies** dialog box.

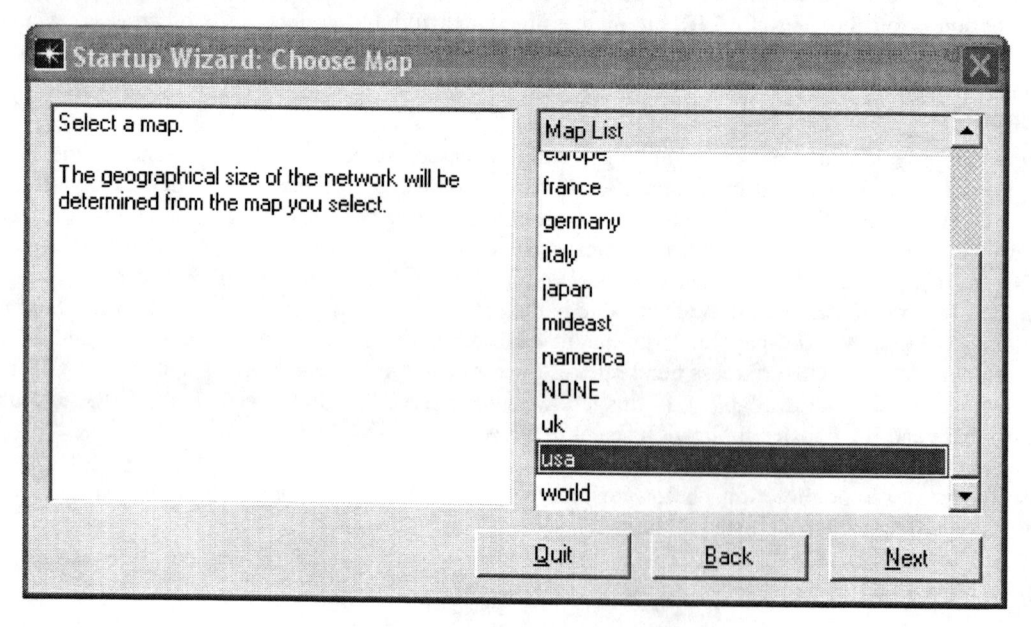

[Screenshot: Startup Wizard: Choose Map dialog. "Select a map. The geographical size of the network will be determined from the map you select." Map List: europe, france, germany, italy, japan, mideast, namerica, NONE, uk, usa (selected), world. Buttons: Quit, Back, Next.]

5. Scroll down until you see **Frame Relay** under the **Model Family** column. Under the **Include column**, click on the box next to **frame relay**. This will change the "no" to "yes" indicating that Frame Relay technologies will be included. Scroll down a bit and click to also include **internet_toolbox** and **routers** in your project. Click **Next**.

104

Lab 6   Frame Relay

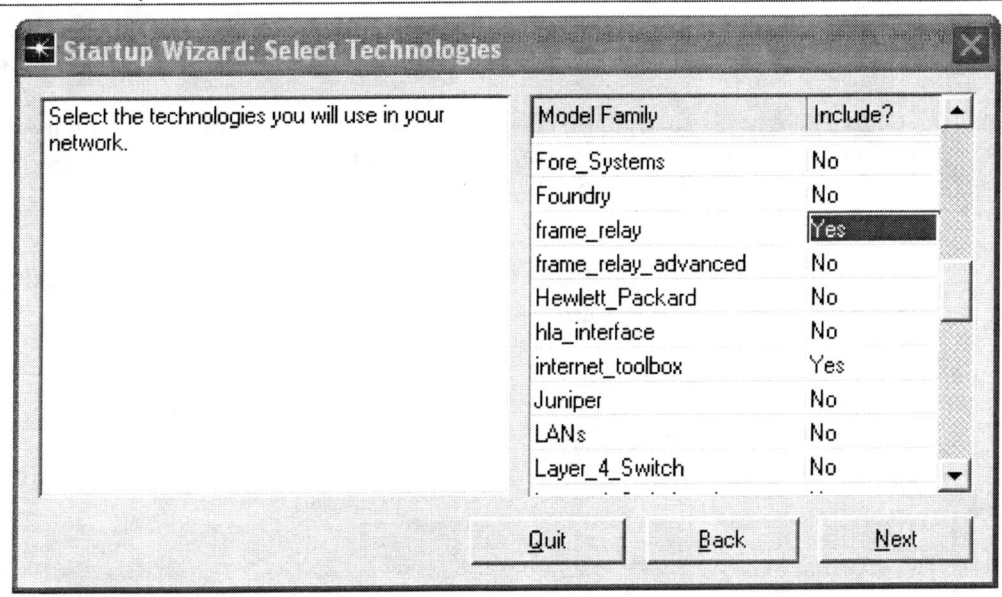

6. In the **Startup Wizard Review** dialog box, verify that the **frame relay, internet_toolbox** and **routers** technologies are chosen and that the scale is World and the map is **usa**. Click **OK**. You should see the **Object Palette** window.

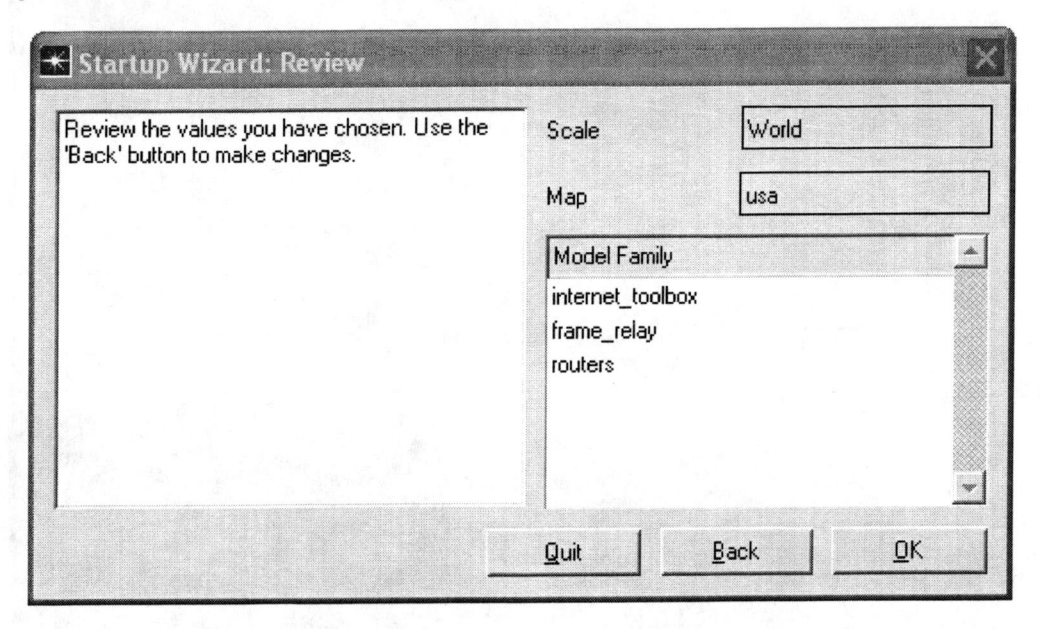

Building the Frame Relay Network

1. We will now create the frame relay network. We will start by placing two subnet objects in the areas of California and South Carolina. From the **Object Palette**, click on the **subnet** object to select it. Click to paste it on the project grid in the area of California. Click again to paste another copy of it in the area of South Carolina. Right click to release the subnet object.

2. Right click on the **CA subnet** and choose **Set Name** from the pull-down menu. Name the subnet **CAProcessingCenter**. Repeat this step and name the other subnet **SCProcessingCenter.**

3. We will begin by creating the SC Processing Center's subnet, which will consist of a 100BaseT LAN and a router that will connect to the frame relay WAN. Double click on the **SCProcessingCenter** subnet object to descend into the subnet. From the **Object Palette**, click on the **100BaseT LAN** object to select

it. Click to paste a copy of it on the map. Right click to release the object. Right click on the **100BaseT LAN** object and choose **Set Name** from the pull-down menu. Name the LAN **SCLAN**.

4. From the **Object Palette**, click on the **fr4_ethernet2_gtwy** object to select it. Drag it to the subnet area, and click to paste a copy of it on the map. Right click to release the object. Right click on the **fr4_ethernet2_gtwy** object and choose **Set Name** from the pull-down menu. Name the router **SCRouter**. Your project should resemble the following figure:

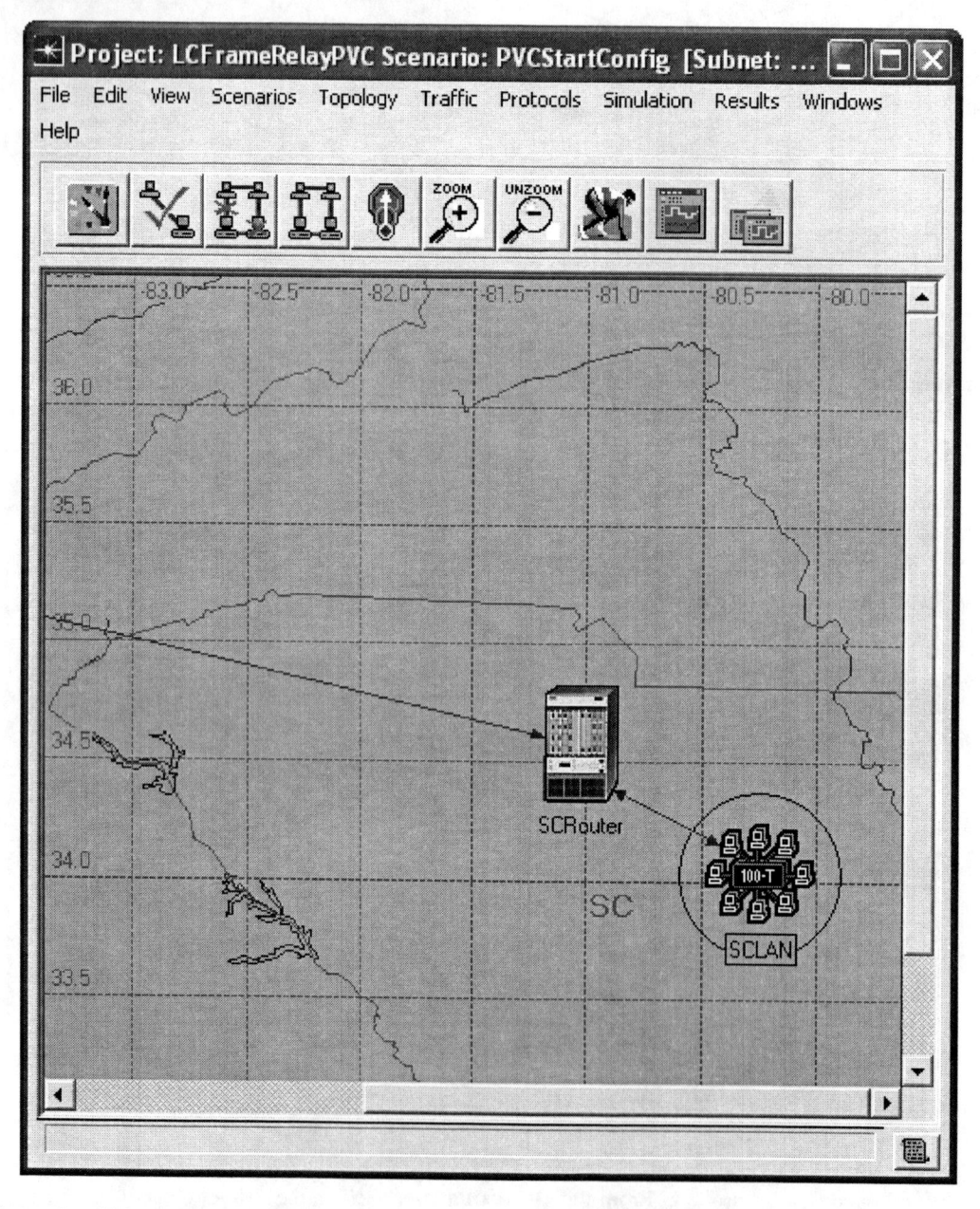

5. From the **Object Palette**, click on the **100BaseT link** object to select it. Connect the 100BaseT LAN object to the **fr4_ethernet2_gtwy** router object. Right click to release the object.

6. Click on the **Arrow Button** to go to the next higher level in the project hierarchy.

7. Now we will build the CA Processing Center's subnet, which will be similar to that of the SC Processing Center. Double click on the **CAProcessingCenter subnet** object to descend into the subnet. From the **Object Palette**, click on the **100BaseT_LAN** object to select it. Click to paste a copy of it on the map.

Lab 6   Frame Relay

Right click to release the object. Right click on the **100BaseT_LAN** object and choose **Set Name** from the pull-down menu. Name the LAN **CALAN**.

8. From the **Object Palette**, click on the **fr4_ethernet2_gtwy** object to select it. Click to paste a copy of it on the map. Right click to release the object. Right click on the **fr4_ethernet2_gtwy** object and choose **Set Name** from the pull-down menu. Name the router **CARouter**.

9. From the **Object Palette**, click on the **100BaseT link** object to select it. Connect the **100BaseT_LAN** object to the **fr4_ethernet2_gtwy** router object. Right click to release the object.

10. Click on the **Arrow Button** to go to the next higher level in the project hierarchy.

11. We will now connect our subnets with a Frame Relay object and some T1 links. From the **Object Palette**, click on the **fr32_cloud** object to select it. Click to paste it on the map between the two subnets. Right click to release the object. Right click on the **fr32_cloud** object and choose **Set Name** from the pull-down menu. Name the object **FrameRelayNetwork**.

12. We will connect our subnets to the frame relay network via T1 links. A T1 is made up of 24 DS0s (64Kbps). A T1, therefore, has a speed of 1.544 Mbps. From the **Object Palette**, click on the **frame relay T1 link** object to select it. Connect the frame relay object to the **SCProcessingCenter**. The **Select Node** window will appear. This allows you to choose which object to link to within the subnet. Choose the **SCProcessingCenter.SCRouter** as shown in the next image. Repeat the step to connect the frame relay object to the **CAProcessingCenter.CARouter**. Right click to release the object.

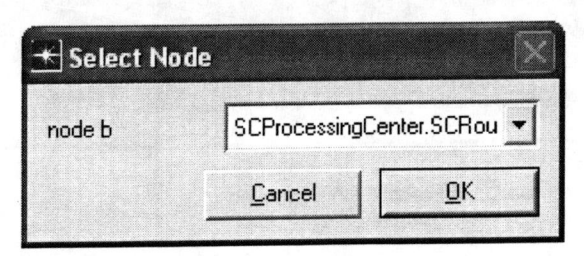

13. Now we will add FR PVC Config, Application Config, and Profile Config objects to the workspace. The FR PVC Configuration object allows you to set characteristics for the PVCs in the frame relay network. We will be interested particularly in CIR and burst rate. The application object will allow us to choose the applications we want to include in our simulation. We will use the database and file transfer applications. The Profile Config object allows us to use default traffic profiles for different types of users or to create new profiles. For this project, we will create our own profile.

14. From the **Object Palette**, click on the **FR PVC Config** object and click to paste it on the project grid. Right click to release the object. Right click on the **FR PVC Config** object and choose **SetName**. Name the object **PVCConfig**. Again, from the **Object Palette**, click on the **Profile Config** object and click to paste it on the project grid. Right click to release the object. Right click on the **Profile Config** object and choose **SetName**. Name the object **ProfileConfig**. Finally, from the **Object Palette**, click on the **Application Config** object and click to paste it on the project grid. Right click to release the object. Right click on the **Application Config** object, and choose **SetName**. Name the object **ApplicationConfig**. Verify that your project looks like the screen image that follows.

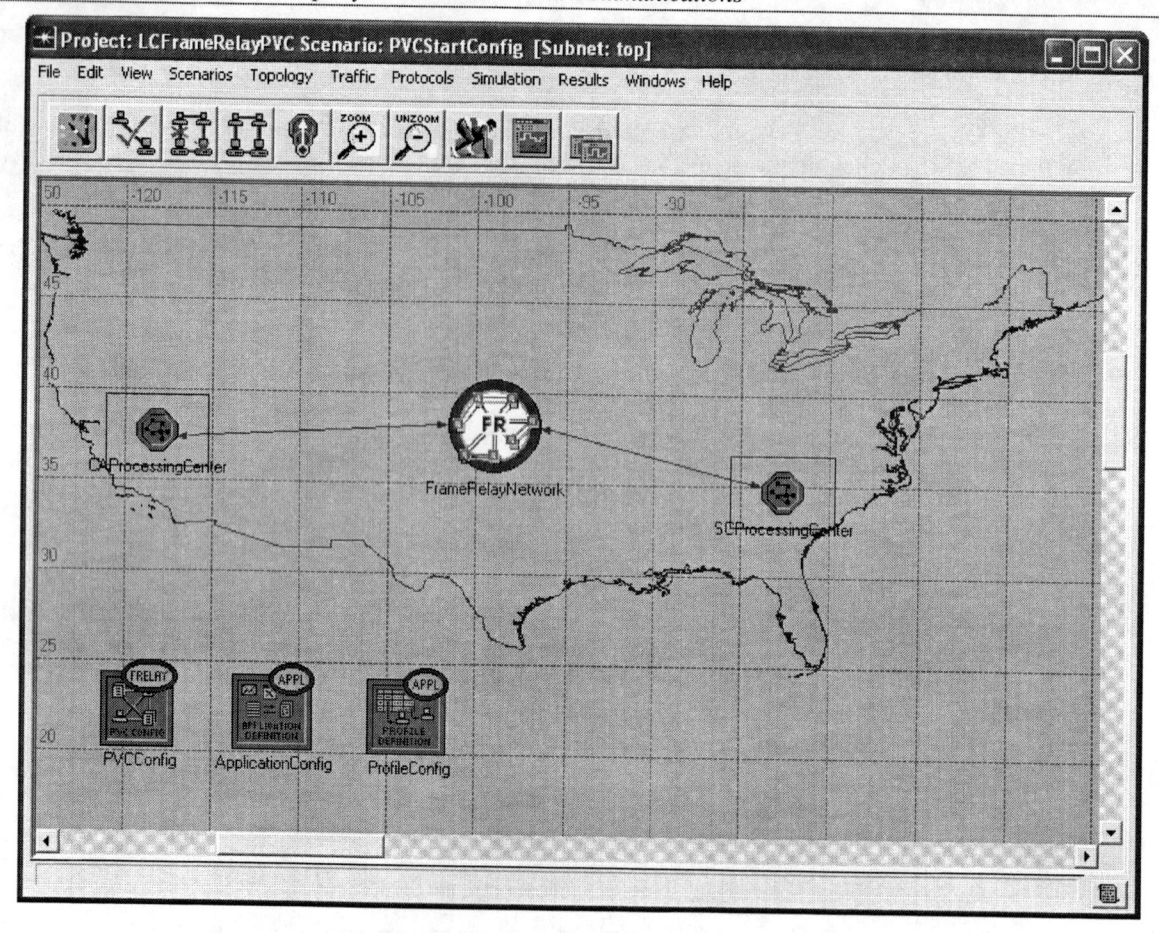

OPNET Lab Manual to accompany Business Data and Communications

[Screenshot of OPNET IT Guru project window showing LCFrameRelayPVC Scenario: PVCStartConfig with a US map containing CAProcessingCenter, FrameRelayNetwork, SCProcessingCenter nodes, and PVCConfig, ApplicationConfig, ProfileConfig objects.]

15. From the Main Menu, choose **File/Save** to save your project.

Configuring the Network

1. We will start by configuring the **ApplicationConfig** object to provide the applications we are interested in using during our simulation. OPNET IT Guru allows you to define new applications or use default descriptions for standard applications. We are interested in database and file transfer applications, which are predefined. Right click on the **ApplicationConfig** object and choose **Edit Attributes**. Click on the value next to **Application Definitions**. Choose **Default** from the pull-down menu. This will allow us to use the default applications during our simulation. Click **OK**.

2. Now we will create a new traffic profile. We will call the profile **CreditUSA,** and it will describe the type of applications and traffic patterns we expect the CreditUSA processing center employees to use. The profile will include **Database (Heavy)** and **File Transfer (Heavy)** applications. Right click on the **ProfileConfig** object, and choose **Edit Attributes**. Click on the value next to **Profile Configuration.** Choose **Edit** from the pull-down menu. Click on the value next to **Rows** and choose **1** from the pull-down menu. Under **Profile Name,** type CreditUSA. Choose **Simultaneous** under **Operation Mode**. This will allow the database and file transfer applications to run at the same time.

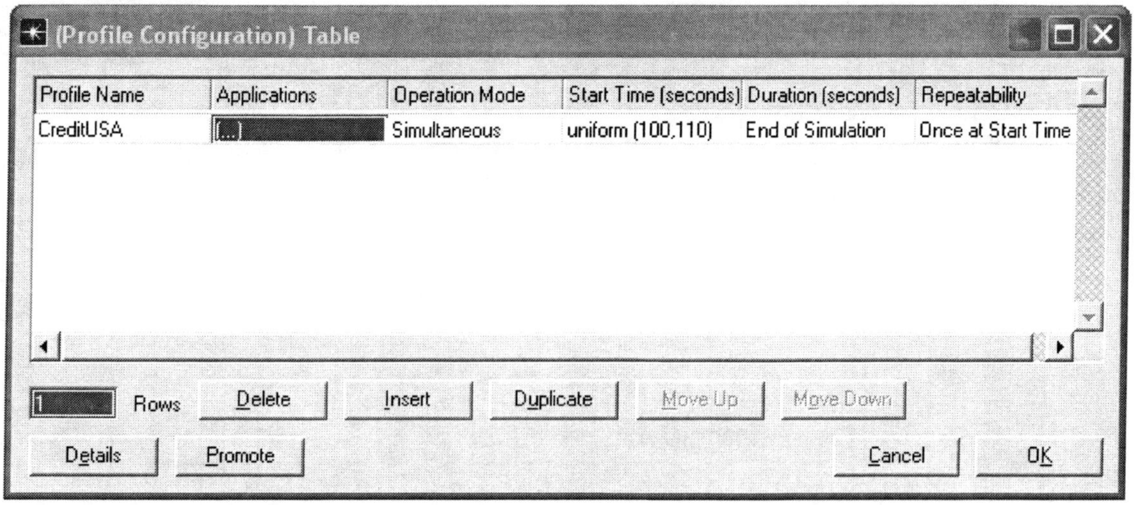
(Profile Configuration) Table
Profile Name
Applications
Operation Mode
Start Time (seconds)
Duration (seconds)
Repeatability
CreditUSA
(...)
Simultaneous
uniform (100,110)
End of Simulation
Once at Start Time
1
Rows
Delete
Insert
Duplicate
Move Up
Move Down
Details
Promote
Cancel
OK

3. Now we will choose the applications that go with this profile. Under **Applications** choose **Edit**. A new window will appear. Click on the value next to **Rows,** and choose **2** from the pull-down menu. In the first row, under **Name**, choose **Database Access (Heavy)** from the pull-down menu. In the second row, under **Name**, choose **File Transfer (Heavy)** from the pull-down menu. Click **OK** to close this window. Click **OK** again to close the **Applications Table** window.

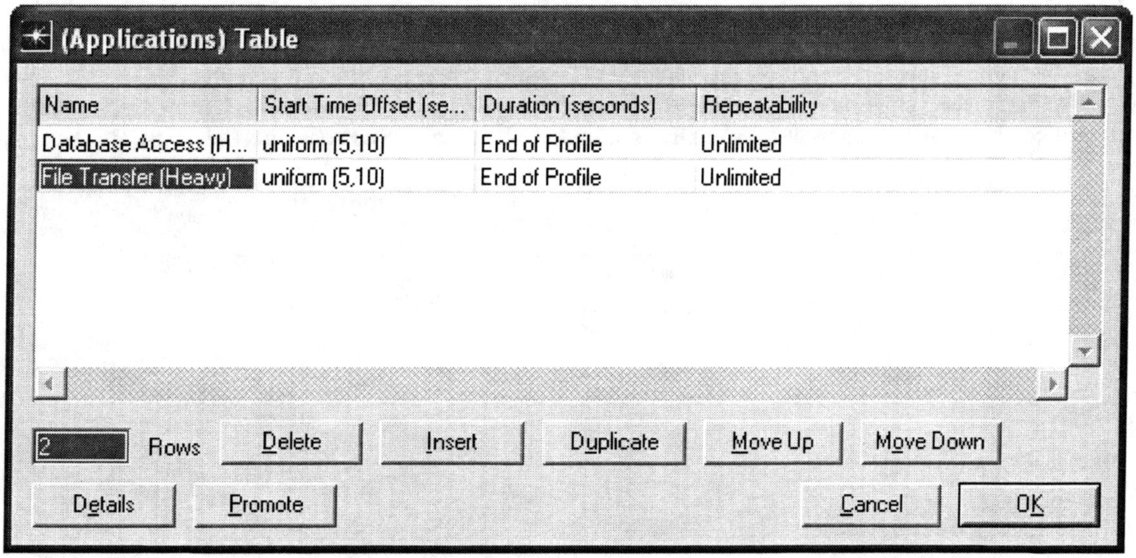
(Applications) Table
Name
Start Time Offset (se...
Duration (seconds)
Repeatability
Database Access (H...
uniform (5,10)
End of Profile
Unlimited
File Transfer (Heavy)
uniform (5,10)
End of Profile
Unlimited
2
Rows
Delete
Insert
Duplicate
Move Up
Move Down
Details
Promote
Cancel
OK

4. Now we will give the CAProcessing Center the CreditUSA profile. Double click on the CA Processing subnet to descend into the subnet. Click on the **LAN** object and choose **Edit Attributes.** Next, click on the + next to **Application: Supported Profiles** to expand the parameters. Click on the value next to **Application: Supported Profiles** and choose **Edit.** A new window will appear.

5. Click on the value listed next to **Rows** (displayed in the bottom left hand corner), and choose **1** from the pull-down menu. You will see a row added to the table. Now click on the value under **Profile Name.** Choose CreditUSA from the pull-down menu. This configures the LANs to create traffic with the CreditUSA profile. Click **OK.** Click **OK** again to close the **Application: Supported Profiles Table.**

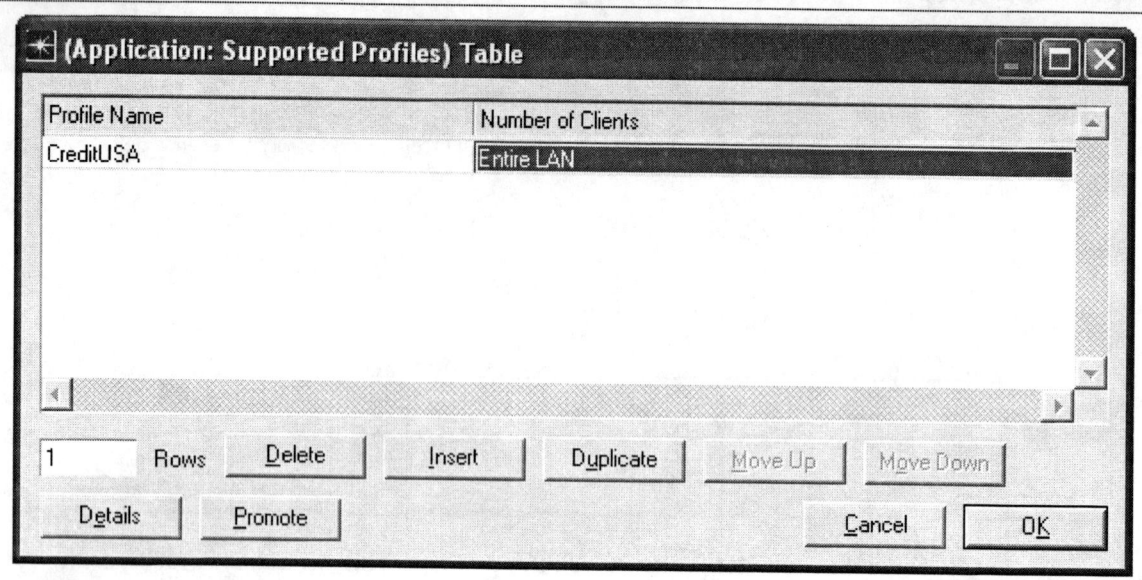

(Application: Supported Profiles) Table

Profile Name	Number of Clients
CreditUSA	Entire LAN

1 Rows

6. We can configure the **PVC Config** object, or we can configure each PVC individually through the router object that is connected to the frame relay network. Since we are in the subnet, we will configure the router PVC. Click on the **CARouter** object and choose **Edit Attributes**. Click on the value next to **Frame Relay Network PVC Configuration** and choose **Edit**. A new window will appear.

7. Click on the value listed next to **Rows**, and choose **1** from the pull-down menu. You will see a row added to the table. Now click on the value under **PVC Name**. Choose **Edit,** and give the PVC a name such as **CAPVC**. Now click on the value under **Source FRAD**. Choose **CAProcessingCenter.CARouter** as the source. Now click on the value under **Destination FRAD**. Choose **SCProcessingCenter.SCRouter**.

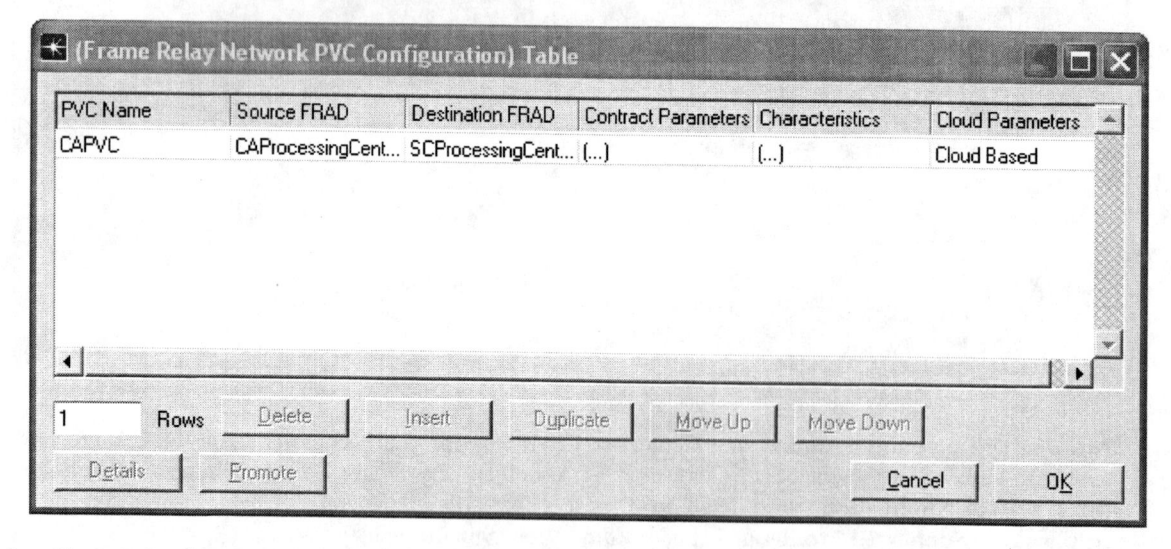

(Frame Relay Network PVC Configuration) Table

PVC Name	Source FRAD	Destination FRAD	Contract Parameters	Characteristics	Cloud Parameters
CAPVC	CAProcessingCent...	SCProcessingCent...	(...)	(...)	Cloud Based

1 Rows

8. Change the **Contract Parameters** to those shown in the following figure. We will ask for an **Outgoing CIR rate (bits/sec)** of **64,000,** an **Outgoing Bc (bits)** of **32,000,** and an **Outgoing Be (bits)** of **32,000**. Bc is the committed burst rate while Be is the excess burst rate. The speed of 64Kbps is a popular CIR choice. This speed is the speed of a DS0, or standard telephone line. High data rates are expensive; therefore, 64Kbps is a popular choice because it is a moderate speed that is very affordable. Click **OK** to close the window.

Lab 6   Frame Relay

Attribute	Value
Outgoing CIR (bits/sec)	64,000
Outgoing Bc (bits)	32,000
Outgoing Be (bits)	32,000
Incoming CIR (bits/sec)	Same
Incoming Bc (bits)	Same
Incoming Be (bits)	Same

(Contract Parameters) Table

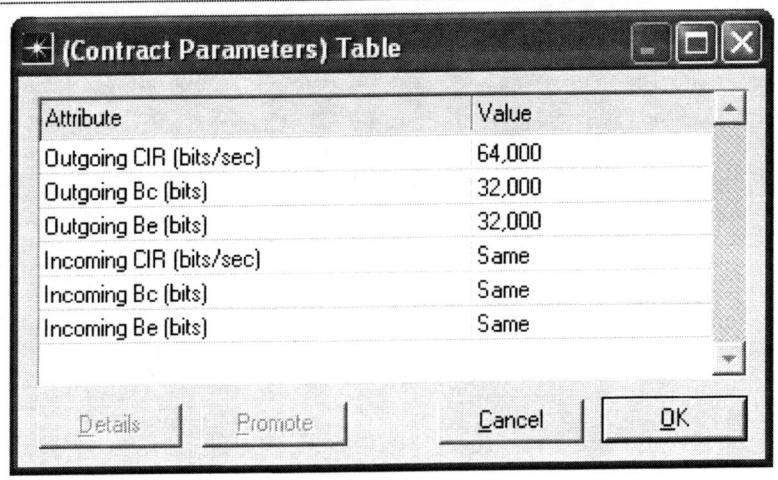

9. Click **OK** to close the PVC Configuration window and then click **OK** again to close the Router Attribute window.

10. Click on the **Arrow Button** to go up to the next higher level of the project hierarchy.

11. Choose **File/Save** from the Main Menu to save your project.

12. Now we will configure the South Carolina Processing Center LAN so that it will provide database and file transfer services. We will also configure the SCRouter PVC. Double click on the **SCProcessing** subnet to descend into the subnet. Click on the **LAN** object, and choose **Edit Attributes**. Next, click on the value next to **Application: Supported Services,** and choose **Edit**. A new window will appear.

13. Click on the value listed next to **Rows,** and choose **2** from the pull-down menu. You will see two rows added to the table. Now click on the value under **Name**. Choose **Database Access (Heavy)** from the pull-down menu. In the second row under **Name**, choose **File Transfer (Heavy)** from the pull-down menu. Now the South Carolina LAN will provide access to these applications. Click **OK**. Click **OK** again to close the Applications: Supported Services Table.

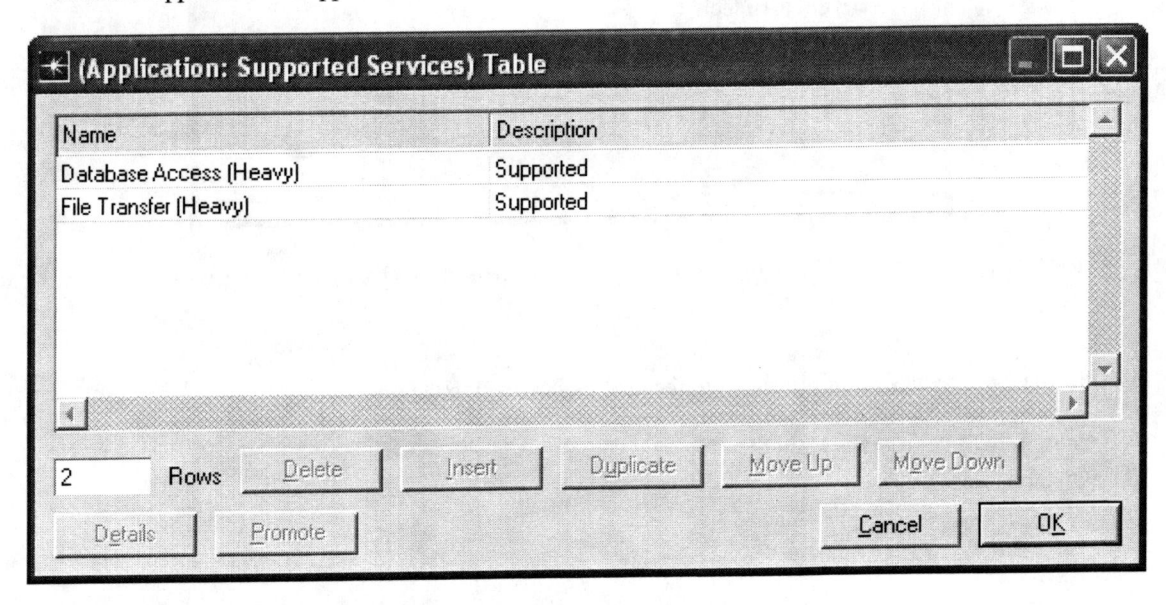

(Application: Supported Services) Table

Name	Description
Database Access (Heavy)	Supported
File Transfer (Heavy)	Supported

2 Rows

14. Now we will configure the SCrouter's PVC. We will use the same values we used for the CARouter PVC. Click on the **SCRouter** object, and choose **Edit Attributes**. Click on the value next to **Frame Relay Network PVC Configuration,** and choose **Edit**. A new window will appear.

111

OPNET Lab Manual to accompany Business Data and Communications

15. Click on the value listed next to **Row,s** and choose **1** from the pull-down menu. You will see a row added to the table. Now click on the value under **PVC Name**. Choose **Edit,** and give the PVC a name such as **SCPVC**. Now click on the value under **Source FRAD**. Choose **SCProcessingCenter.SCRouter** as the source. Now click on the value under **Destination FRAD**. Choose **CAProcessingCenter.CARouter**.

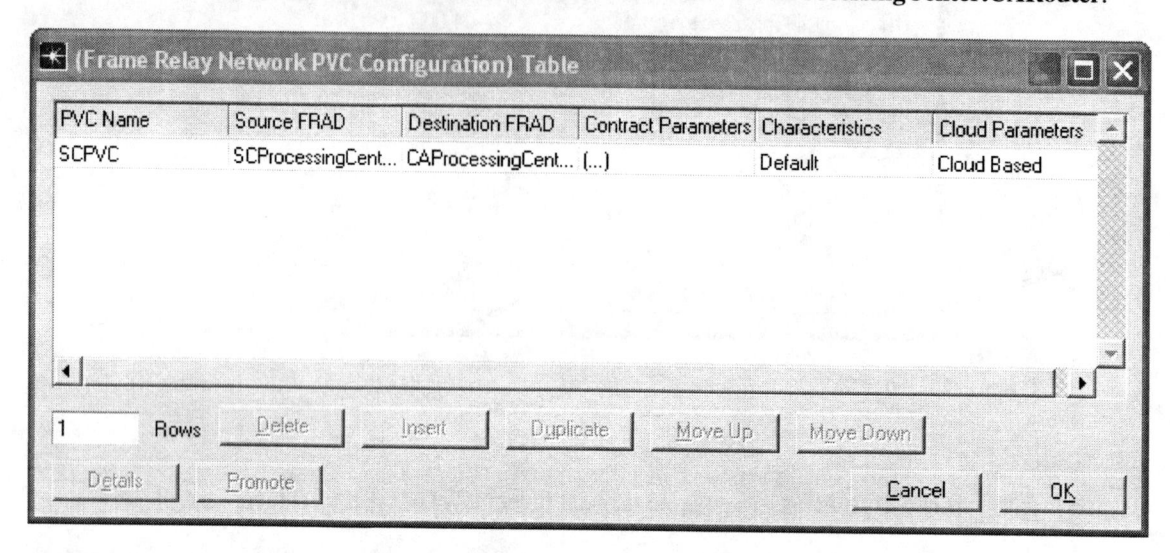

16. Change the **Contract Parameters** to those shown in the figure below. Again we will use an **Outgoing CIR (bits/sec)** rate of **64,000**, an **Outgoing Bc** of **32,000**, and an **Outgoing Be** of **32,000**. Click **OK** to close the Contract Parameters Table.

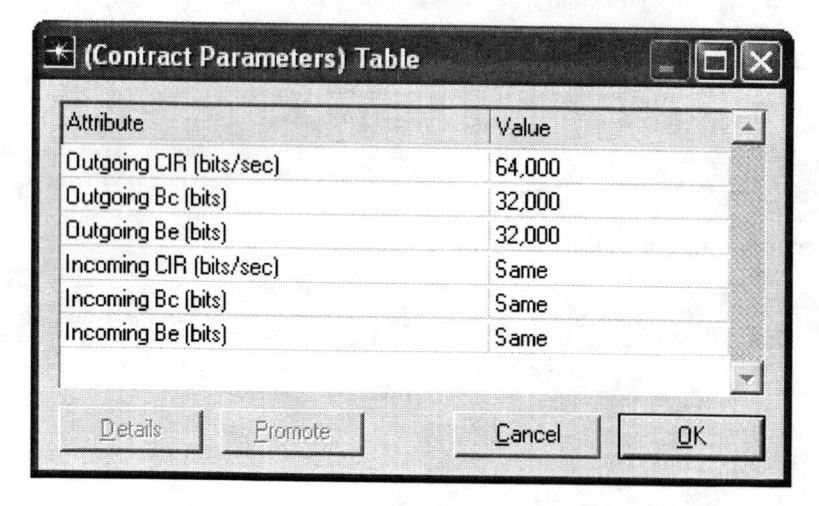

17. Click **OK** to close the PVC Configuration window and then click **OK** again to close the Router attribute window.

18. Click on the **Arrow Button** to go to the next higher level in the project hierarchy.

19. Choose **File/Save** from the Main Menu to save your project.

Lab 6   Frame Relay

Configuring the Simulation and Choosing Statistics

Now we would like to choose the statistics that will reflect response time for our applications. We want to be able to evaluate our choices for improving response time. Our choices are to pay for a higher CIR or to pay for faster links to the frame relay network.

1. Right click somewhere on the project workspace and select **Choose Individual Statistics** from the pull-down menu. Click on the + next to **Global** to expand the parameters. First, we will select statistics for the database application. Click on the + next to **DB Entry** to expand the options. Click on the box next to **Response Time (sec)**, **Traffic Received (bytes/sec)**, and **Traffic Sent (bytes/sec)**. Click on the + next to **DB Query** to expand the options. Again, choose **Response Time (sec)**, **Traffic Received (bytes/sec,)** and **Traffic Sent (bytes/sec)** statistics.

2. Next we will select the statistics for the File Transfer (FTP) application. Click on the + next to **FTP** to expand the options. Click on the box next to **Download Response Time (sec)**, **Upload Response Time (sec)**, **Traffic Received (bytes/sec)**, and **Traffic Sent (bytes/sec)**. Click **OK** to close the window.

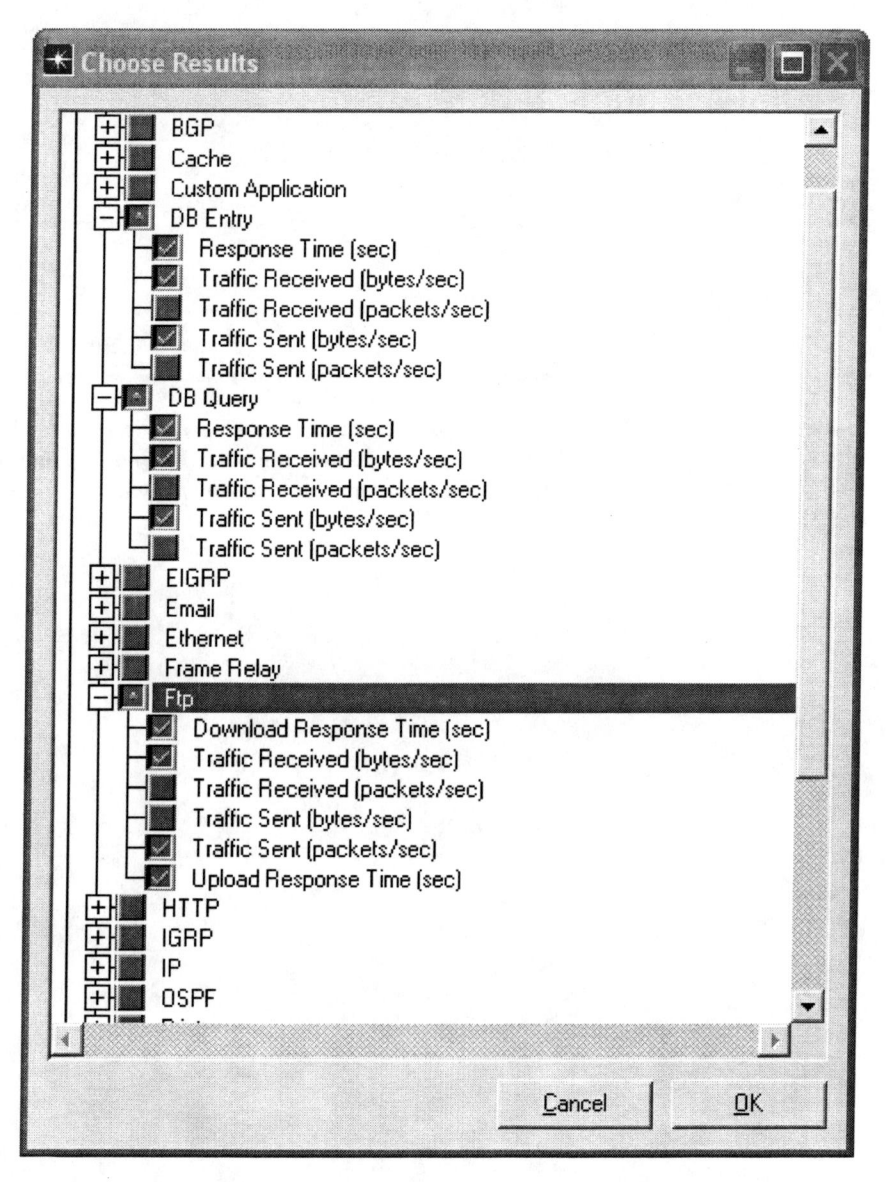

113

OPNET Lab Manual to accompany Business Data and Communications

Running the Simulation

1. Click on the **Configure and Run** Button. Change the **Duration** of the simulation to **10**. Choose **Minute(s)** from the pull-down menu. Click the **Run** button to run the simulation. Depending on your processor it should take several minutes.

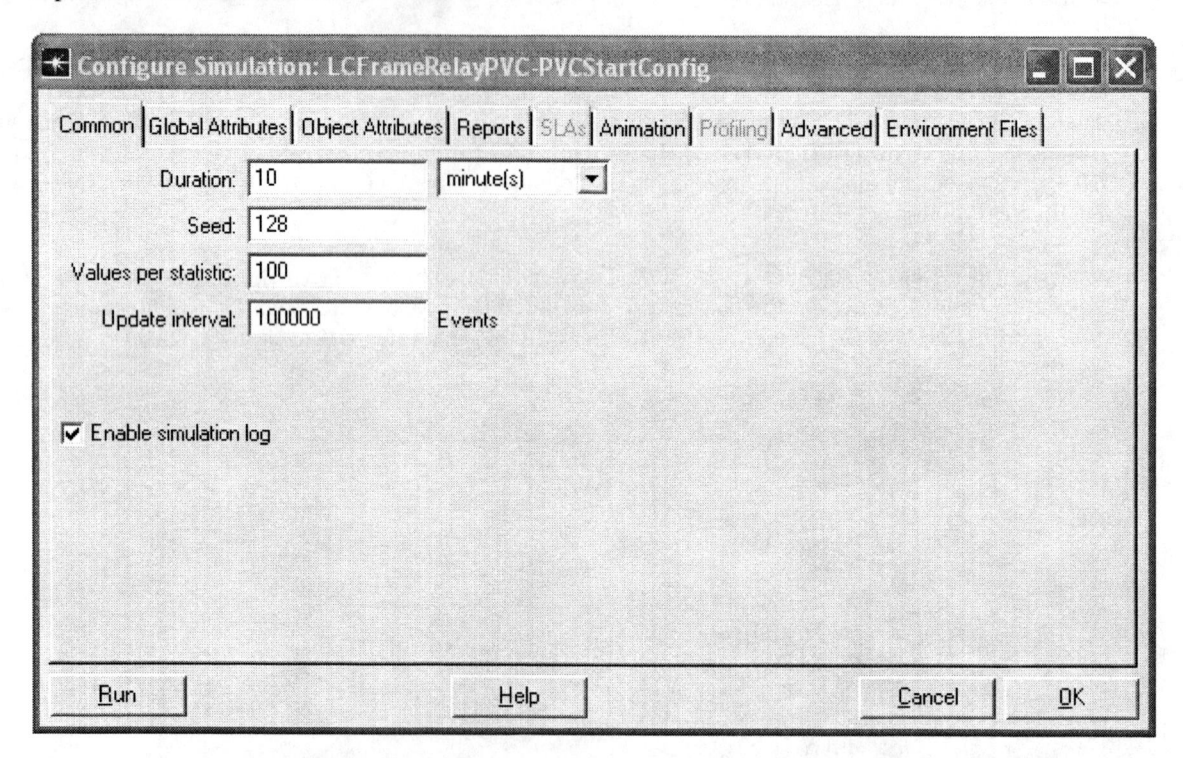

2. To view results, choose **Results/View Results** from the Main Menu. To view results for **Database Query** response time, click on the **+** next to **Database Query,** and click next to **Response Time (seconds).** You can view the results **As Is** or select **Time Average** from the pull-down menu. Choose **time_average.** Click on the **Show** button to display the graph in a new window. Your graph should resemble the next figure shown.

Lab 6   Frame Relay

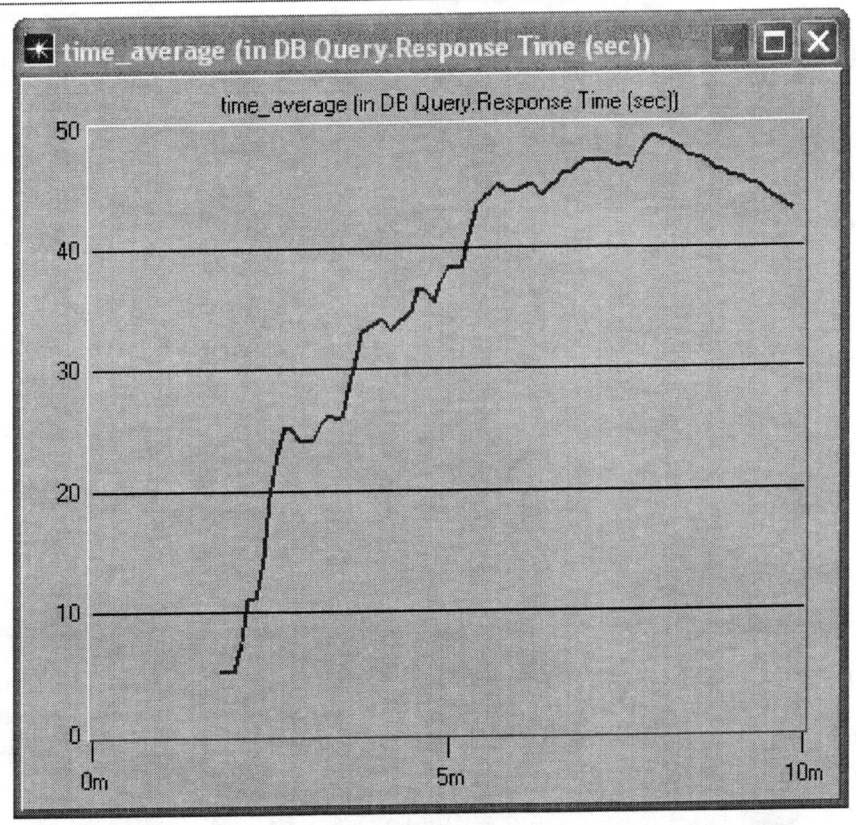

[Figure: time_average (in DB Query.Response Time (sec)) graph showing response time rising from near 0 at 0m to about 45-50 at 10m]

Notice that the response time is fairly high and is increasing over time. Click on the **X** to close the window. The **Close Analysis** window will appear asking you if you want to hide the panel or permanently delete it. Click on the **Delete** button to delete the window.

Before we look at all the statistics for our original Start Configuration, we will create a second scenario with a higher Committed Information Rate (CIR), and a third scenario with faster links. We will then compare results.

Creating Two New Scenarios

1. Our purpose is to evaluate two options for increasing application response time. We will begin by creating a scenario in which the link from the processing center routers to the frame relay network is faster. We will upgrade the links to T3 lines that run at 1.544 Mbps. From the Main Menu, choose **Scenarios/Duplicate Scenario**. Give the new scenario a name such as **PVCLinkT3**.

2. Click on the **T1 link** to select it. Now, right click on the **T1 link** and choose **Select Similar Links**. Right click on one of the selected links and select **Edit Attributes** from the pull-down menu. Click on **Apply Changes to Selected Objects**. Click on the value next to **model** and change the value from **FR_T1** to **FR_T3**.

3. From the Main Menu, choose **Simulation/Run Discrete Event Simulation** to gather results for the new scenario.

4. Choose **File/Save** from the Main Menu to save your project.

5. Now, we will create a third scenario with T1 links and a CIR rate that is twice what we specified previously. We will also increase burst rate. From the Main Menu choose **Scenarios/Previous Scenario**. You should see the name of your first scenario (**PVCStart**) at the top of the window. Now, from the Main Menu, choose **Scenarios/Duplicate Scenario**. Give the new scenario a name, such as **PVC128**.

6. Double click on the SCProcessing Center subnet. Click on the **SCRouter** object and choose **Edit Attributes**. Click on the value next to **Frame Relay Network PVC Configuration** and choose **Edit**. A new window will appear.

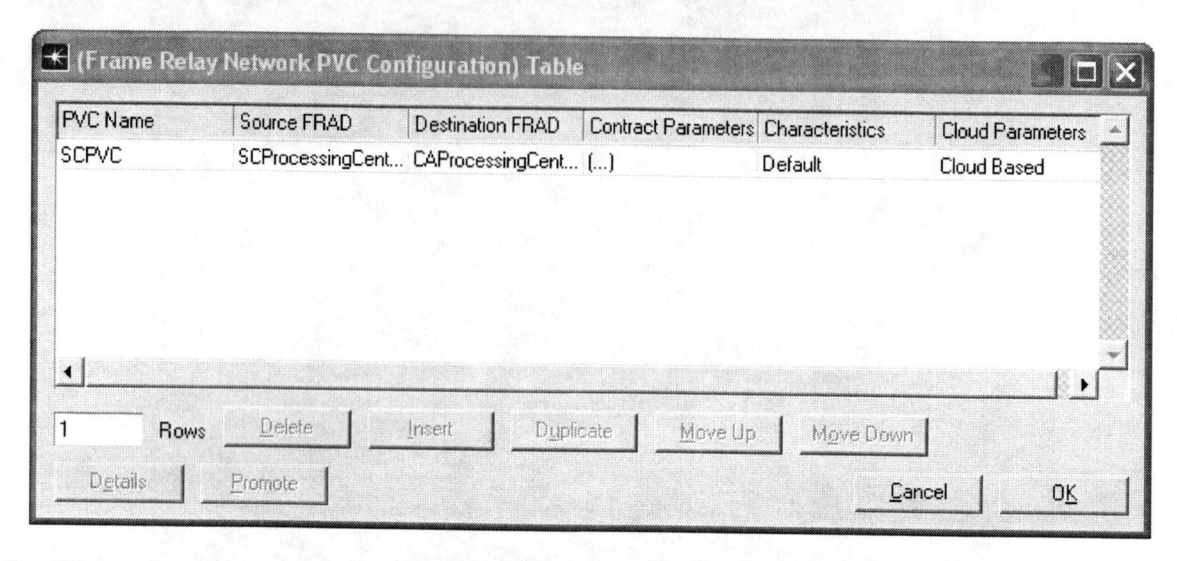

```
(Frame Relay Network PVC Configuration) Table

PVC Name   Source FRAD        Destination FRAD   Contract Parameters  Characteristics   Cloud Parameters
SCPVC      SCProcessingCent...  CAProcessingCent...  (...)              Default           Cloud Based

1 Rows   Delete   Insert   Duplicate   Move Up   Move Down
Details   Promote                                  Cancel   OK
```

7. Click on the value under **Contract Parameters** and choose **Edit**. Change the parameters to those shown in the next figure. We will double the **Outgoing CIR (bits/sec)** rate to **128,000** and double the burst rate **Outgoing Bc** and **Outgoing Be** to **64,000**. Click **OK** to close the **Contract Parameters Table**.

```
(Contract Parameters) Table

Attribute                 Value
Outgoing CIR (bits/sec)   128,000
Outgoing Bc (bits)        64,000
Outgoing Be (bits)        64,000
Incoming CIR (bits/sec)   Same
Incoming Bc (bits)        Same
Incoming Be (bits)        Same

Details   Promote          Cancel   OK
```

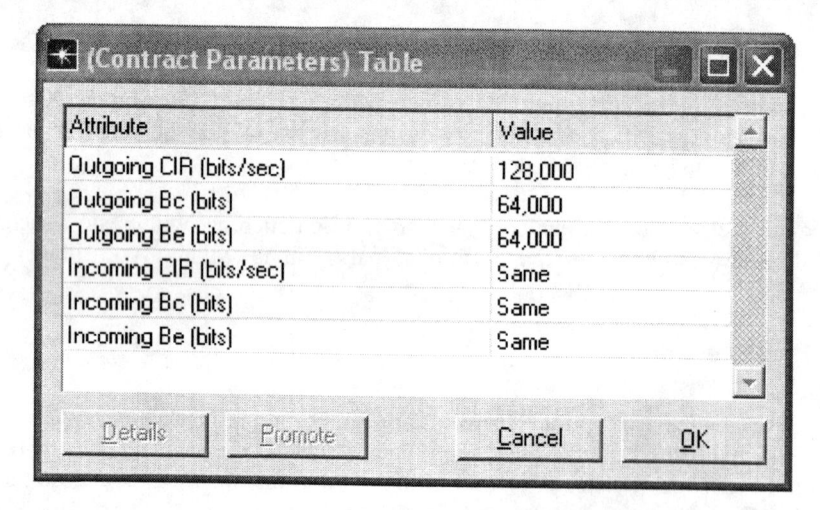

8. Click **OK** to close the PVC Configuration window and then click **OK** again to close the Router attribute window. Repeat the same steps to increase the **CARouter** PVC Outgoing CIR (bits/sec) to **128,000** and the **Outgoing Bc** and **Outgoing Be** to **64,000**.

9. Choose **Simulation/Run Discrete Simulation** to gather results for the new scenario.

10. Choose **File/Save** from the Main Menu to save your project.

Comparing Results

1. We can now compare results for all three scenarios. To do this, choose **Results/Compare Results** from the Main Menu. To compare results for **Database Query Response Time**, click on the + next to **Database**, and click next to **Query Response Time (seconds)**. Choose **time_average** from the pull-down menu and click **Show**. This will give you a graph with data from all three scenarios, as well as a legend for the graph. Your graph should resemble the one shown in the next window image.

Lab 6   Frame Relay

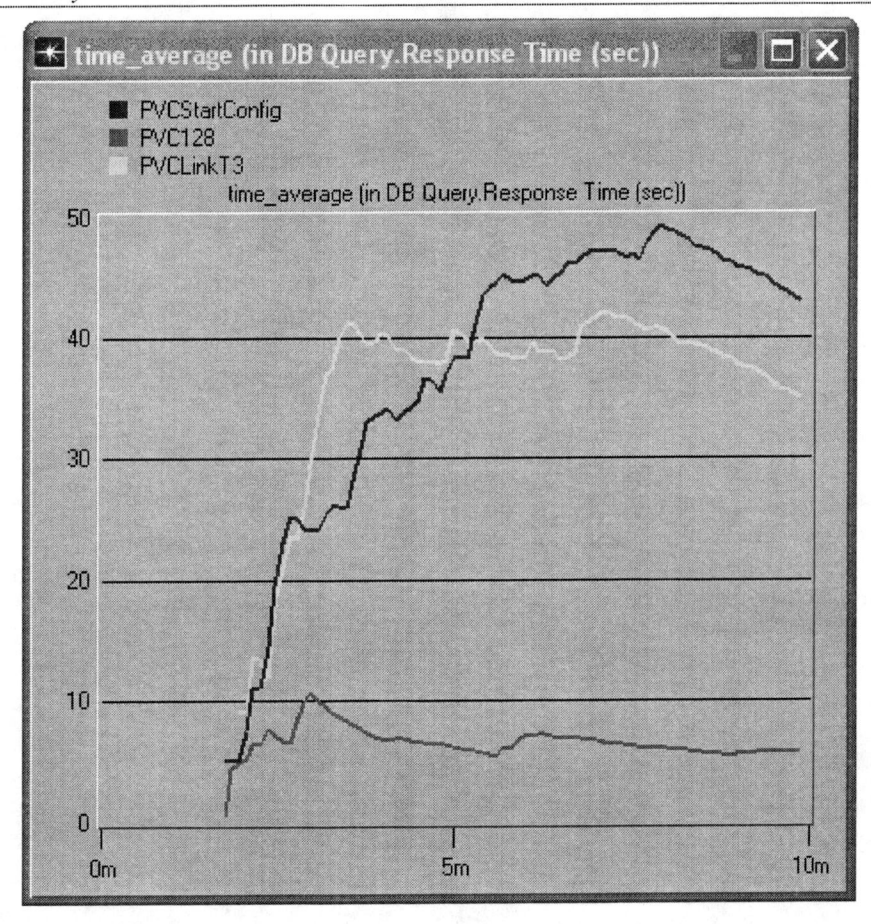

[Figure: time_average (in DB Query.Response Time (sec)) graph showing PVCStartConfig, PVC128, and PVCLinkT3 curves from 0m to 10m]

If you would like to **Zoom In** on an area of the graph, you may left click on the graph and drag a square around the area that you would like to enlarge.

Click on the **X** to close the window. The **Close Analysis** window will appear asking you if you want to hide the panel or permanently delete it. Click on the **Delete** button to delete the window.

You may discard the results of your simulations, rerun them, and manage your scenarios by doing the following:

Choose **Scenarios/Manage Scenarios** from the Main Menu. A window will appear with the names of your scenarios, the length of time that the simulation ran, and the status of the data.

By clicking on a scenario name, you may choose to **Discard Results** or **Collect Results**. By clicking on the values under the **Results** column, you may choose **recollect** or **collect** from the pull-down menu. Click **OK** to close the window.

Using the results you obtained in your simulations, answer the following questions.

Questions

1. In the first scenario, when the CIR was 64Kbps, what was the maximum response time for a database update? What was the minimum response time for a database update?
2. In the first scenario, when the CIR was 64Kbps, what was the maximum response time for downloading a file? What was maximum response time for uploading a file? What was the difference between the maximum and the minimum response times for uploading and downloading?

3. Based on what you observed, what would you advise CreditUSA to do in order to decrease application response time? Explain your answer.
4. How does burst speed affect your results? Change the PVC configuration of the scenario to use a B_e value of 0 in the scenario where the CIR rate was 128 Kbps. Does this change the response times? How do response times compare to those you created with the slower PVC value?
5. We saw that the database query response time did not improve when T3 links were used instead of T1 links. Explain why faster links did not help response time.
6. Change the attributes for the **Frame Relay** object. Click on the object and choose **Edit Attributes**. Change the **Packet Discard Ratio %** to **10%**. Now choose **Individual Statistics** for the **Frame Relay** Object. Choose **Packet Drops, Traffic Forwarded (packets/second),** and **Traffic Received (packets/second)**. Run the simulation and describe the results you observed.

Summary Questions

1. Do you think results would vary if you used different applications? Try creating a new profile called **HomeUser** that uses Web Browsing (Light HTTP1.1) and **Email (Heavy)**. Run your three simulations again with this traffic profile. What is your advice for improving response time based on this profile?
2. Locate a carrier and price out the cost for PVCs of various speeds. Does doubling the speed result in doubling the cost? How much more does it cost to upgrade from a T1 line to a T3 line? Note that the length of the link may be a factor.

LAB 7   ETHERNET HUB VERSUS SWITCH

Objectives

1. To compare the performance of 10BaseT hub, 10BaseT switch, and 100BaseT hub LAN connection devices.

Motivation

PopMag is a small company that publishes a local magazine containing articles that describe the local art and music scene. PopMag's circulation is growing steadily. They have recently hired several new employees and expect to hire more by the end of the year. PopMag currently uses a shared hub, 10BaseT LAN, for all of their desktop publishing and communication needs. They are currently considering migrating to a new technology to support their expansion plans. PopMag's options are to keep the line speed at 10 Mbps and replace the hub with a switch or to increase line speed to 100 Mbps and replace the 10BaseT hub with a 100BaseT hub. Satchel Wong has taken over the system administration duties and has borrowed equipment to run tests in order to choose the best and most cost-efficient solution for the company.

Description

Most Local Area Networks (LANs) are broadcast networks, meaning that all messages sent on the LAN are "heard" by all hosts on the LAN. Only the host for whom the message is addressed will receive the message. 10BaseT Ethernet LANs operate at 10 Mbps and utilize the Carrier Sense Multiple Access Collision Detection (CSMA/CD) protocol. Using CSMA/CD, a host must sense the line to discover if it is free before sending a packet. Collisions may occur in this environment if more than one host senses the line at the same time, finds it free, and tries to send a packet. 100BaseT Ethernet LANs operate at 100 Mbps and also use the CSMA/CD protocol. Due to their faster speed, 100BaseT LANs have a shorter maximum link length than 10BaseT LANs. Ethernet LANs traditionally connect hosts through a hub. Alternatively, a switch may connect hosts. Switches differ from hubs in that they treat each host to port connection as a separate CSMA/CD domain. The result is fewer collisions and greater throughput.

In this lab, we will examine throughput and delay statistics for an Ethernet LAN using these 10BaseT hub, 100BaseT hub, and 10BaseT switch connection devices.

Follow the directions presented next to create an OPNET model, run simulations, and gather statistics. Questions regarding the lab are listed at the end.

Create a New Project

1. Start the **OPNET IT Guru Academic Edition Application.** Create a **New Project** by selecting **File/New/Project** from the main menu. Click **OK**.

2. Give the project a name such as **your_initialsHubSwitch**. Give the scenario a name such as **Hub10**. Click **OK**. You should see the **Startup Wizard Initial Topology** dialog box.

3. Verify that **Create Empty Scenario** is selected in the **Initial Topology Dialog Box**. Click **Next**. You will now see the **Network Scale** dialog box.

4. Choose **Office** from the **Network Scale** list. Click **Next**.

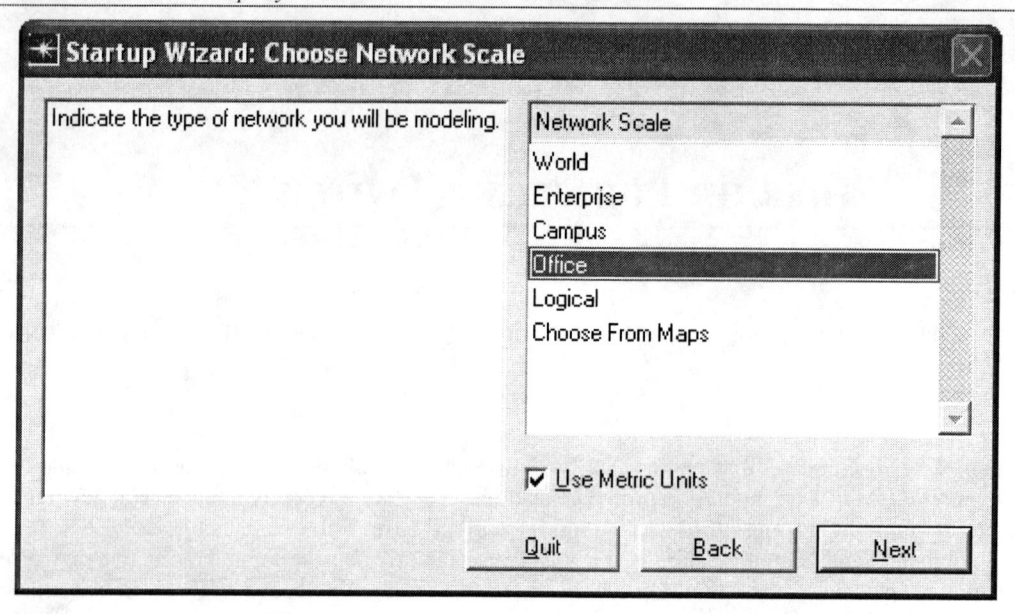

5. Choose **Meters** in the **Size** drop-down box. Type **100** for the **X span** and **100** for the **Y span.** Click **Next**. You should see the **Select Technologies** dialog box.

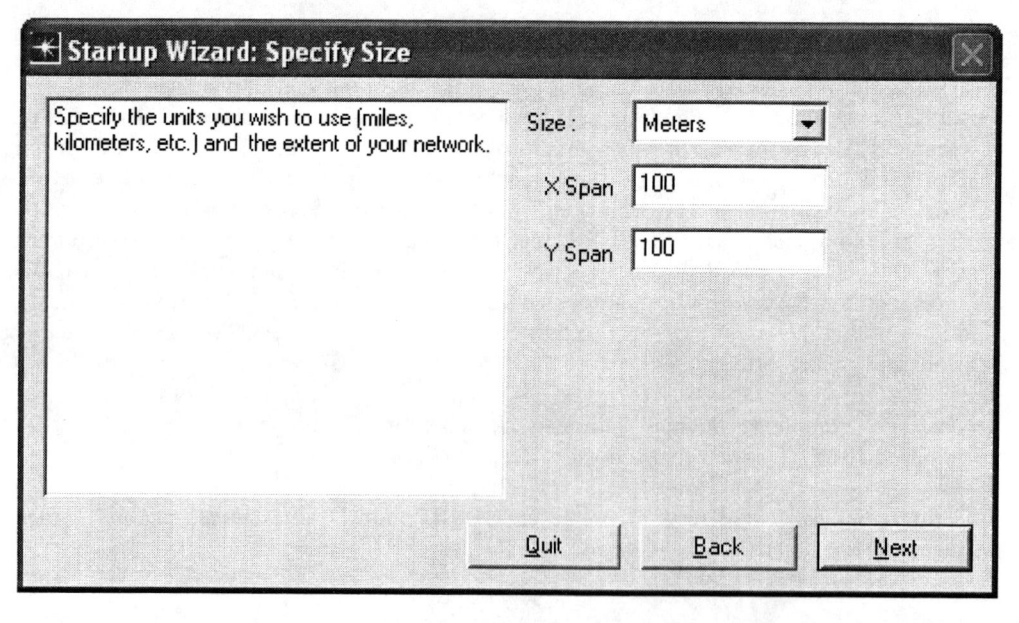

6. Scroll down until you see **Ethernet** under the **Model Family** column. Under the **Include column**, click on the box next to **Ethernet**. This will change the "no" to "yes," indicating that Ethernet technologies will be included. Click **Next.**

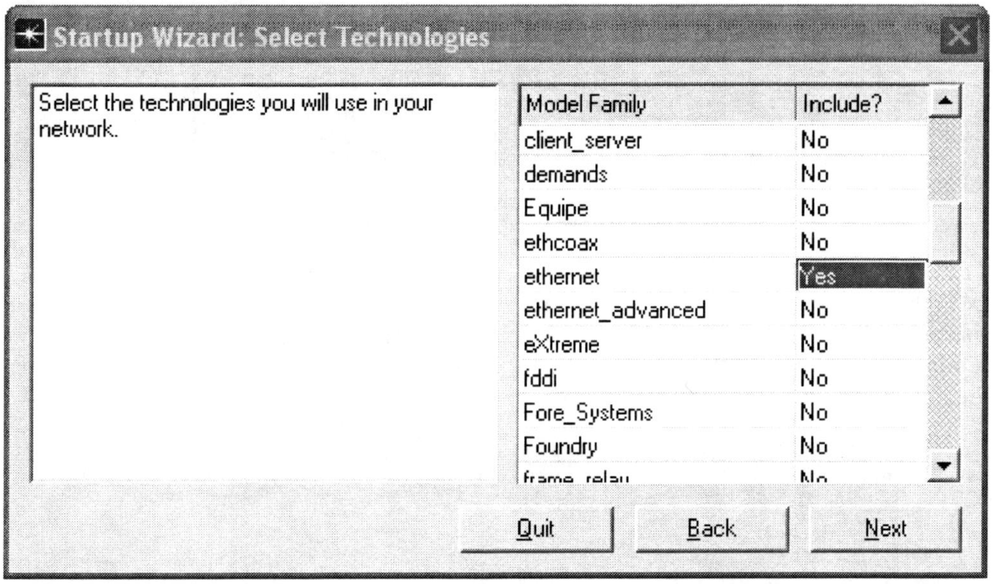

7. In the **Start of Wizard Review** dialog box, verify that the Ethernet technology was chosen and that the scale is Office 100m x 100m. Click **OK**. You should see the **Object Palette** window.

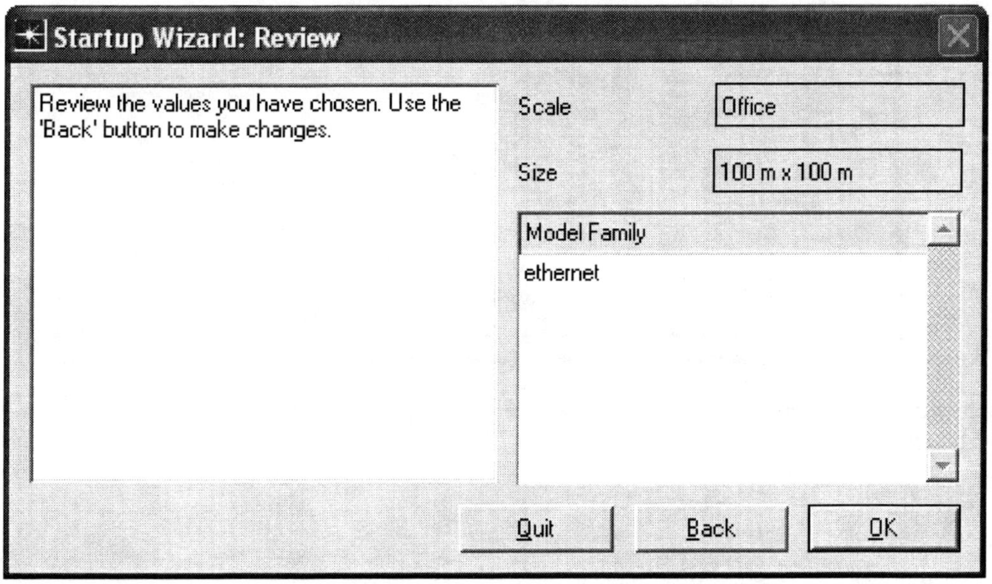

Building the Ethernet Network

1. We will begin by using Rapid Configuration to create the hub-based 10BaseT LAN. We will create a star network with 16 Ethernet stations connected to an Ethernet hub. 10BaseT links will be used to connect the hosts to the hub. From the Main Menu, choose **Topology/Rapid Configuration**. A window will appear prompting you to choose the topology or network shape. Choose **Star** from the pull-down menu.

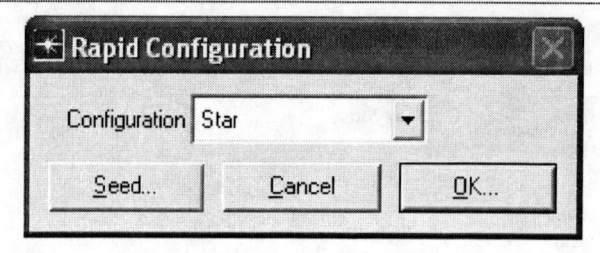

2. A new window will appear which will allow you to choose the nodes in the network. For the **Center node,** choose **ethernet16_hub**from the pull-down menu. For the **Periphery node** model, choose **ethernet_station**. Change the number of nodes to **16**. This will create 16 stations on the network. For the **Link Model**, choose **10BaseT**. Set the **X** coordinate to **50**, the **Y** coordinate to **50** and the **Radius** to **24**. This will put the LAN in the center of the workspace. Click **OK**.

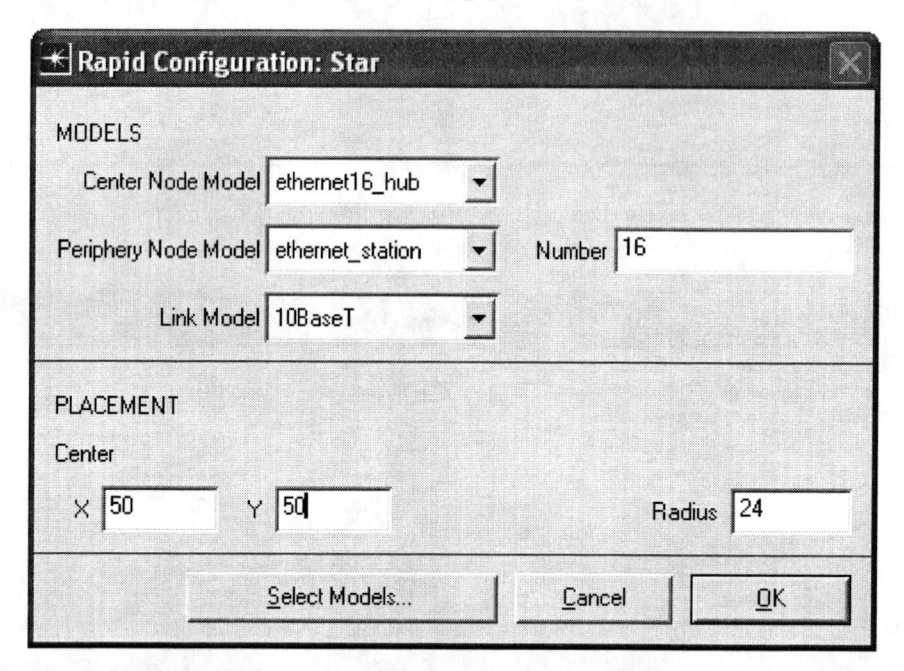

3. Right click on the hub object and choose **Set Name**. Name the hub **Hub10**. Your project should resemble the following:

Lab 7   Ethernet Hub versus Switch

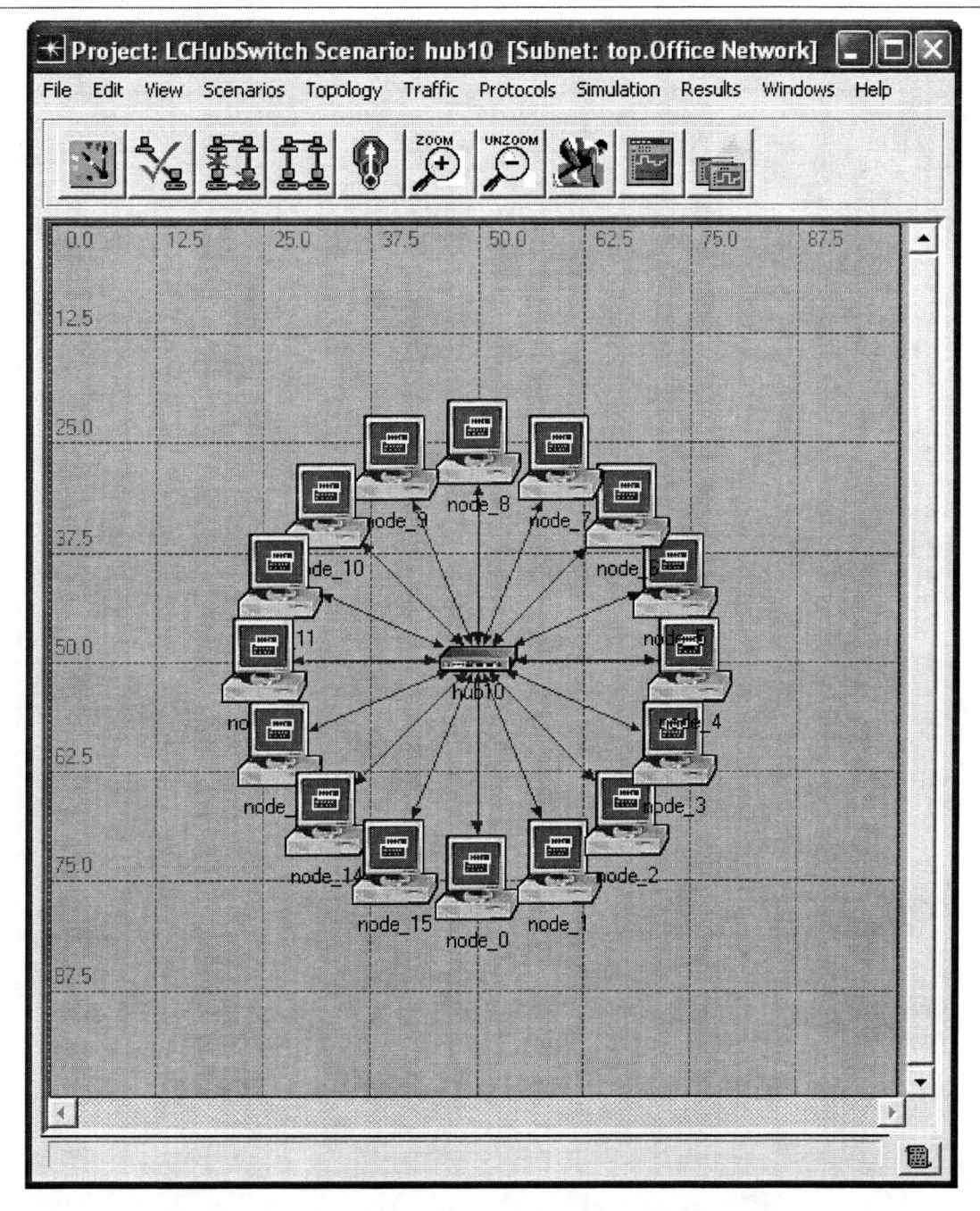

4. We will now configure the **ethernet_stations** so that they send a large amount of traffic. We will set the packet interarrival time and the packet size. Right click on one of the **ethernet_station** objects and choose **Select Similar Nodes** from the pull-down menu. Right click again on one of the stations and choose **Edit Attributes.**

5. Click on the box to **Apply Changes to Selected Objects.** Click on + next to **Traffic generation parameters** to expand the parameter list. You should see the following screen image:

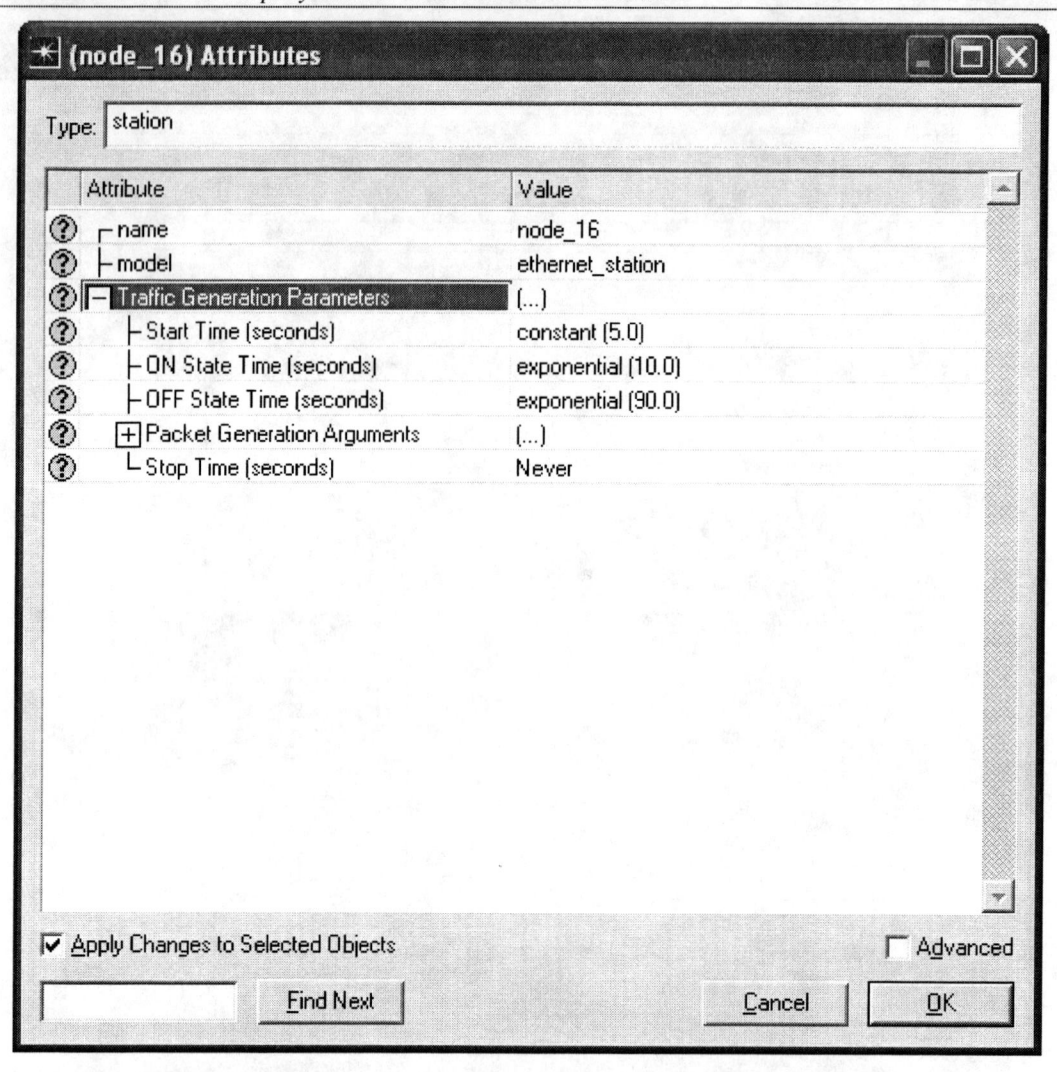

6. We will first configure the simulation start time. We would like our simulation to start up immediately. Click on the value next to **Start Time (Seconds)**. A new window will appear. Change the value to **constant (1)**. This will cause our simulation to start after 1 second. Click **OK**.

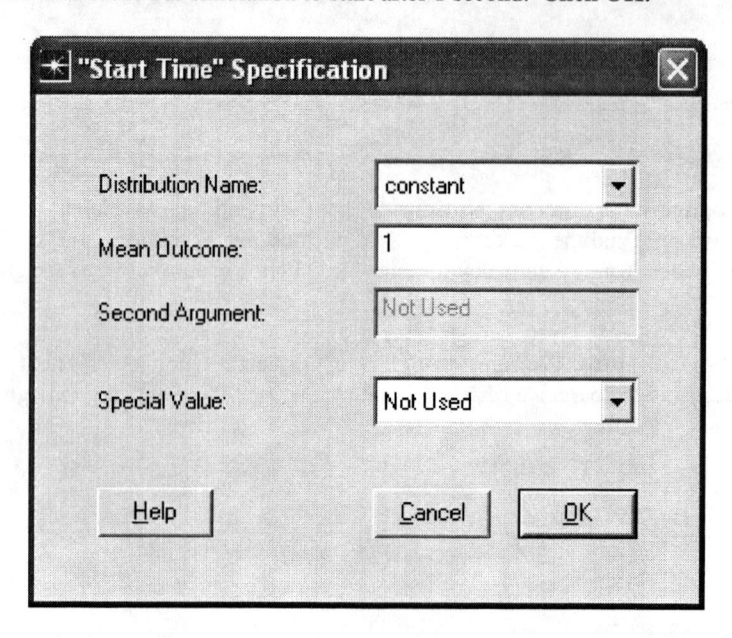

Lab 7   Ethernet Hub versus Switch

7. Next click on the value next to **ON State Time**. This will bring up a new window. Change the **Mean Outcome** to **100**. Click **OK**.

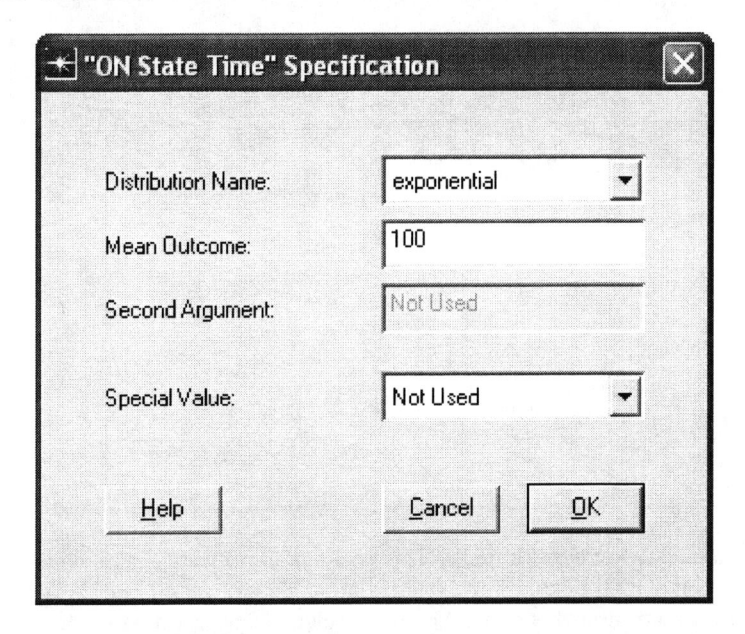

8. Click on the value next to **OFF State Time**. A new window will appear. Change the **Mean outcome time** to **0** and click **OK**. This causes our simulation to generate traffic 100% of the time.

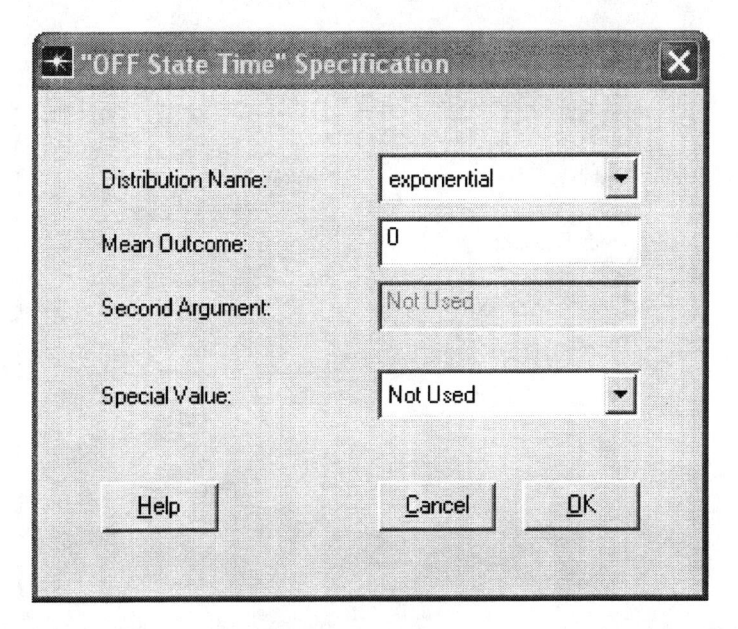

9. Now click on the **+** next to **Packet Generation Arguments**. We will change the interarrival time and the packet size. Click on the value next to **Interarrival time (seconds)**. This will bring up a new window. Change the **Mean outcome time** to **.01** and click **OK**.

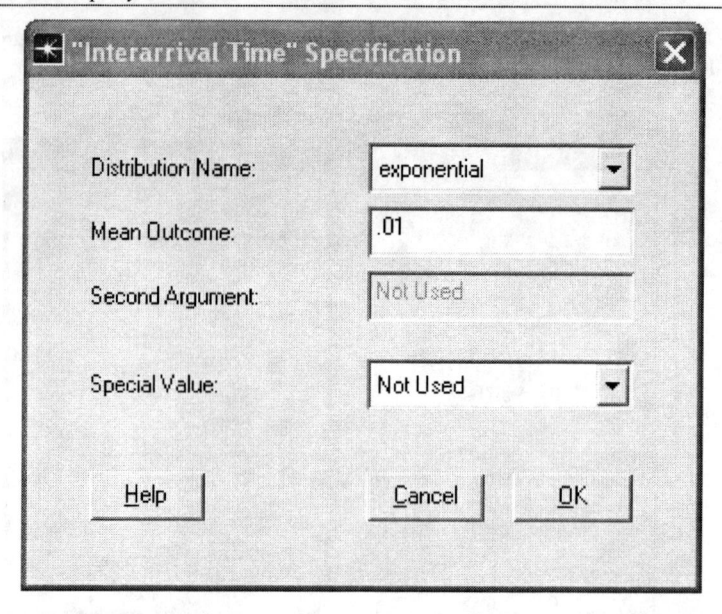

OPNET Lab Manual to accompany Business Data and Communications

[Dialog box: "Interarrival Time" Specification — Distribution Name: exponential; Mean Outcome: .01; Second Argument: Not Used; Special Value: Not Used]

10. Click on value after **Packet Size (bytes).** This will again bring up a new window. Ethernet packets can range between 46 and 1500 bytes. For this simulation we will set all our packets to be a constant size of 1000 bytes to make our throughput calculations simpler. Choose **Constant** from the **Distribution Name** pull-down menu. Next, change the **Mean Outcome** to **1000**. This will cause all packets generated to be 1000 bytes in size. Click **OK**.

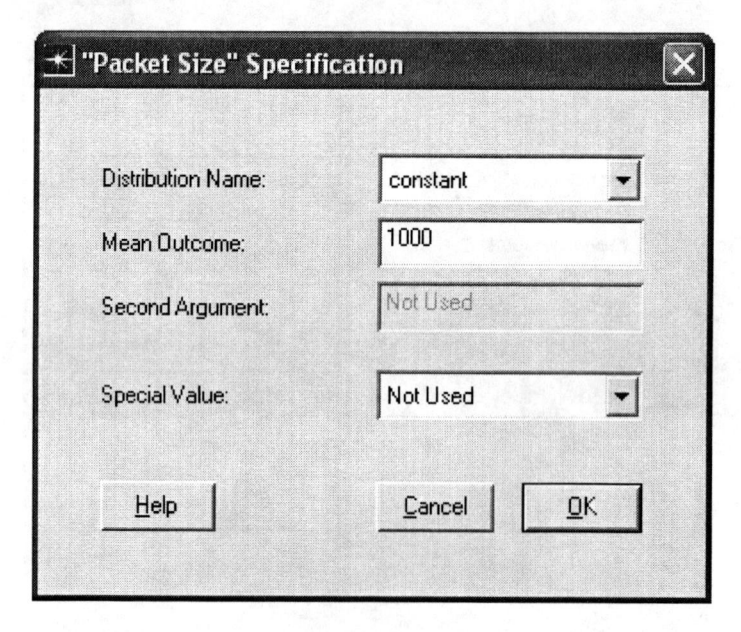

[Dialog box: "Packet Size" Specification — Distribution Name: constant; Mean Outcome: 1000; Second Argument: Not Used; Special Value: Not Used]

11. You should now have the values listed in the **Attribute** window below. Click **OK** to close the window.

(node_0) Attributes

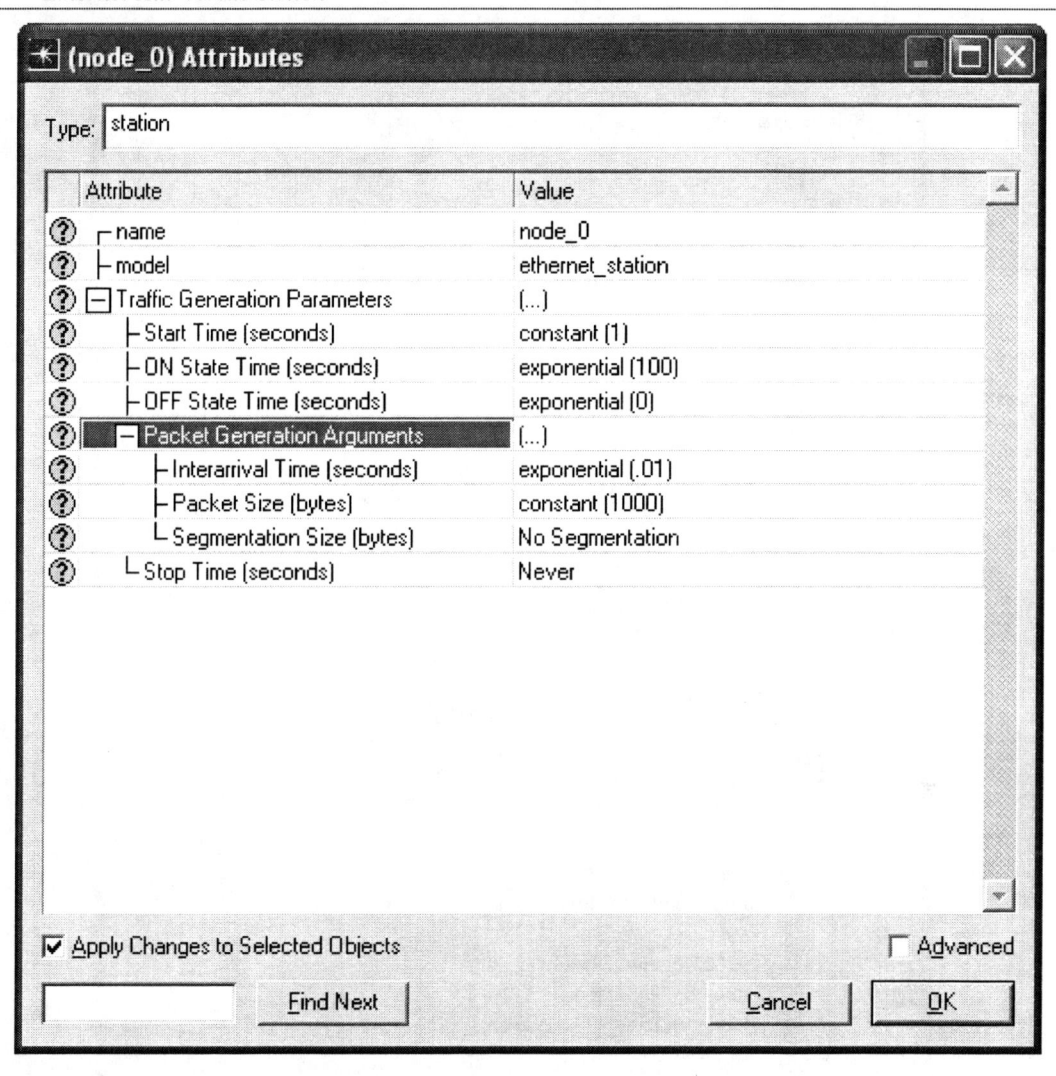

Type: station

Attribute	Value
⊢ name	node_0
⊢ model	ethernet_station
⊟ Traffic Generation Parameters	(...)
⊢ Start Time (seconds)	constant (1)
⊢ ON State Time (seconds)	exponential (100)
⊢ OFF State Time (seconds)	exponential (0)
⊟ **Packet Generation Arguments**	(...)
⊢ Interarrival Time (seconds)	exponential (.01)
⊢ Packet Size (bytes)	constant (1000)
⊔ Segmentation Size (bytes)	No Segmentation
⊔ Stop Time (seconds)	Never

☑ Apply Changes to Selected Objects   ☐ Advanced

[Find Next]   [Cancel] [OK]

12. From the Main Menu, choose **File/Save** to save your project.

Configuring the 10BaseT Ethernet simulation

1. Now we will choose the statistics that we want to evaluate after our simulation runs. We will be most interested in delay, throughput, and collisions. Right click on the project workspace and select **Choose Individual Statistics** from the pull-down menu. Click on the + next to **Global** to expand the options. Click on **Ethernet** and **Delay**.

2. Also under Global, click on the + next to **Traffic Sink** to expand the options. Click on **End-to-End Delay (Sec)** and **Traffic Received (bits/sec)**. Click on the + next to **Traffic Source** to expand the options. Click on **Traffic Sent (bits/sec)**.

3. Click on the + next to **Node Statistics** to expand the options. Click on the + next to **Ethernet.** Click on **Collision Count, Delay, Load (bits/sec), Traffic Forwarded (bits/sec), Traffic Received (bits/sec), Transmission Attempts,** and **Utilization.**

OPNET Lab Manual to accompany Business Data and Communications

```
Choose Results
├─ Global Statistics
│  ├─ Ethernet
│  │  └─ ☑ Delay (sec)
│  ├─ Traffic Sink
│  │  ├─ ☑ End-to-End Delay (seconds)
│  │  ├─ ☐ Traffic Received (bits)
│  │  ├─ ☑ Traffic Received (bits/sec)
│  │  ├─ ☐ Traffic Received (packets)
│  │  └─ ☐ Traffic Received (packets/sec)
│  └─ Traffic Source
│     ├─ ☐ Traffic Sent (bits)
│     ├─ ☑ Traffic Sent (bits/sec)
│     ├─ ☐ Traffic Sent (packets)
│     └─ ☐ Traffic Sent (packets/sec)
└─ Node Statistics
   └─ Ethernet
      ├─ ☐ Burst Duration (sec)
      ├─ ☐ Burst ON/OFF
      ├─ ☐ Burst Size (packets)
      ├─ ☑ Collision Count
      ├─ ☑ Delay (sec)
      ├─ ☐ Load (bits)
      ├─ ☑ Load (bits/sec)
      ├─ ☐ Load (packets)
      ├─ ☐ Load (packets/sec)
      ├─ ☑ Traffic Forwarded (bits/sec)
      ├─ ☐ Traffic Forwarded (packets/sec)
      ├─ ☐ Traffic Received (bits)
      ├─ ☑ Traffic Received (bits/sec)
      ├─ ☐ Traffic Received (packets)
      ├─ ☐ Traffic Received (packets/sec)
      ├─ ☑ Transmission Attempts
      └─ ☑ Utilization

                        [ Cancel ]   [ OK ]
```

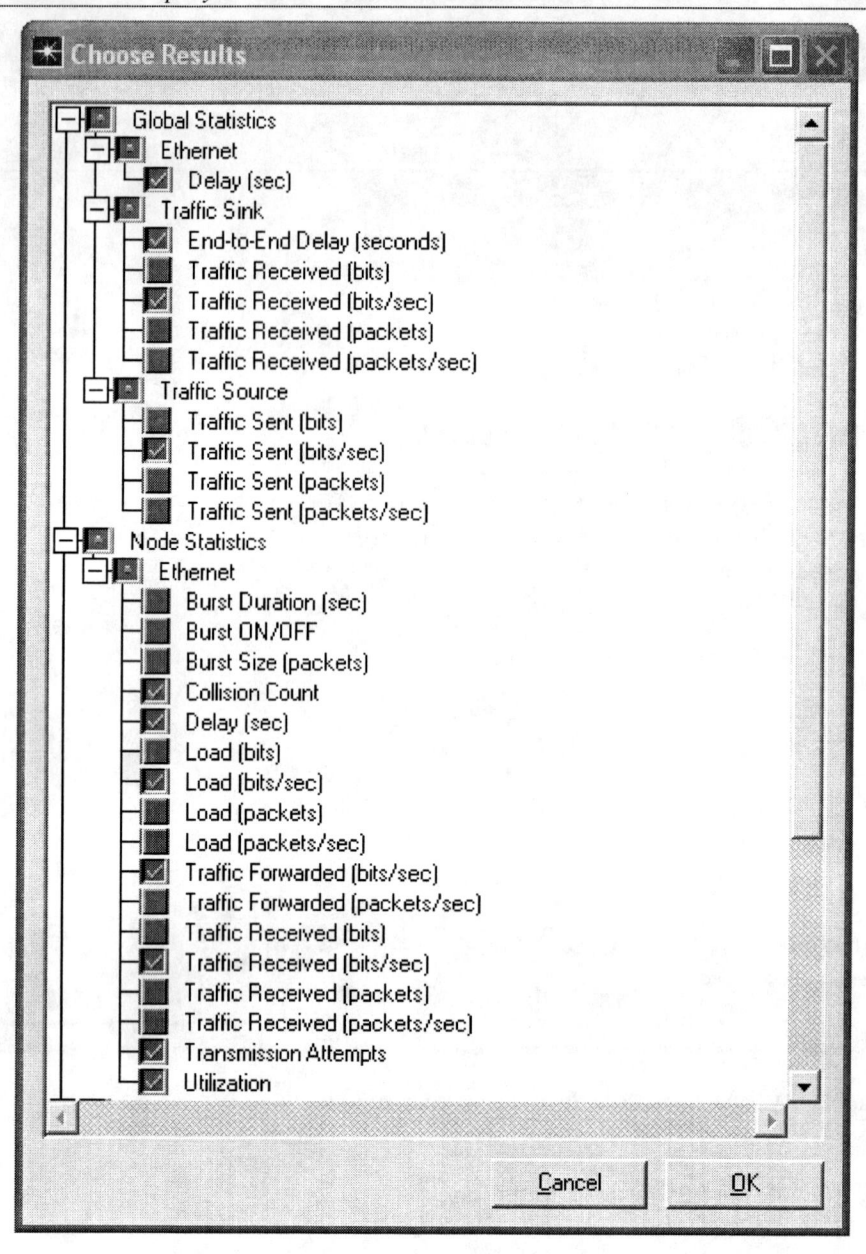

4. Finally, click on the + next to **Link Statistics** to expand the options. Click on **point-to-point.** Click on **throughput (bits/sec) →, throughput (bits/sec) ←, utilization →, and utilization ←.** Verify that your parameters match those shown next and click **OK** to save these parameters.

Lab 7 Ethernet Hub versus Switch

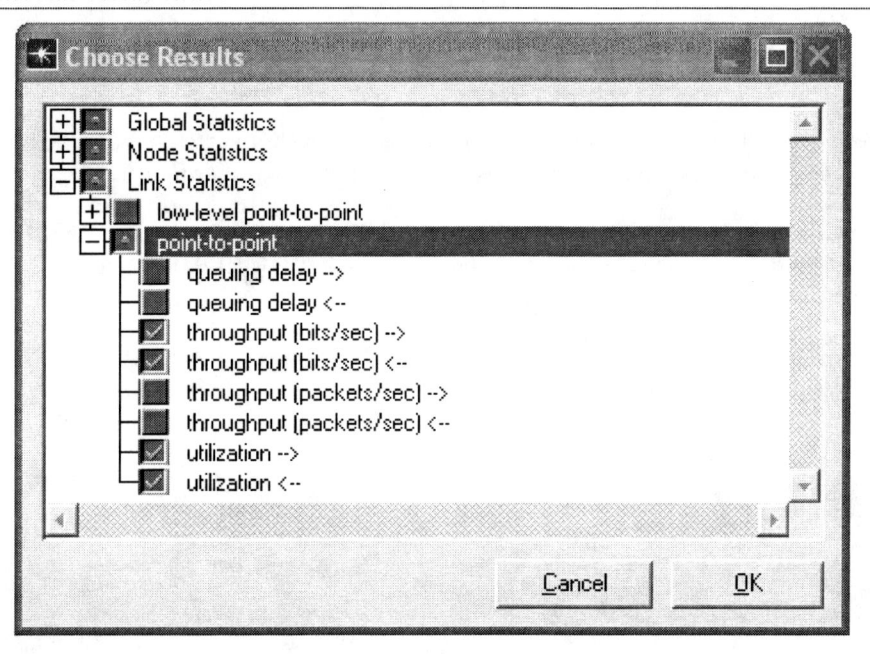

Running the Simulation

1. Click on the **Configure and Run** Button. Set the **Duration** value to **1** and the time unit to **Minutes**. Click the **Run** button to run the simulation. Depending on your processor, the simulation should take several minutes to run.

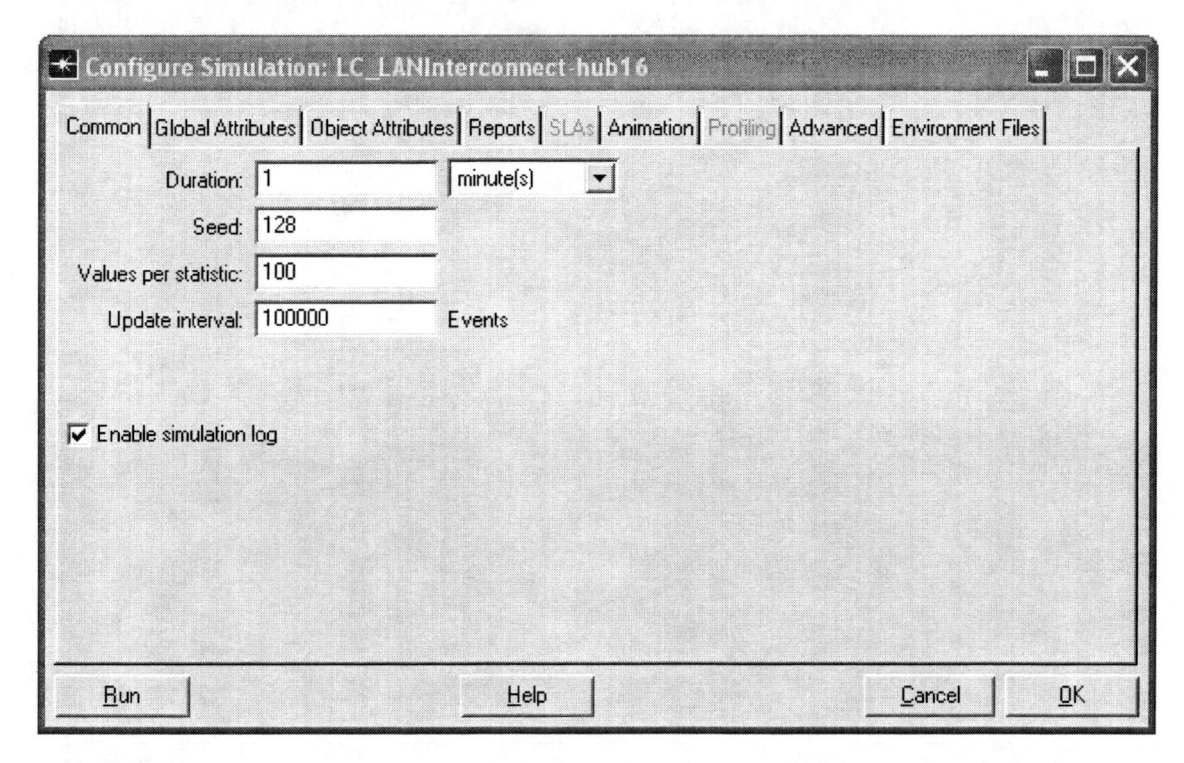

2. We will look at our results after simulating two other scenarios. Choose **File/Save** from the Main Menu to save your project.

OPNET Lab Manual to accompany Business Data and Communications

Creating the Switched Scenario

1. Now we will duplicate the scenario, but replace the hub with a switch. From the Main Menu choose **Scenarios/Duplicate Scenario**. Give the new scenario a name such as **switch10.**

2. Right click on the hub to select it and choose **Edit Attributes**. Change the value for the **Model** attribute to **Ethernet16_switch**. Change the **Name** attribute to **switch10**. Click **OK** to close the Attribute window. Your project should resemble the following screen image:

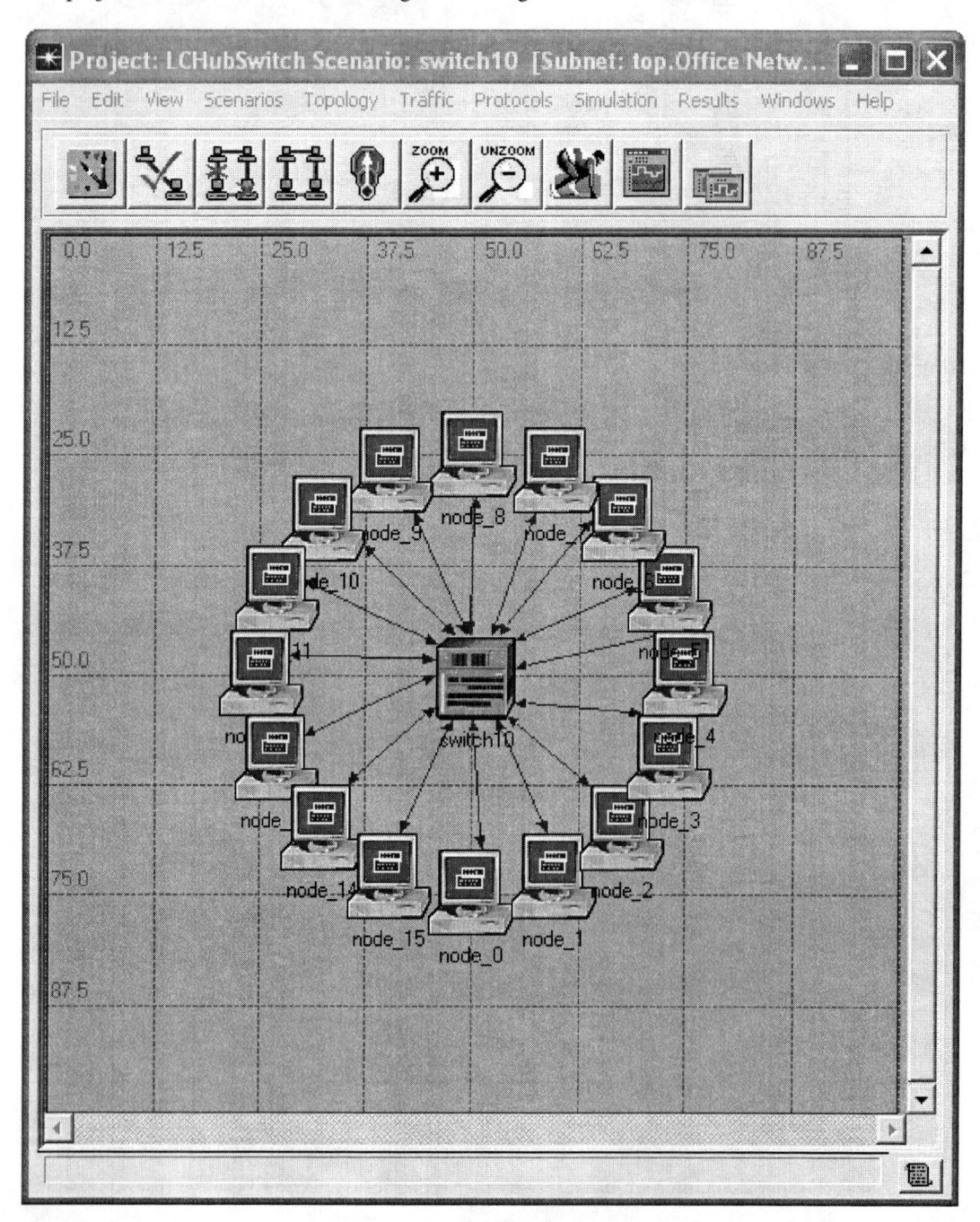

3. Right click on the project workspace and select **Choose Individual Statistics** from the pull-down menu. We would like to capture the switch statistics. Click on the **+** next to **Node** to expand the options. Click on

130

Lab 7   Ethernet Hub versus Switch

---

the + next to **Switch** and choose **Traffic Forwarded (bits/sec)** and **Traffic Received (bits/sec)**. Click **OK** to close the window.

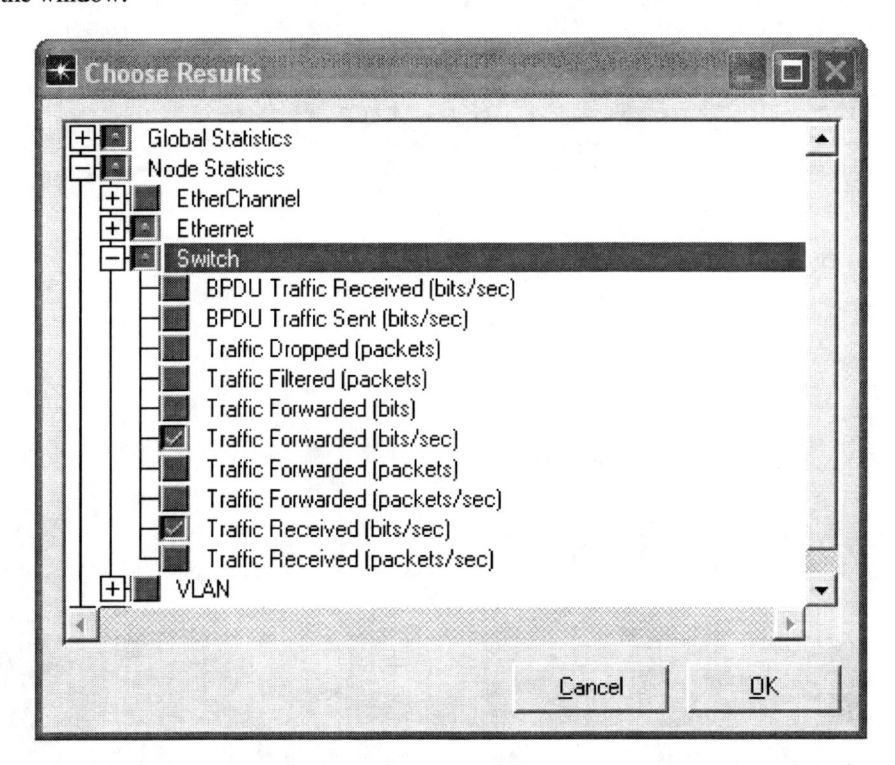

4. From the Main Menu, choose **Simulation/Run Discrete Time Simulation**. We will look at the results after we run our third simulation.

5. From the Main Menu, choose **File/Save** to save your project.

Creating the 100BaseT Scenario

1. At this time, we will create a third scenario that uses a 100BaseT hub and 100BaseT links. From the Main, Menu choose **Scenarios/ Switch to Scenario → hub 10.** We will duplicate this scenario. Again from the Main Menu choose **Scenarios/Duplicate Scenario**. Give the new scenario a name such as **hub100**.

2. Right click on the hub and choose **Set Name**. Name the hub, **Hub100,** to indicate that it will operate at 100 Mbps. The hub in our simulation has auto port sensing, meaning that it can operate at either 10 Mbps or 100 Mbps depending on the type of link attached to it. Right click on one of the links and choose **Edit Attributes**. Check the **Apply Changes to Selected Objects** checkbox and change the model to **100BaseT**. Click **OK** to close the Attribute window. Your scenario should resemble the following one:

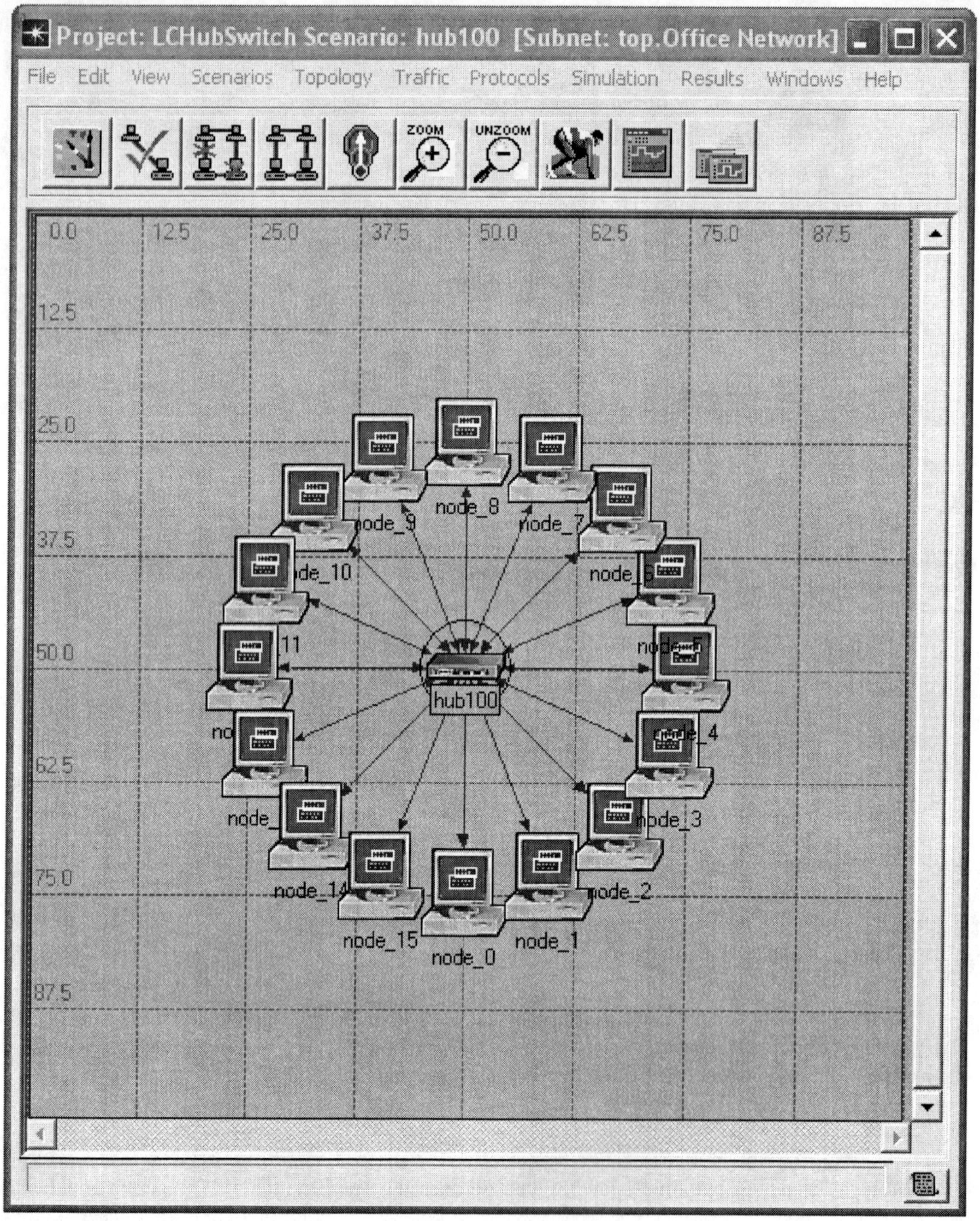
Project: LCHubSwitch Scenario: hub100 [Subnet: top.Office Network]
File Edit View Scenarios Topology Traffic Protocols Simulation Results Windows Help
ZOOM
UNZOOM
0.0
12.5
25.0
37.5
50.0
62.5
75.0
87.5
hub100
node_0
node_1
node_2
node_3
node_4
node_7
node_8
node_15

3. From the Main Menu, choose **Simulation/Run Discrete Event Simulation**.

4. Choose **File/Save** from the Main Menu to save your project.

Comparing Results

To view results, choose **Results/Compare Results** from the Main Menu. To view results for **Traffic Sink Traffic Received (bits/sec),** click on the **+** next to **Traffic Sink** and click next to **Traffic Received (bits/sec).** Select **Time Average** from the pull-down menu**.** Click on the **Show** button to display the graph in a new window. Your graph should resemble the following figure:

Lab 7   Ethernet Hub versus Switch

[Figure: time_average (in Traffic Sink.Traffic Received (bits/sec)) graph showing hub100, hub10, and switch10 curves over 0s–60s. hub100 reaches ~12,500,000; switch10 is slightly below hub100; hub10 reaches ~9,000,000.]

If you would like to **Zoom In** on an area of the graph, you may left click on the graph and drag a square around the area that you would like to enlarge.

Click on the **X** to close the window. The **Close Analysis** window will appear asking you if you want to hide the panel or permanently delete it. Click on the **Delete** button to delete the window.

You may discard the results of your simulations, rerun them, and manage your scenarios by doing the following:

Choose **Scenarios/Manage Scenarios** from the Main Menu. A window will appear with the names of your scenarios, the length of time that the simulation ran, and the status of the data.

By clicking on a scenario name, you may choose to **Discard Results** or **Collect Results**. By clicking on the values under the **Results** column, you may choose **recollect** or **collect** from the pull-down menu. Click **OK** to close the window.

Use the results you obtained in your simulations to answer the questions below.

Questions

1. Based on the results you observed in the Traffic Sink, Traffic Sent graph, which interconnection technology (10BaseT hub, 10BaseT switch, 100BaseT hub) provides the highest throughput? Is this what you expected? Comment on the relationship between throughput for the 100BaseT hub and the switch.
2. Based on the packet size (100 bytes) and interarrival time (.01), how many bits per second is your simulation sending? How does the number you calculated compare to the Traffic Source, Traffic Sent statistics for the 10BaseT scenario?

3. Look at the delay value for the 10BaseT scenario. Explain why delay is increasing. How does this delay value compare to the 100BaseT scenario?
4. Look at the 10BaseT hub statistics. What is the utilization value? Based on this value, what would you expect in regards to collisions? Look at the collision statistics for this scenario. What is the maximum number of collisions?
5. Are there any collisions in the 100Base T hub scenario? Are there any collisions in the 10BaseT switch scenario? Explain how a switch might have collisions.
6. What are the utilization statistics for the 10BaseT hub and the 100BaseT hub? Does the difference in utilization make sense based on your calculation of the number of bits being sent? Explain your answer.
7. Create a new scenario that uses 100 Mbps links and a switch. What is the difference in throughput between this new scenario and the 100 Mbps hub scenario?

LAB 8   PACKET SIZE, THROUGHPUT, AND DELAY IN 10BaseT ETHERNET LANs

Objectives
1. To build a shared 10BaseT Ethernet LAN.
2. To compare the effect of various packet sizes on throughput and delay in 10BaseT LANs.

Motivation
John Garcia is a support technician for EtherLAN, a company that builds and installs Ethernet LANs. EtherLAN has a client that has heard that packet size can affect throughput and delay. The client is interested in maximizing throughput and minimizing delay. She has asked John to do a study to determine exactly how packet size affects these statistics and to make a recommendation regarding the optimal packet size.

Description
Ethernet LANs utilize the Carrier Sense Multiple Access/Collision Detection (CSMA/CD) protocol. Using CSMA/CD, a host must sense the line to discover if it is free before sending a packet. Collisions may occur in this environment if more than one host senses the line, finds it free, and tries to send a packet. Collisions will decrease throughput and lead to poor performance. Packet size also affects throughput. Small packets generally have less transmission delay, but invoke more overhead when considering protocol header size versus data payload size. Bigger packets, despite having longer transmission delays, may be more efficient because header overhead will be minimized. In this lab, we will explore the relationship between packet size, throughput, and delay in 10BaseT Ethernet LANs.

Follow the directions listed next to create an OPNET model, run simulations, and gather statistics. Questions regarding the lab appear at the end.

Create a New Project

1. Start the **OPNET IT Guru Academic Edition Application.** Create a **New Project** by selecting **File/New/Project** from the Main Menu. Click **OK**.

2. Give the project a name such as **your_initialsEthPkt**. Give the scenario a name such as **Pkt512**. Click **OK**. You should see the **Startup Wizard Initial Topology** dialog box.

3. Verify that **Create Empty Scenario** is selected in the **Initial Topology Dialog Box**. Click **Next**. You will now see the **Network Scale** dialog box.

4. Choose **Office** from the **Network Scale** list. Click **Next**. Choose **Meters** in the **size** drop down box. Type **200** for the **X span** and **200** for the **Y span**. Click **Next**. You should see the **Select Technologies** dialog box

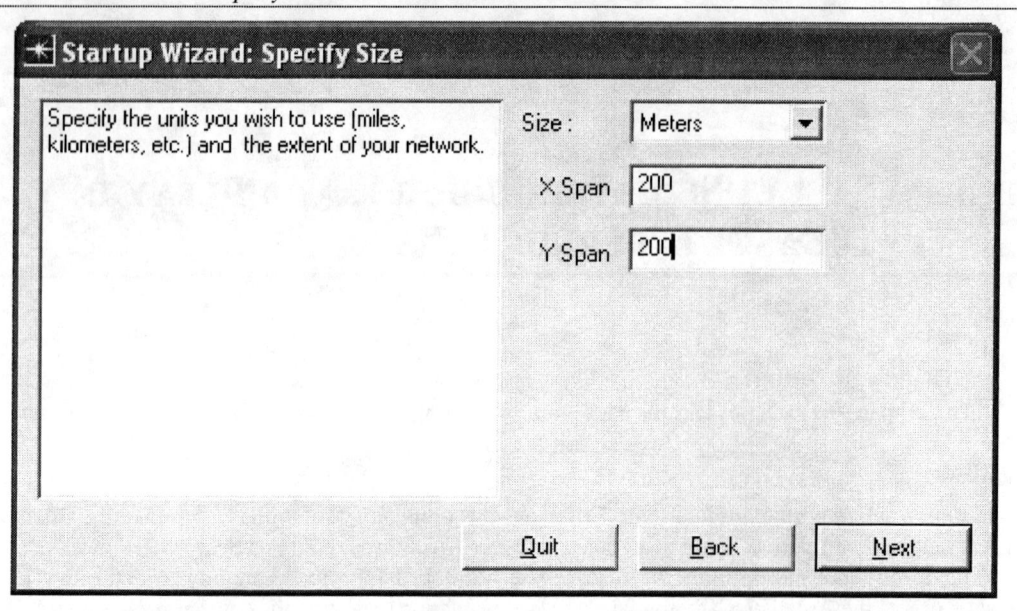

5. Scroll down until you see **ethernet** under the **Model Family** column. Under the **Include column**, click on the box next to **ethernet**. This will change the "no" to "yes," indicating that Ethernet technologies will be included. Click **Next**.

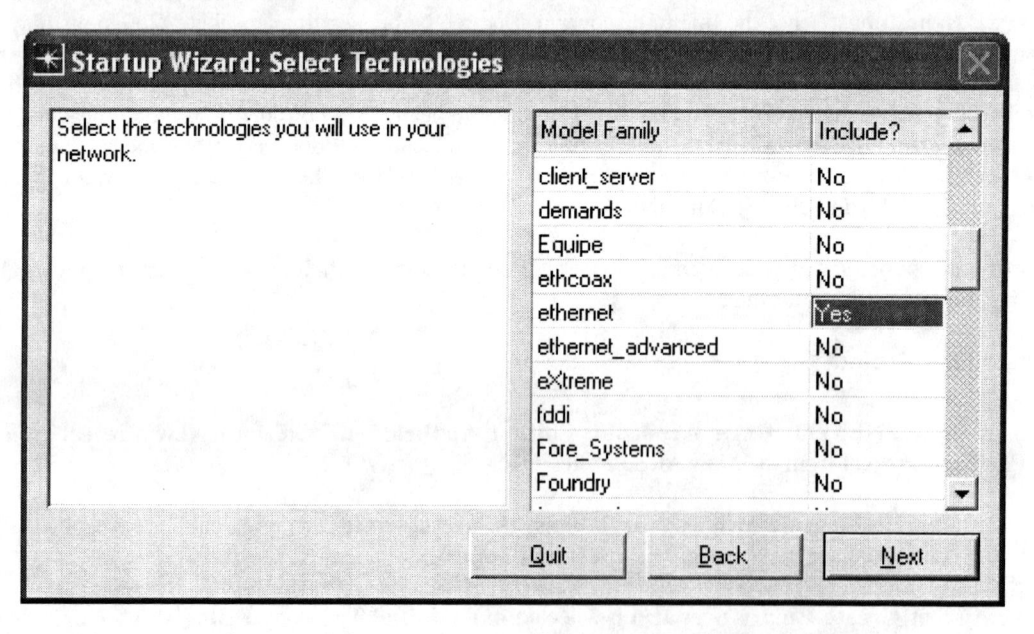

6. In the **Startup Wizard Review** dialog box, verify that the ethernet technology was chosen and that the scale is Office 200m x 200m. Click **OK**. You should see the **Object Palette** window.

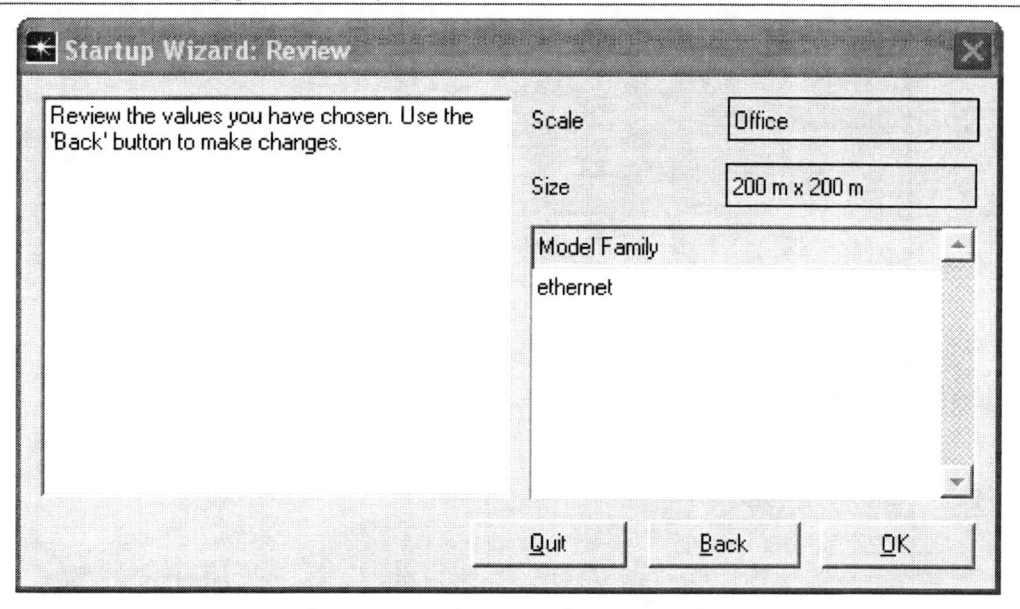

Lab 8   Packet Size, Throughput, and Delay in 10BaseT Ethernet LANs

Building the Ethernet Network

1. We will begin by using Rapid Configuration to create the hub-based 10BaseT LAN. We will create a star network with 16 Ethernet stations connected to an Ethernet hub. We will choose 10BaseT links to connect the hosts to the hub. From the Main Menu, choose **Topology/Rapid Configuration**. A window will appear prompting you to choose the topology or network shape. Choose **Star** from the pull-down menu.

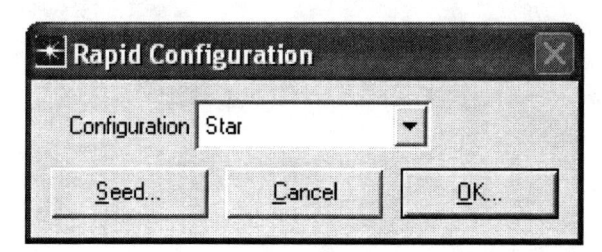

2. A new window will appear which will allow you to choose the nodes in the network. For the **Center Node Model,** choose **ethernet16_hub** from the pull-down menu. For the **Periphery Node Model,** choose **ethernet_station.** Change the **Number** to **16**. This will create 16 stations on the network. For the **Link Model**, choose **10BaseT**. Set the **X** coordinate to **100**, the **Y** coordinate to **100,** and the **Radius** to **50**. This will put the LAN in the center of the workspace. Click **OK**.

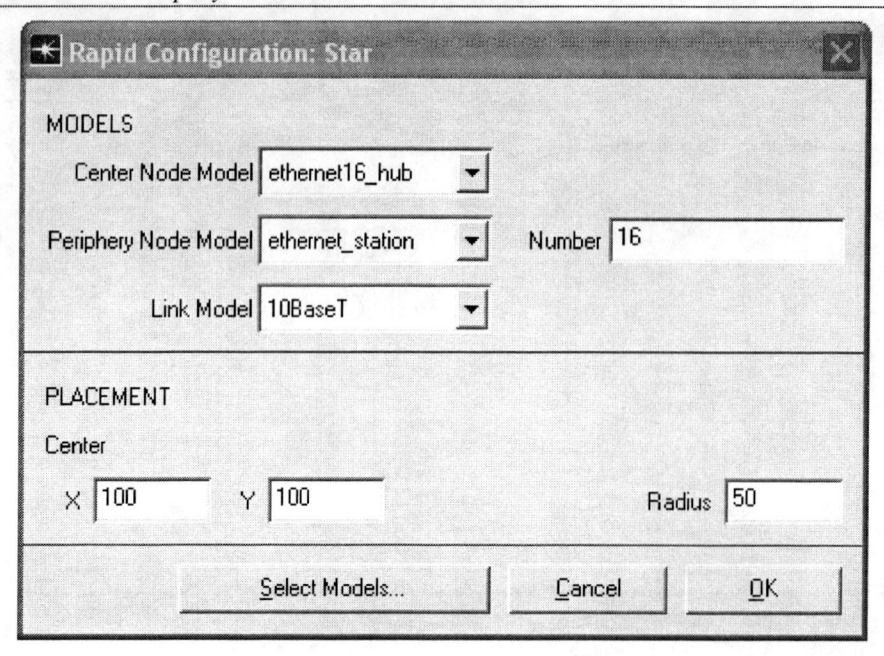

Rapid Configuration: Star

MODELS
- Center Node Model: ethernet16_hub
- Periphery Node Model: ethernet_station   Number: 16
- Link Model: 10BaseT

PLACEMENT
Center
X: 100   Y: 100   Radius: 50

[Select Models...]   [Cancel]   [OK]

3. Right click on the hub object, and choose **Set Name**. Name the hub **Hub10**. Your project should resemble the following one:

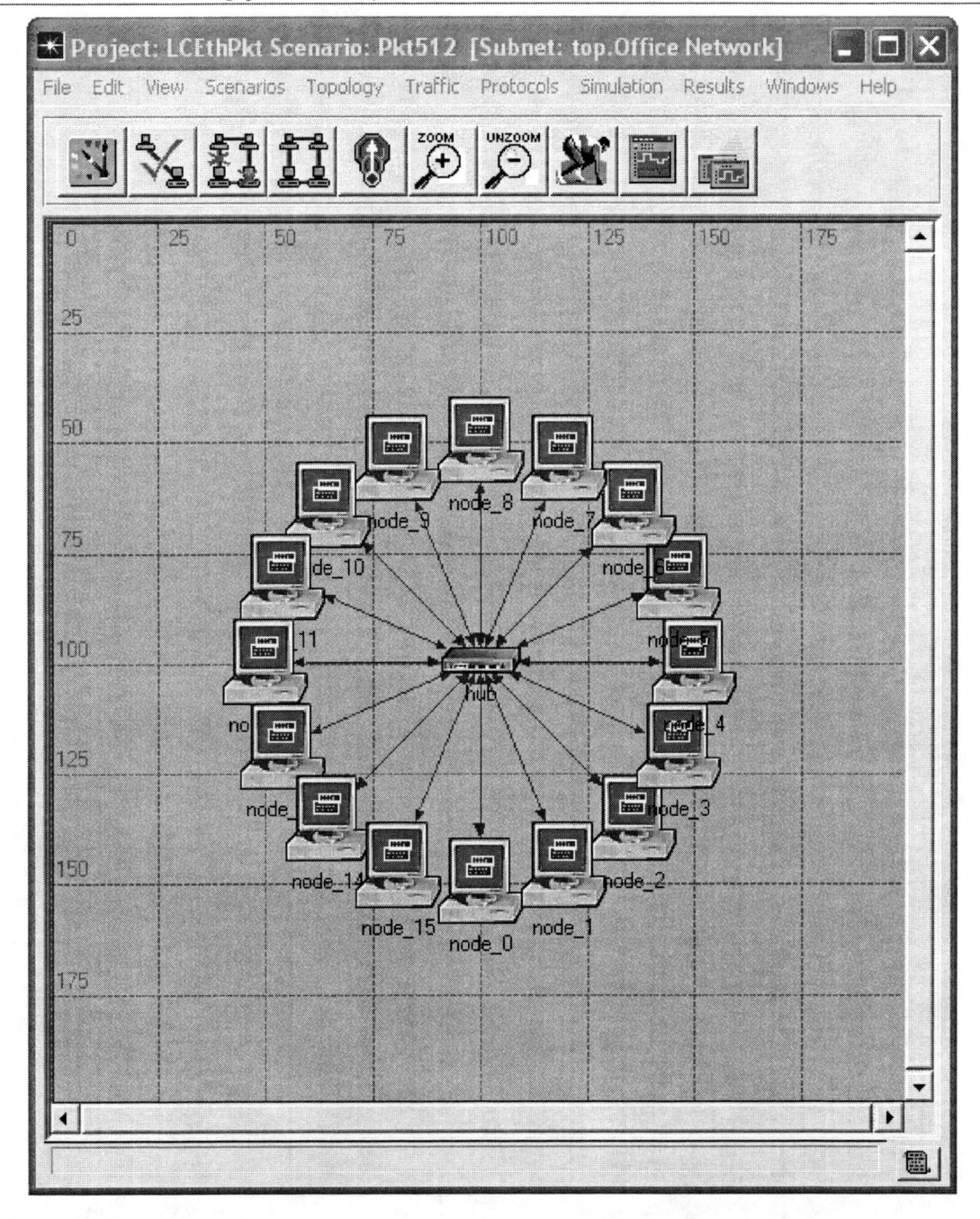

Lab 8   Packet Size, Throughput, and Delay in 10BaseT Ethernet LANs

4. We will now configure the **ethernet_stations**. Right click on one of the **ethernet_station** objects and choose **Select Similar Nodes** from the pull-down menu. Right click again on one of the stations and choose **Edit attributes.**

5. Click on the box to **Apply Changes to Selected Objects**. Click on + next to **Traffic Generation Parameters** to expand the parameter list. You should see the following:

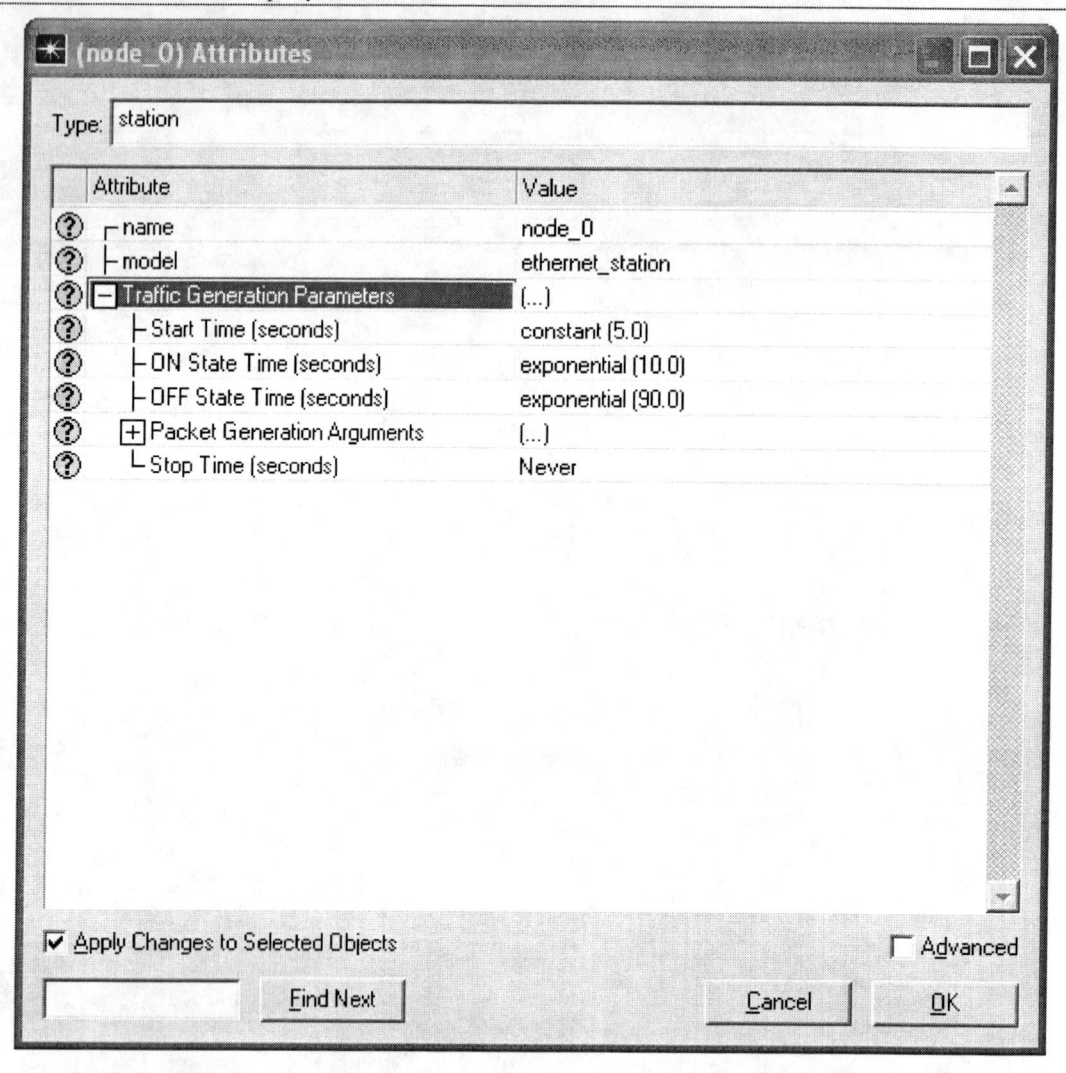

![node_0 Attributes dialog]

Attribute	Value
name	node_0
model	ethernet_station
Traffic Generation Parameters	(...)
├ Start Time (seconds)	constant (5.0)
├ ON State Time (seconds)	exponential (10.0)
├ OFF State Time (seconds)	exponential (90.0)
⊞ Packet Generation Arguments	(...)
└ Stop Time (seconds)	Never

6. We will first configure the simulation start time. We would like our simulation to start up immediately. Click on the value next to **Start Time (seconds)**. A new window will appear. Change the value to **constant (1)**. This will cause our traffic to be generated after 1 second. Click **OK**.

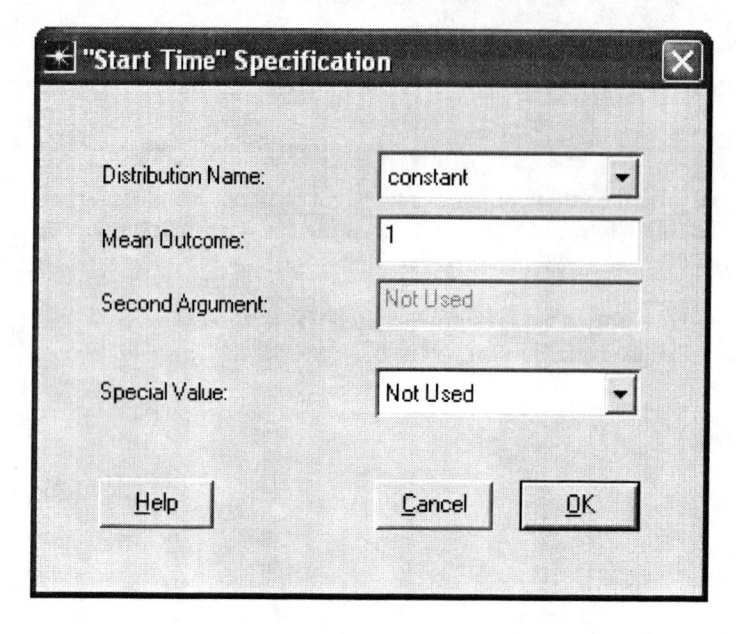

"Start Time" Specification

- Distribution Name: constant
- Mean Outcome: 1
- Second Argument: Not Used
- Special Value: Not Used

Lab 8   Packet Size, Throughput, and Delay in 10BaseT Ethernet LANs

7. Next, click on the value next to **ON State Time**. This will bring up a new window. Change the **Mean Outcome** to **100**.

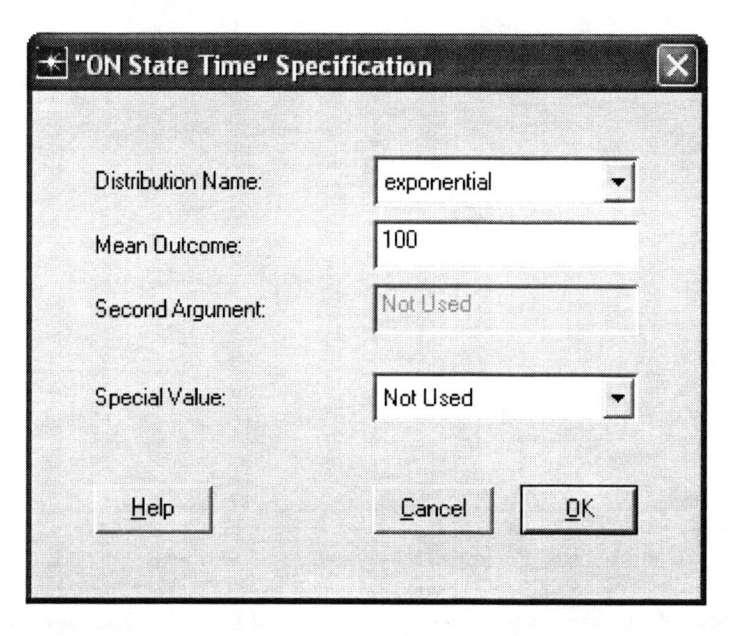

8. Click on the value next to **OFF State Time**. A new window will appear. Change the **Mean outcome time** to **0** and click **OK**. This causes our simulation to generate traffic 100% of the time. Click **OK**.

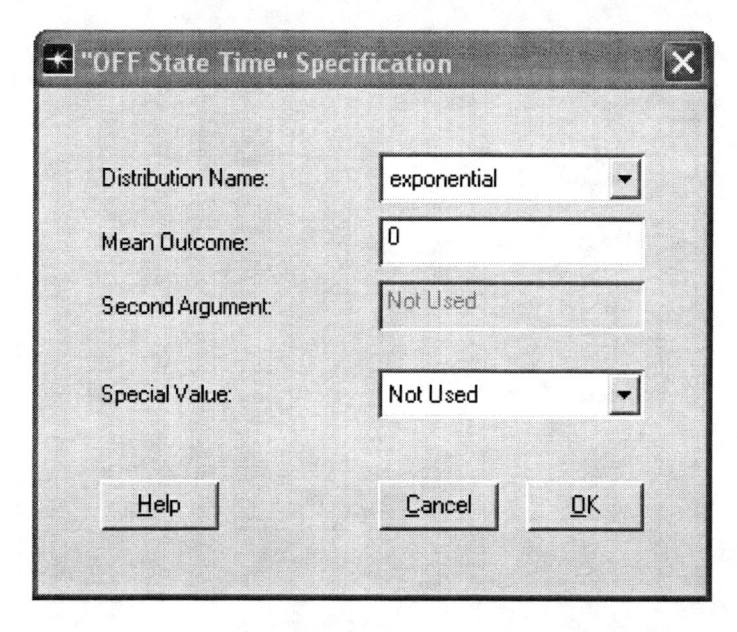

9. Now click on the + next to **Packet Generation Arguments**. We will change the interarrival time, which is the time between each packet, and the packet size. Click on the value next to **Interarrival Time (seconds)**. This will bring up a new window. Change the **Mean outcome time** to **.02** and click **OK**.

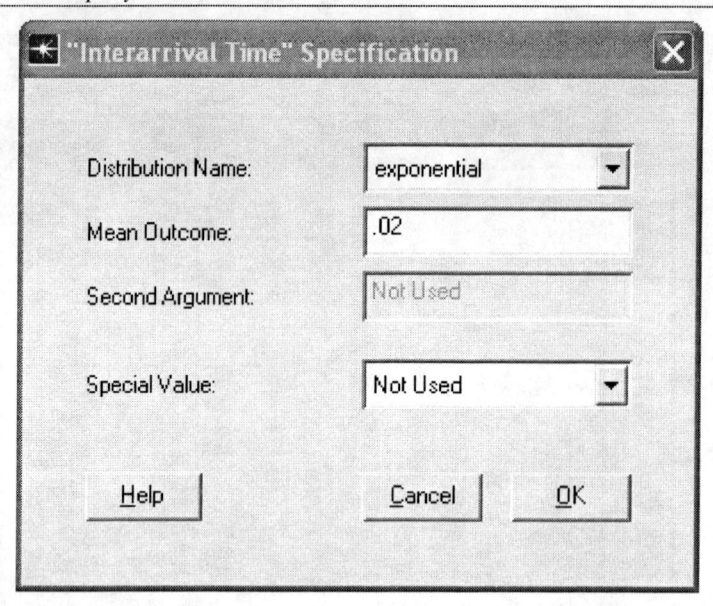

10. Next, click on the value next to **Packet Size (bytes).** This will again bring up a new window. Ethernet packets can range between 46 and 1500 bytes. We will begin by choosing a packet size that is a power of 2 near the middle of these two values. Choose **Constant** from the **Distribution** drop-down box. Next change the **Mean Outcome** to **512**. This will cause all packets generated to be 512 bytes in size. Click the **OK** button.

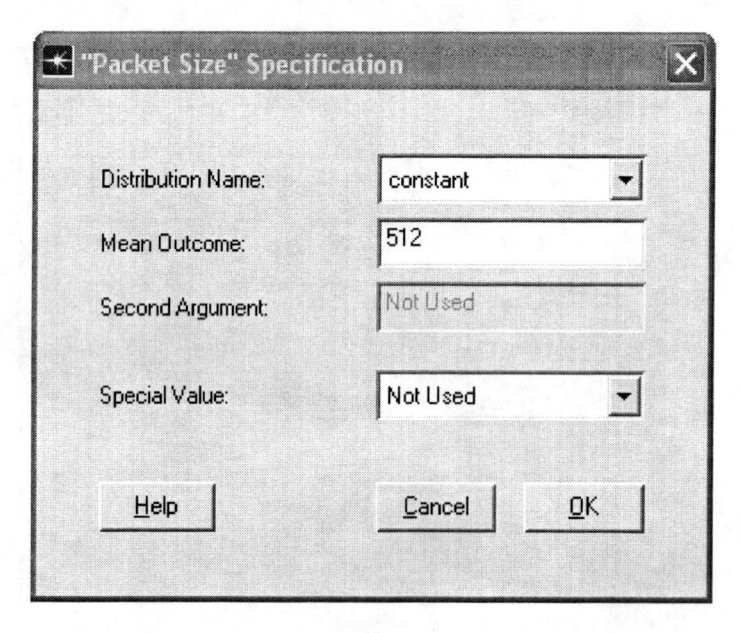

11. You should now have the values listed on the attribute screen image that follows. Note that node numbers may differ. Here node_0 is being edited. Click **OK** to close the attribute window.

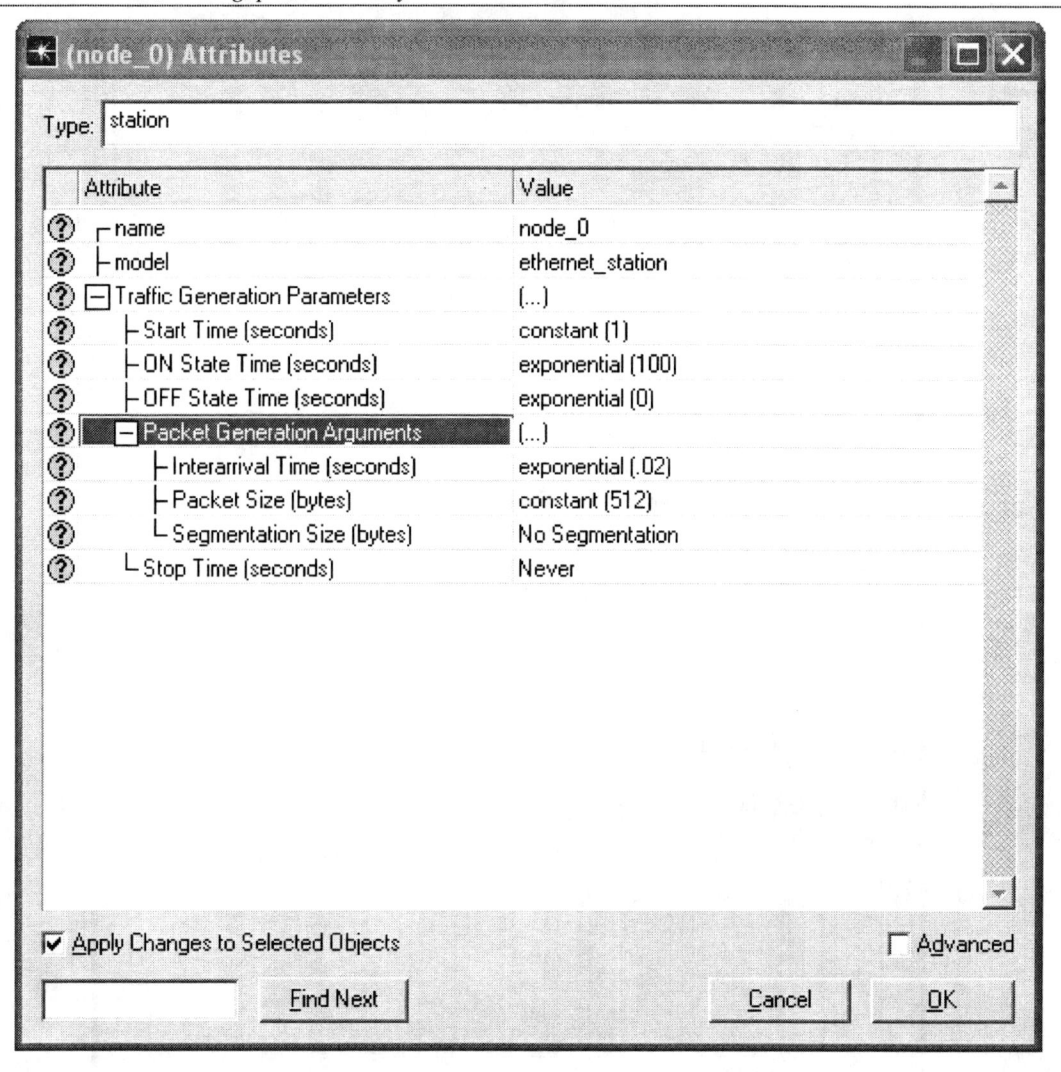

Lab 8 *Packet Size, Throughput, and Delay in 10BaseT Ethernet LANs*

```
┌─ (node_0) Attributes ──────────────────────────────── _ □ X ┐
│                                                              │
│  Type: station                                               │
│  ┌──────────────────────────────────────────────────────┐   │
│  │ Attribute                    │ Value                │▲  │
│  ├──────────────────────────────┼──────────────────────┤   │
│  │ ⓘ ┌name                      │ node_0               │   │
│  │ ⓘ ├model                     │ ethernet_station     │   │
│  │ ⓘ ├─Traffic Generation Parameters │ (...)           │   │
│  │ ⓘ │  ├Start Time (seconds)   │ constant (1)         │   │
│  │ ⓘ │  ├ON State Time (seconds)│ exponential (100)    │   │
│  │ ⓘ │  ├OFF State Time (seconds)│ exponential (0)     │   │
│  │ ⓘ │  ├─Packet Generation Arguments │ (...)          │   │
│  │ ⓘ │  │  ├Interarrival Time (seconds) │ exponential (.02) │   │
│  │ ⓘ │  │  ├Packet Size (bytes) │ constant (512)       │   │
│  │ ⓘ │  │  └Segmentation Size (bytes) │ No Segmentation │   │
│  │ ⓘ │  └Stop Time (seconds)    │ Never                │   │
│  │                                                      │▼  │
│  └──────────────────────────────────────────────────────┘   │
│  ☑ Apply Changes to Selected Objects          ☐ Advanced    │
│  ┌─────────┐  ┌─Find Next─┐       ┌─Cancel─┐  ┌──OK──┐     │
│  └─────────┘  └───────────┘       └────────┘  └──────┘     │
└──────────────────────────────────────────────────────────────┘
```

12. From the Main Menu, choose **File/Save** to save your project.

Configuring the Ethernet simulation

1. Now we will choose the statistics that we want to evaluate after our simulation runs. We will be most interested in delay, throughput, and collisions. Right click on the project workspace and select **Choose Individual Statistics** from the pull-down menu. Click on the + next to **Global** to expand the options. Click on **Ethernet** and **Delay**.

2. Also, under Global, click on the + next to **Traffic Sink** to expand the options. Click on **End-toEnd Delay (sec)** and **Traffic Received (bits/sec), Traffic Received (packets)**. Click on the + next to **Traffic Source** to expand the options. Click on **Traffic Sent (bits/sec)** and **Traffic Sent (packets)**.

3. Click on the + next to **Node Statistics** to expand the options. Click on the + next to **Ethernet.** Click on **Collision Count, Delay, Traffic Forwarded (bits/sec), Traffic Received (bits/sec),** and **Utilization.**

4. Finally, click on the + next to **Link Statistics** to expand the options. Click on **point-to-point.** Click on **throughput (bits/sec)** →, **throughput (bits/sec)** ←, **utilization** →, **and utilization** ← Verify that your parameters match those shown next and click **OK** to save these parameters.

143

Choose Results

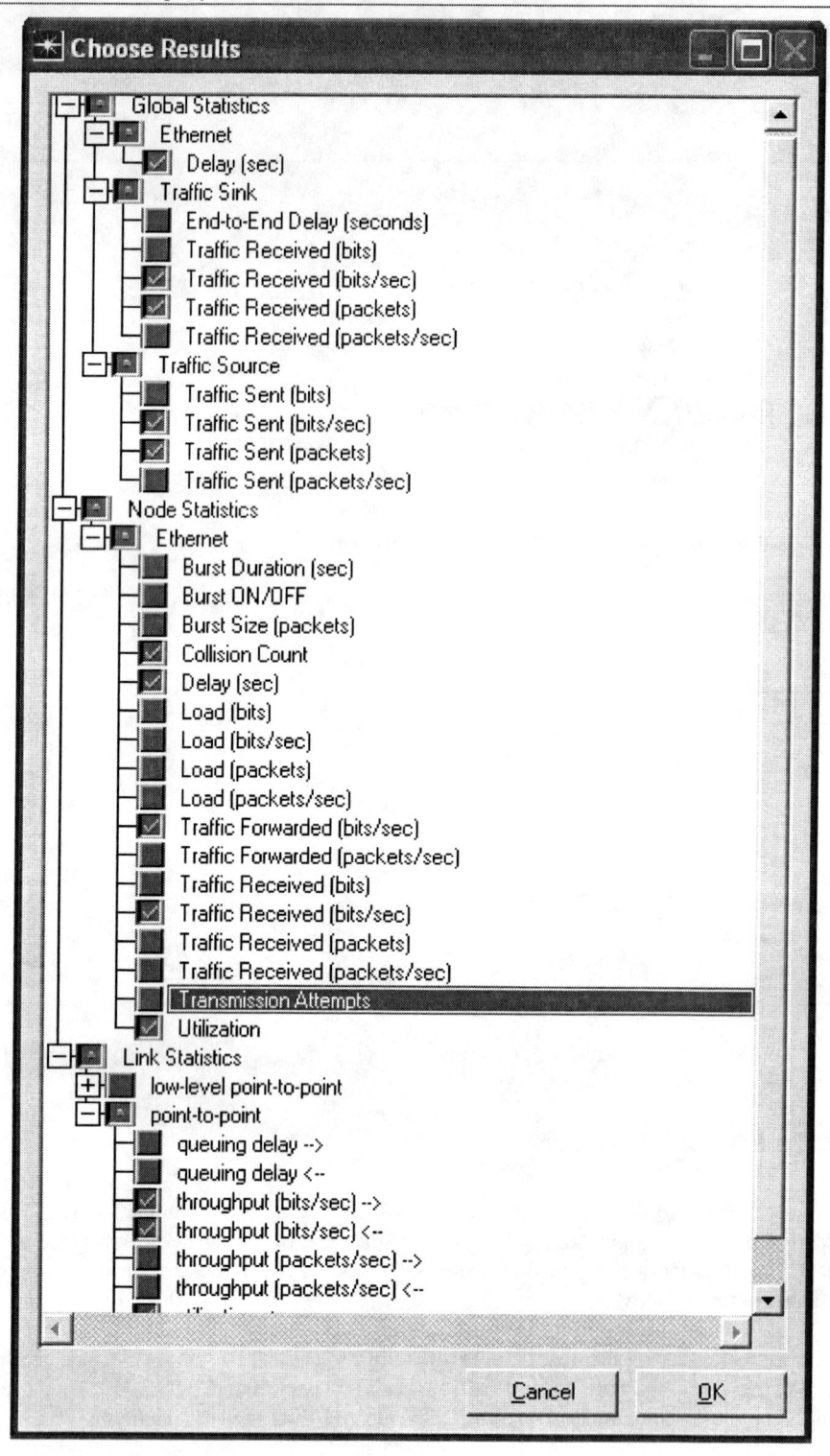

- Global Statistics
  - Ethernet
    - ☑ Delay (sec)
  - Traffic Sink
    - ☐ End-to-End Delay (seconds)
    - ☐ Traffic Received (bits)
    - ☑ Traffic Received (bits/sec)
    - ☑ Traffic Received (packets)
    - ☐ Traffic Received (packets/sec)
  - Traffic Source
    - ☐ Traffic Sent (bits)
    - ☑ Traffic Sent (bits/sec)
    - ☑ Traffic Sent (packets)
    - ☐ Traffic Sent (packets/sec)
- Node Statistics
  - Ethernet
    - ☐ Burst Duration (sec)
    - ☐ Burst ON/OFF
    - ☐ Burst Size (packets)
    - ☑ Collision Count
    - ☑ Delay (sec)
    - ☐ Load (bits)
    - ☐ Load (bits/sec)
    - ☐ Load (packets)
    - ☐ Load (packets/sec)
    - ☑ Traffic Forwarded (bits/sec)
    - ☐ Traffic Forwarded (packets/sec)
    - ☐ Traffic Received (bits)
    - ☑ Traffic Received (bits/sec)
    - ☐ Traffic Received (packets)
    - ☐ Traffic Received (packets/sec)
    - **Transmission Attempts**
    - ☑ Utilization
- Link Statistics
  - [+] low-level point-to-point
  - [−] point-to-point
    - ☐ queuing delay -->
    - ☐ queuing delay <--
    - ☑ throughput (bits/sec) -->
    - ☑ throughput (bits/sec) <--
    - ☐ throughput (packets/sec) -->
    - ☐ throughput (packets/sec) <--

[Cancel]   [OK]

Running the Simulation

1. Click on the **Configure and Run** Button. Set the **Duration** value to **1** and the time unit to **Minute(s)**. Click the **Run** button to run the simulation. Depending on your processor, the simulation should take several minutes to run.

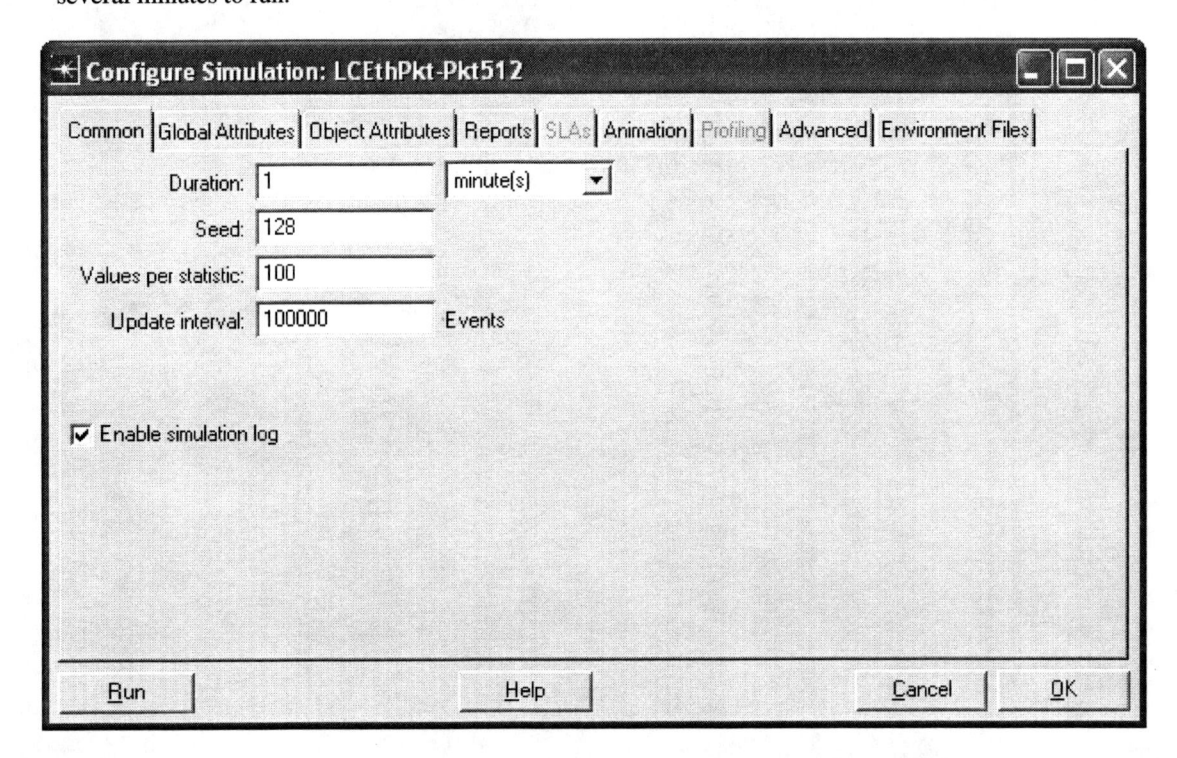

2. We will look at our results after creating a second scenario. Choose **File/Save** from the Main Menu to save your project.

Changing the Packet Size for the Ethernet Simulation

1. Now we want to see how things change when a larger packet size is chosen. We will choose the largest possible packet size that a 10BaseT Ethernet can transmit (1500 bytes). From the main menu, choose **Scenarios/Duplicate** the scenario. This will bring up a dialog box. Give your scenario a name, such as **Pkt1500,** and click **OK**.

2. Right click on one of the **ethernet stations** and choose **Select Similar Nodes**. Right click on one of the selected stations and choose **Edit Attributes**. Click in the check box next to **Apply Changes to Selected Objects**.

3. Now click on the **+** next to **Packet Generation Parameters**. Click on the value next to **Packet Size (bytes)**. This will bring up a new window. Change the **Mean Outcome** to **1500**. This will cause all packets generated to be 1500 bytes in size. Click **OK.** Verify that all 16 highlighted workstations had their attributes changed by observing the message on the status bar. It should read, "16 objects changed."

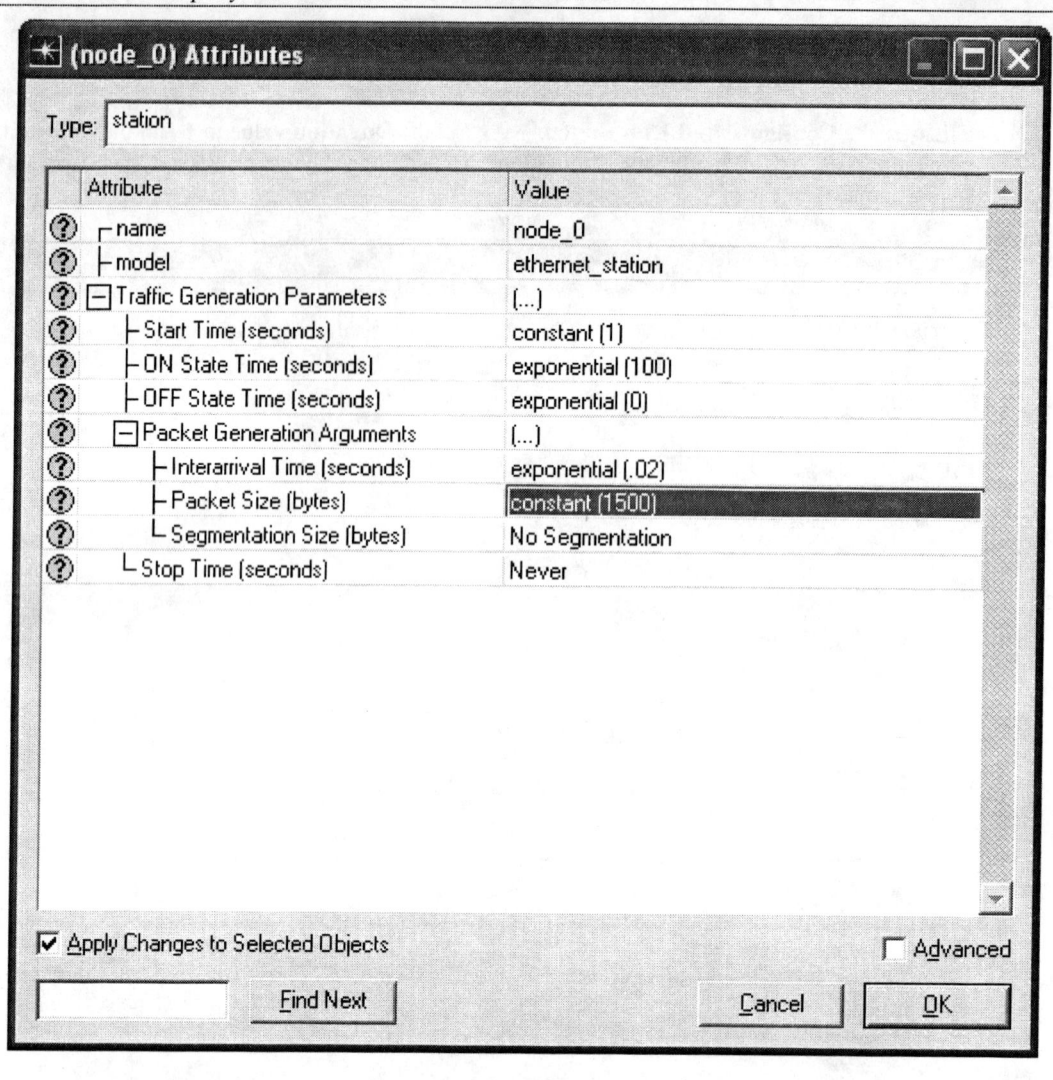

OPNET Lab Manual to accompany Business Data and Communications

[node_0 Attributes dialog box showing:
- Type: station
- name: node_0
- model: ethernet_station
- Traffic Generation Parameters: (...)
  - Start Time (seconds): constant (1)
  - ON State Time (seconds): exponential (100)
  - OFF State Time (seconds): exponential (0)
  - Packet Generation Arguments: (...)
    - Interarrival Time (seconds): exponential (.02)
    - Packet Size (bytes): constant (1500)
    - Segmentation Size (bytes): No Segmentation
  - Stop Time (seconds): Never]

4. From the Main Menu, choose **Simulation/ Run Discrete Event Simulation**. The simulation will take several minutes to run, depending on your processor. We will now compare results to those in the previous run.

5. From the Main Menu, choose **Results/Compare Results**. Choose **All Scenarios** from the drop down box at the lower right, below the graph area. Choose **time_average** from the drop-down box at the lower left, below the graph area. Following is a graph of the **Traffic Sink, Traffic Received (bits/sec)** for both scenarios:

146

Lab 8   Packet Size, Throughput, and Delay in 10BaseT Ethernet LANs

[Figure: time_average (in Traffic Sink.Traffic Received (bits/sec)) showing two curves, Pkt512 leveling near 3,000,000 and Pkt1500 leveling near 9,000,000, over 0–60s]

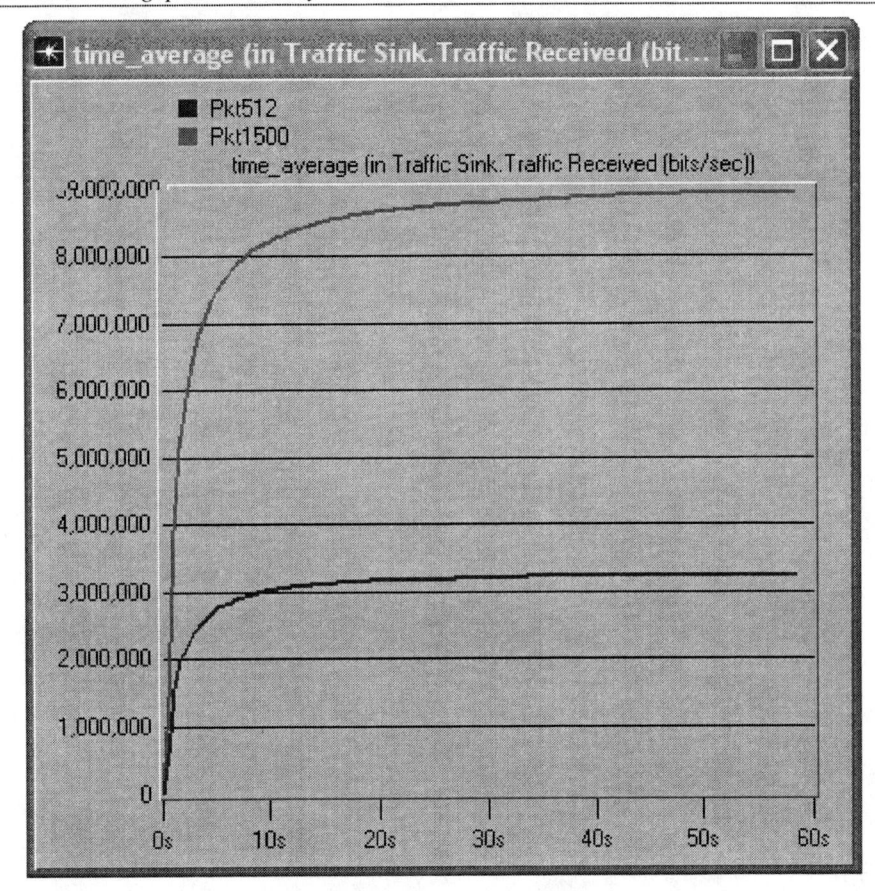

Click on the **X** to close the window. The **Close Analysis** window will appear asking you if you want to hide the panel or permanently delete it. Click on the **Delete** button to delete the window.

You may discard the results of your simulations, rerun them, and manage your scenarios by doing the following:

Choose **Scenarios/Manage Scenarios** from the Main Menu. A window will appear with the names of your scenarios, the length of time that the simulation ran, and the status of the data.

By clicking on a scenario name, you may choose to **Discard Results** or **Collect Results**. By clicking on the values under the **Results** column, you may choose **recollect** or **collect** from the pull-down menu. Click **OK** to close the window.

Using the results you obtained in your simulations, answer the following questions.

Questions for the Scenario with Packet Size 512

1. What were the Global statistics (delay, traffic sent and received in bits/sec) for the Ethernet LAN when the packet size was 512? Were the same number of packets sent as received? List the values and explain your answer.
2. Look at the Object statistics, Office Network. Click on the node number that represents the hub. What was the average collision count? What was the utilization? Is this what you expected?
3. Based on the interarrival time (.02), the packet size, and the speed of the Ethernet LAN, how many bits would you expect to be sent? Show your calculations. Does this match the results that you recorded? How does this compare with the number of bits sent?

Questions for the Scenario with Packet Size 1500

1. What were the Global statistics (delay, traffic sent and received in bits/sec) for the Ethernet LAN when the packet size was 1500? Were the same number of packets sent as received? List the values and explain your answer.
2. Look at the Object statistics, Office Network. Click on the node number that represents the hub. What was the average collision count? What was the utilization? Is this what you expected?
3. Based on the interarrival time (.02), the packet size, and the speed of the Ethernet LAN, how many bits would you expect to be sent? Show your calculations. Does this match the results that you recorded? How does this compare with the number of bits sent?

Summary Questions

1. Based on your observations, does packet size affect delay and throughput? Explain your answer based on what you observed in your simulation. Which packet size seems to be the "better" size based on throughput and delay?
2. How was the number of collisions affected when the packet size increased from 512 to 1500?

LAB 9   100 MBPS ETHERNET VERSUS 100 MBPS FDDI

Objectives
1. To create a shared 100BaseT Ethernet LAN and a 100 Mbps FDDI LAN.
2. To compare throughput and delay for each of these two LAN technologies.

Motivation

Stacey Yamamoto is a system administrator for the Admissions and Records Department at CyberSpy University. A new budget has provided her with some extra funds that she would like to use to upgrade the department's 10 Mbps Ethernet LAN. She has heard that fiber optic cable is more secure than unshielded twisted pair, however, migrating to a faster Ethernet LAN would most likely be easier and less expensive. She would like to evaluate the performance of a 100 Mbps Ethernet LAN and a Fiber Distributed Data Interface (FDDI) LAN. She is particularly interested in comparing throughput and delay statistics for these two LAN technologies.

Description

Ethernet Local Area Networks (LANs) utilize the Carrier Sense Multiple Access/Collision Detection (CSMA/CD) protocol. Using this protocol, each station must sense the line to discover whether it is free before sending a packet. Collisions may occur in this environment if more than one station senses the line, finds it free, and tries to send a packet.

FDDI LANs, however, utilize a token passing protocol. A token is passed from station to station. A station may not send a packet unless it holds the token. This protocol is collision free. We would expect throughput to be higher for LANs that are collision free.

Following the directions to create an OPNET model, run simulations, and gather statistics. Questions regarding the lab are listed at the end.

Create a New Project

1. Start the **OPNET IT Guru Academic Edition Application.** Create a **New Project** by selecting **File/New/Project** from the Main Menu. Click **OK**.

2. Give the project a name such as **your_initials_EthvsFDDI**. Give the scenario a name such as **100BaseT**. Click **OK**. You should see the **Startup Wizard Initial Topology** dialog box.

3. Verify that **Create Empty Scenario** is selected in the **Initial Topology Dialog Box**. Click **Next**. You will now see the **Network Scale** dialog box.

4. Choose **Office** from the **Network Scale** list. Click **Next**.

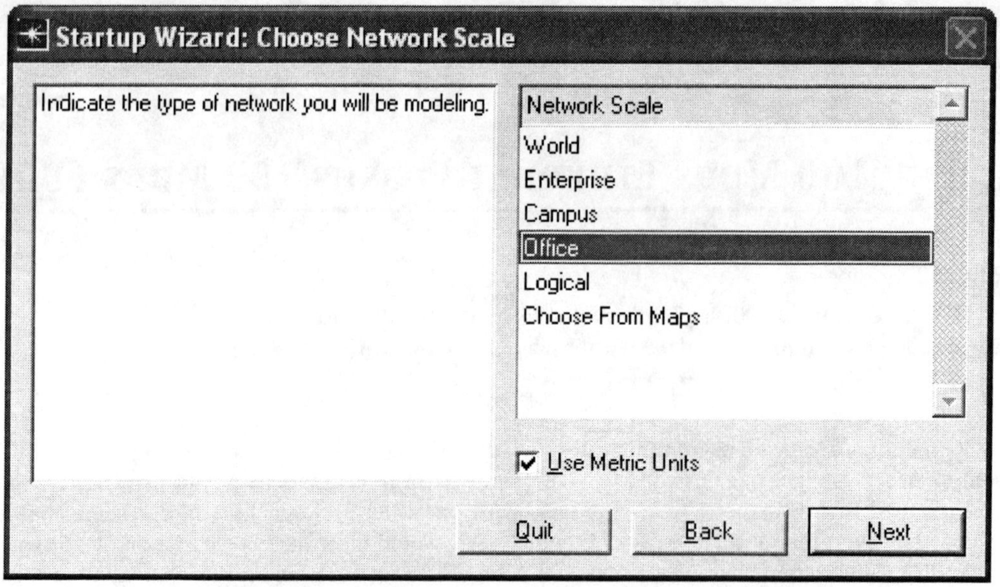
Startup Wizard: Choose Network Scale
Indicate the type of network you will be modeling.
Network Scale
World
Enterprise
Campus
Office
Logical
Choose From Maps
Use Metric Units
Quit
Back
Next

5. Choose **Meters** in the **size** drop-down box. Type **200** for the **X span** and **200** for the **Y span.** Click **Next.** You should see the **Select Technologies** dialog box.

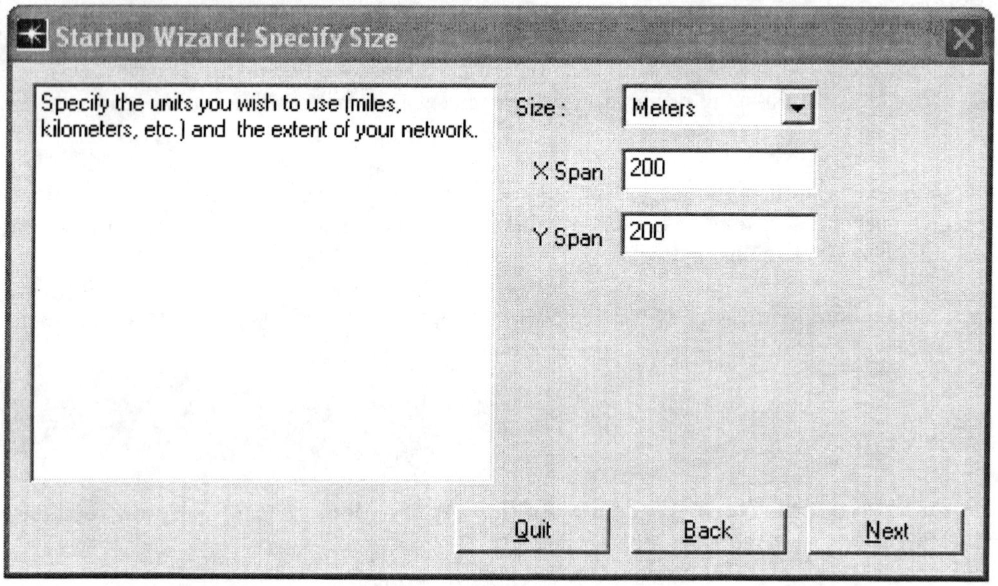
Startup Wizard: Specify Size
Specify the units you wish to use (miles, kilometers, etc.) and the extent of your network.
Size :
Meters
X Span
200
Y Span
200
Quit
Back
Next

6. Scroll down until you see **ethernet** under the **Model Family** column. Under the **Include column**, click on the box next to **ethernet**. This will change the "no" to "yes" indicating that Ethernet technologies will be included. Click **Next.**

Lab 9   100 Mbps Ethernet versus 100 Mbps FDDI

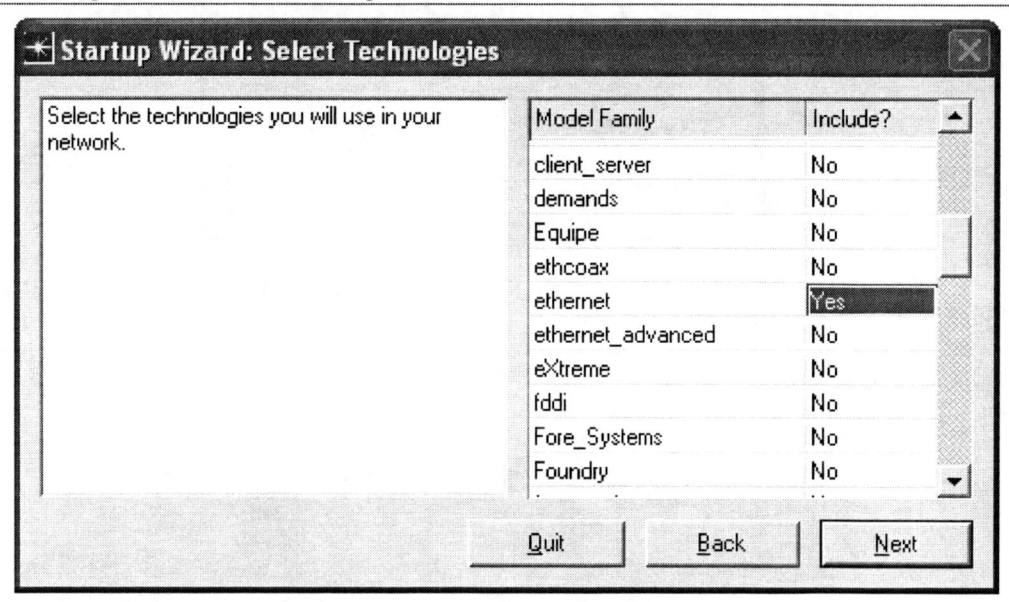

7. In the **Startup Wizard Review** dialog box, verify that the Ethernet technology was chosen and that the scale is Office 200m x 200m. Click **OK**. You should see the **Object Palette** window.

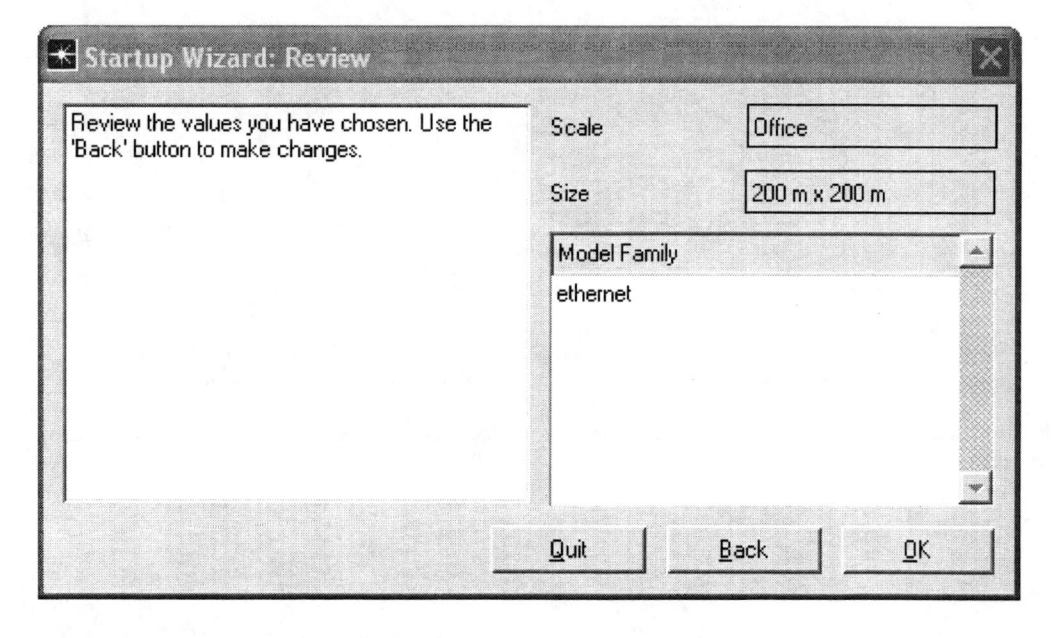

Building the Ethernet Network

1. We will begin by using Rapid Configuration to create the hub-based 100BaseT LAN. We will create a star network with eight Ethernet stations connected to an Ethernet hub. We will choose 100BaseT links to connect the hosts to the hub. From the Main Menu choose **Topology/Rapid Configuration**. A window will appear prompting you to choose the topology or network shape. Choose **Star** from the pull-down menu.

151

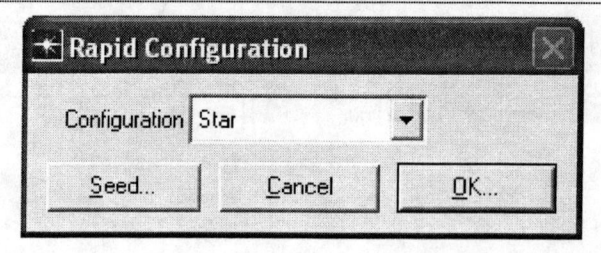

OPNET Lab Manual to accompany Business Data and Communications

2. A new window will appear that will allow you to choose the nodes in the network. For the **Center Node Model** choose **ethernet16_hub** from the pull-down menu. For the **Periphery Node Model** choose **ethernet_station.** Change the **Number** to **8**. This will create eight stations on the network. For the **Link Model**, choose **100BaseT**. Set the **X** coordinate to **100**, the **Y** coordinate to **100,** and the **Radius** to **50**. This will put the LAN in the center of the workspace. Click **OK**.

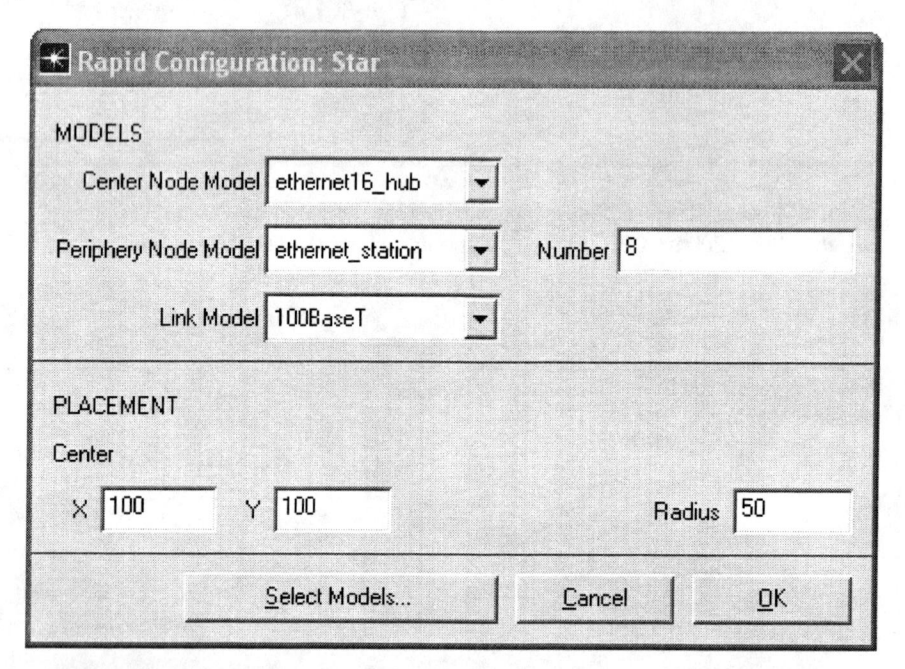

3. Right click on the hub object and choose **Set Name**. Name the hub **Hub100**. Your project should resemble the following Project screen image:

152

Lab 9   100 Mbps Ethernet versus 100 Mbps FDDI

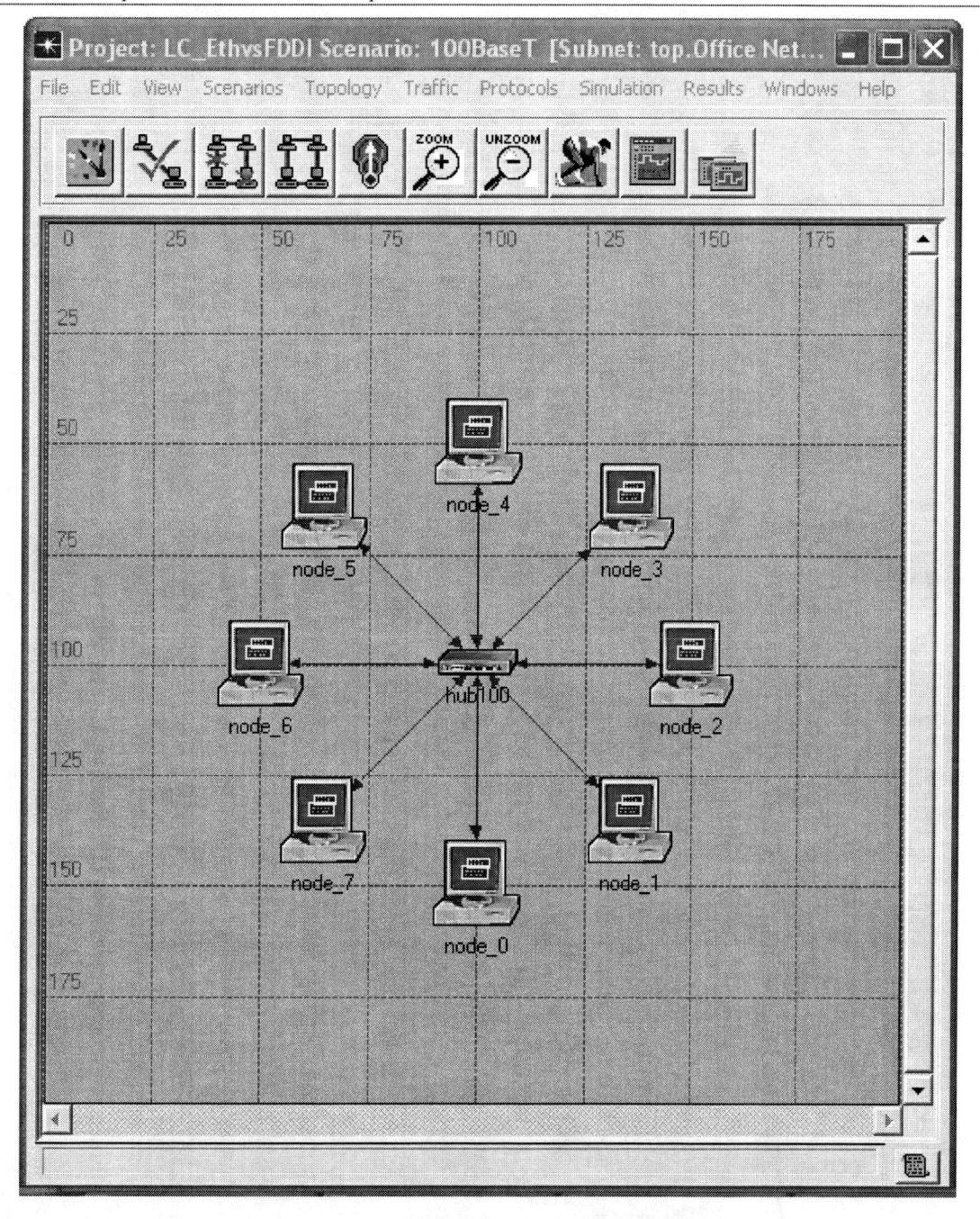

4. Right click on one of the **ethernet_stations** and choose **Select Similar Nodes.** You will see a circle around the station, indicating that it is selected. Right click on one of the selected **ethernet_station** objects and choose **Edit Attributes.** A new window will appear.

5. Click on the box next to **Apply Changes to Selected Objects** at the bottom of the window. This will ensure that all eight stations are configured in the same manner. Click on the + next to **Traffic generation parameters.** (Note: Your node numbers may vary. In this figure, node_2 is being configured.)

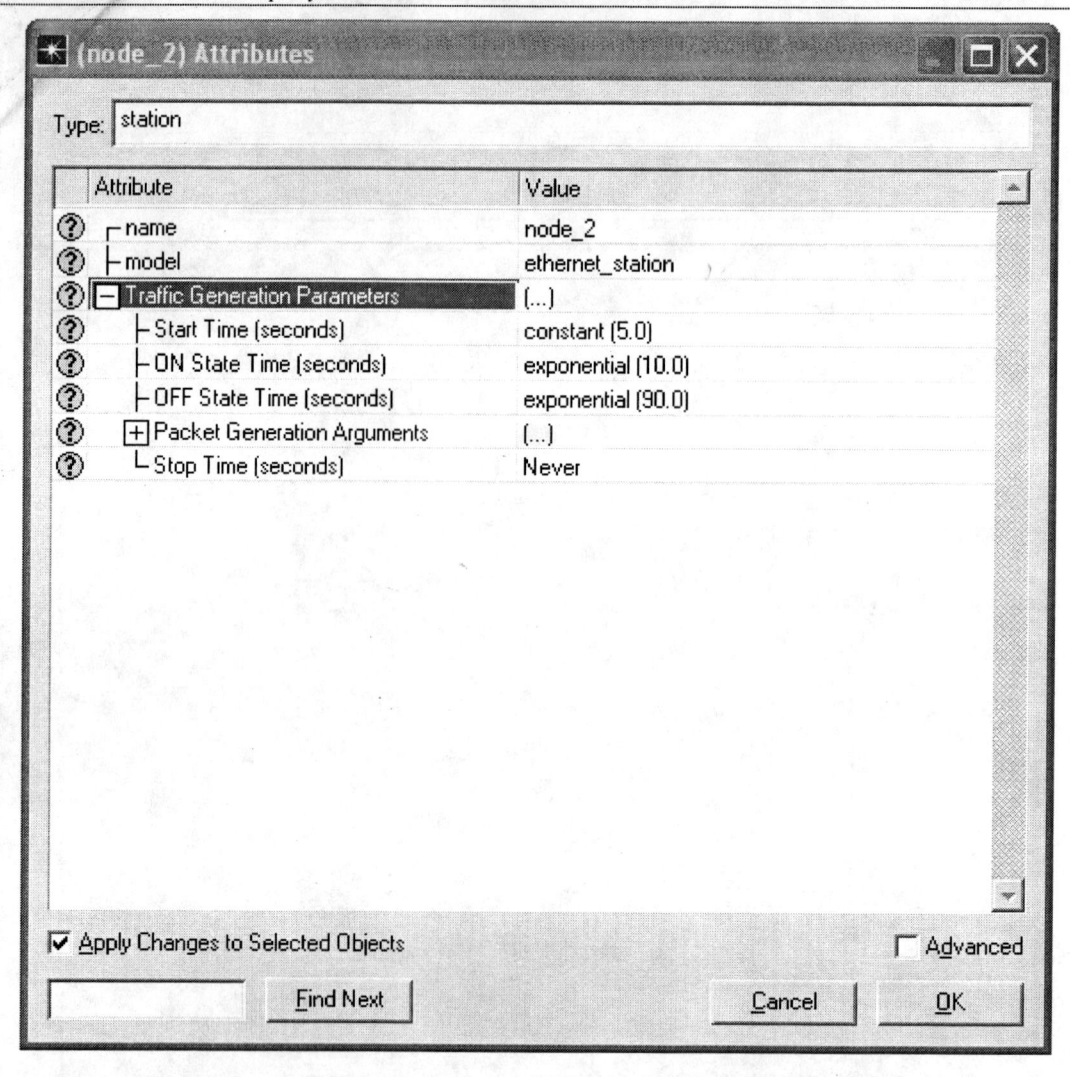

(node_2) Attributes

Type: station

Attribute	Value
⌐ name	node_2
├ model	ethernet_station
⊟ Traffic Generation Parameters	(...)
├ Start Time (seconds)	constant (5.0)
├ ON State Time (seconds)	exponential (10.0)
├ OFF State Time (seconds)	exponential (90.0)
⊞ Packet Generation Arguments	(...)
⌐ Stop Time (seconds)	Never

☑ Apply Changes to Selected Objects   ☐ Advanced

[Find Next]   [Cancel]   [OK]

6. We will first configure the simulation start time. We would like our simulation to start up immediately. Click on the value next to **Start Time (Seconds)**. A new window will appear. Change the value to **constant (1)**. This will cause our simulation to generate traffic after 1 second. Click **OK**.

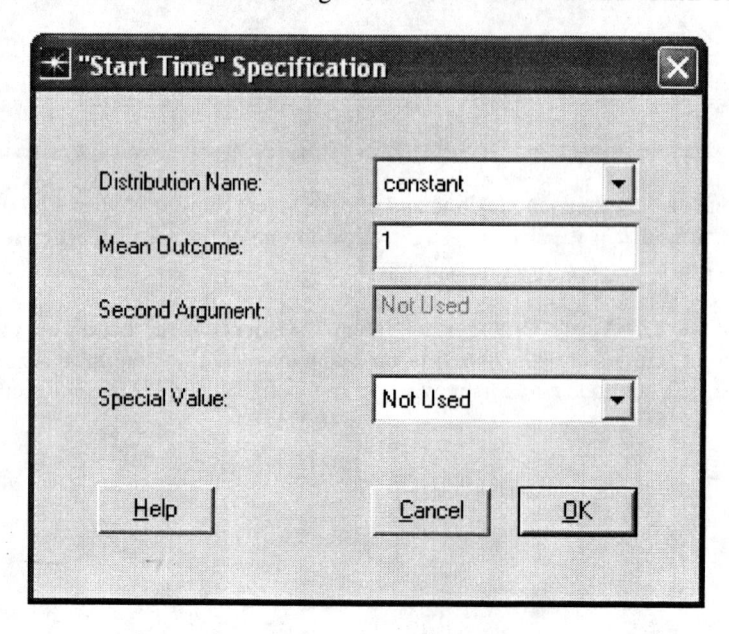

"Start Time" Specification

Distribution Name: constant

Mean Outcome: 1

Second Argument: Not Used

Special Value: Not Used

[Help]   [Cancel]   [OK]

Lab 9   100 Mbps Ethernet versus 100 Mbps FDDI

7. Next, click on the value next to **ON State Time**. This will bring up a new window. Change the **Mean Outcome** to **100**. Click **OK**.

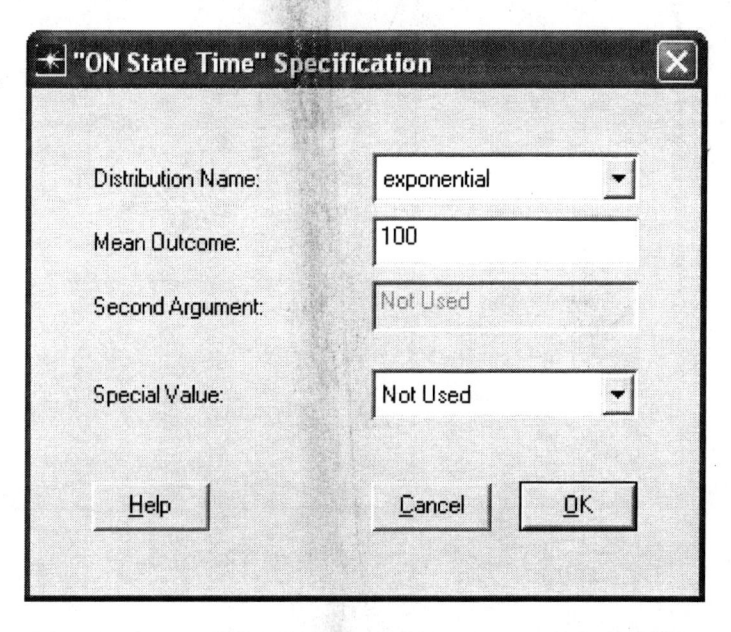

8. Click on the value next to **OFF State Time**. A new window will appear. Change the **Mean outcome time** to **0**. This causes our simulation to generate traffic 100% of the time. Click **OK**.

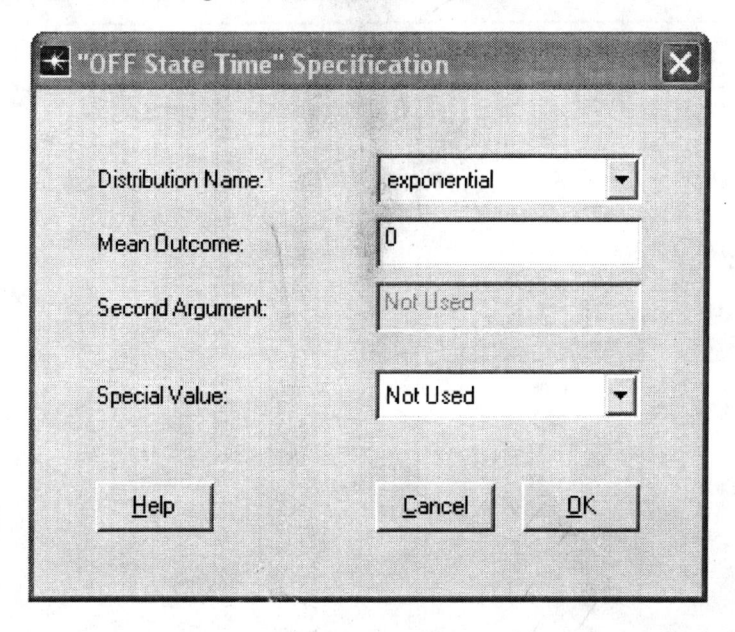

9. Now click on the + next to **Packet Generation Arguments**. We will change the interarrival time and the packet size. Click on the value next to **Interarrival Time (seconds)**. This will bring up a new window. Change the **Mean Outcome** time to **.02** and click **OK**.

155

"Interarrival Time" Specification

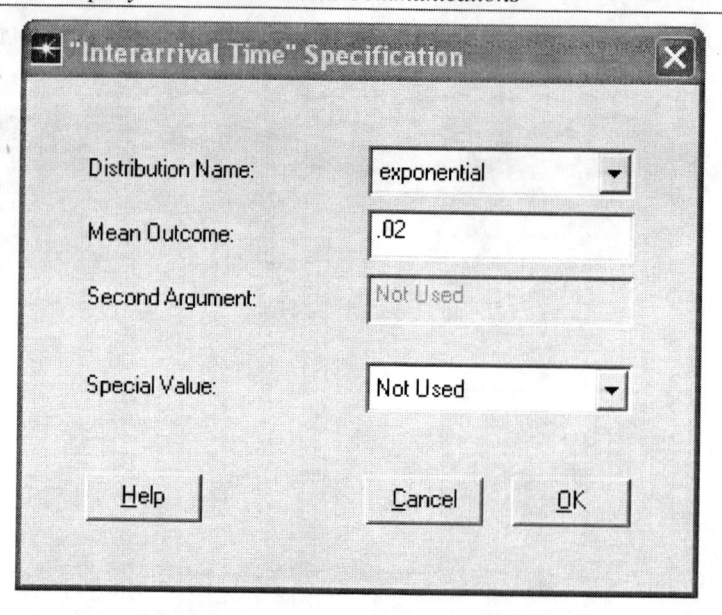

Distribution Name:	exponential
Mean Outcome:	.02
Second Argument:	Not Used
Special Value:	Not Used

10. You should now have the values listed in the Attribute window that follows. Click **OK** to close the window.

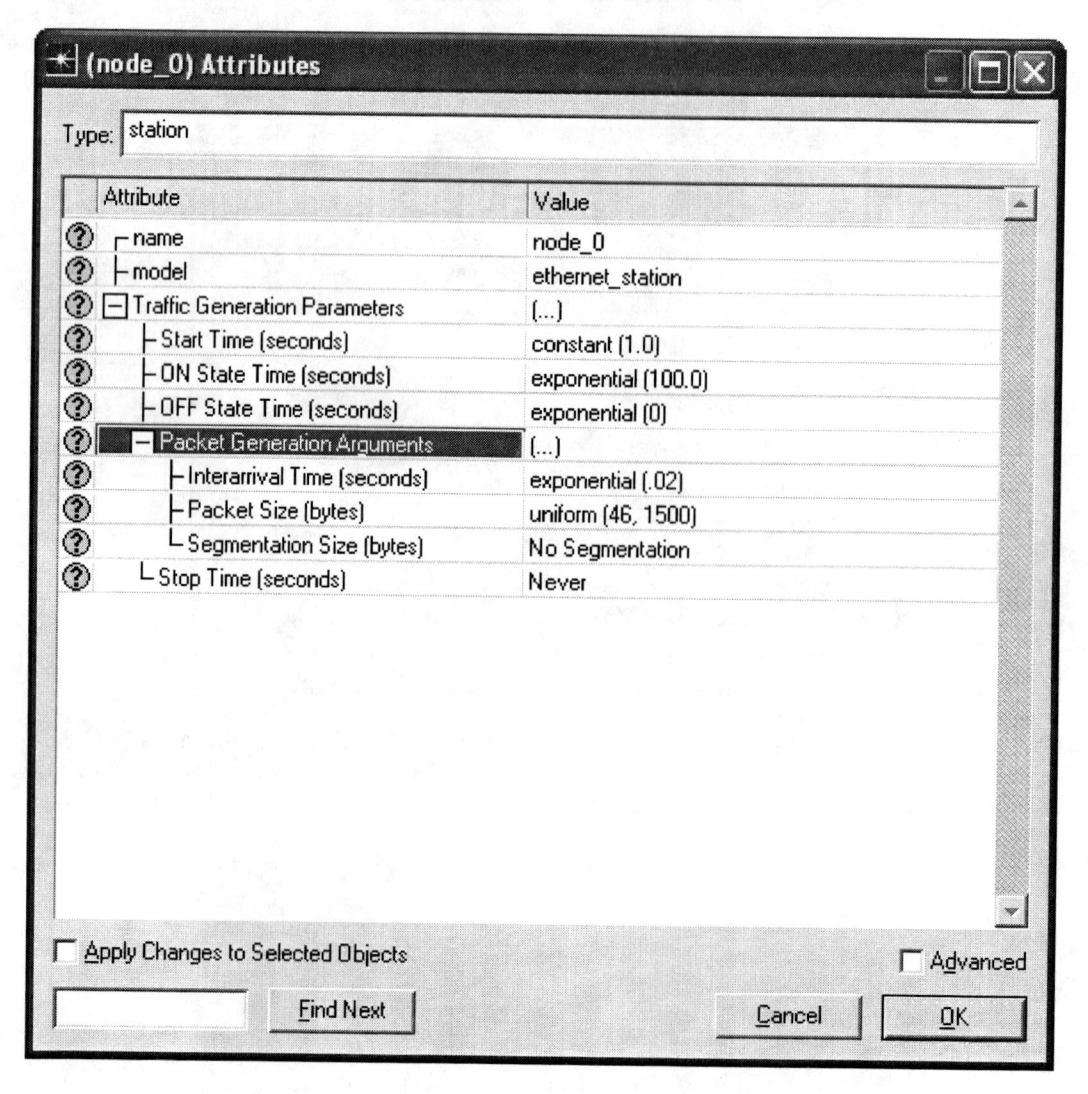

(node_0) Attributes

Type: station

Attribute	Value
name	node_0
model	ethernet_station
⊟ Traffic Generation Parameters	(...)
Start Time (seconds)	constant (1.0)
ON State Time (seconds)	exponential (100.0)
OFF State Time (seconds)	exponential (0)
⊟ Packet Generation Arguments	(...)
Interarrival Time (seconds)	exponential (.02)
Packet Size (bytes)	uniform (46, 1500)
Segmentation Size (bytes)	No Segmentation
Stop Time (seconds)	Never

☐ Apply Changes to Selected Objects   ☐ Advanced

11. From the Main Menu, choose **File/Save** to save your project.

Configuring the Ethernet simulation

1. Right click on the project workspace and select **Choose Individual Statistics** from the pull-down menu. Click on the **+** next to **Global** to expand the parameters. Click on **Ethernet** and **Delay**.

2. Also under Global, click on the **+** next to **Traffic Sink** to expand the parameters. Click on **End-to-End Delay (seconds)** and **Traffic Received (bits/sec)**. Click on the **+** next to **Traffic Source** to expand the parameters. Click on **Traffic Sent (bits/sec)**.

3. Click on the **+** next to **Node Statistics** to expand the statistic options. Click on **Ethernet.** Click on **Collision Count, Delay (sec), Traffic Forwarded (bits/sec), Traffic Received (bits/sec),** and **Utilization.**

4. Finally, click on the **+** next to **Link Statistics** to expand the parameters. Click on **point-to-point.** Click on **throughput (bits/sec)→, throughput (bits/sec) ←, utilization →, and utilization ←.** Verify that your parameters match those chcked in the Choose Results window image that appears next. Click **OK** to save these parameters.

Choose Results

- Global Statistics
  - Ethernet
    - ☑ Delay (sec)
  - Traffic Sink
    - ☑ End-to-End Delay (seconds)
    - ☐ Traffic Received (bits)
    - ☑ Traffic Received (bits/sec)
    - ☐ Traffic Received (packets)
    - ☐ Traffic Received (packets/sec)
  - Traffic Source
    - ☐ Traffic Sent (bits)
    - ☑ Traffic Sent (bits/sec)
    - ☐ Traffic Sent (packets)
    - ☐ Traffic Sent (packets/sec)
- Node Statistics
  - Ethernet
    - ☐ Burst Duration (sec)
    - ☐ Burst ON/OFF
    - ☐ Burst Size (packets)
    - ☑ Collision Count
    - ☐ Delay (sec)
    - ☐ Load (bits)
    - ☐ Load (bits/sec)
    - ☐ Load (packets)
    - ☐ Load (packets/sec)
    - ☑ Traffic Forwarded (bits/sec)
    - ☐ Traffic Forwarded (packets/sec)
    - ☐ Traffic Received (bits)
    - ☑ Traffic Received (bits/sec)
    - ☐ Traffic Received (packets)
    - ☐ Traffic Received (packets/sec)
    - ☐ Transmission Attempts
    - ☑ Utilization
- Link Statistics
  - low-level point-to-point
  - point-to-point
    - ☐ queuing delay -->
    - ☐ queuing delay <--
    - ☑ throughput (bits/sec) -->
    - ☑ throughput (bits/sec) <--
    - ☐ throughput (packets/sec) -->

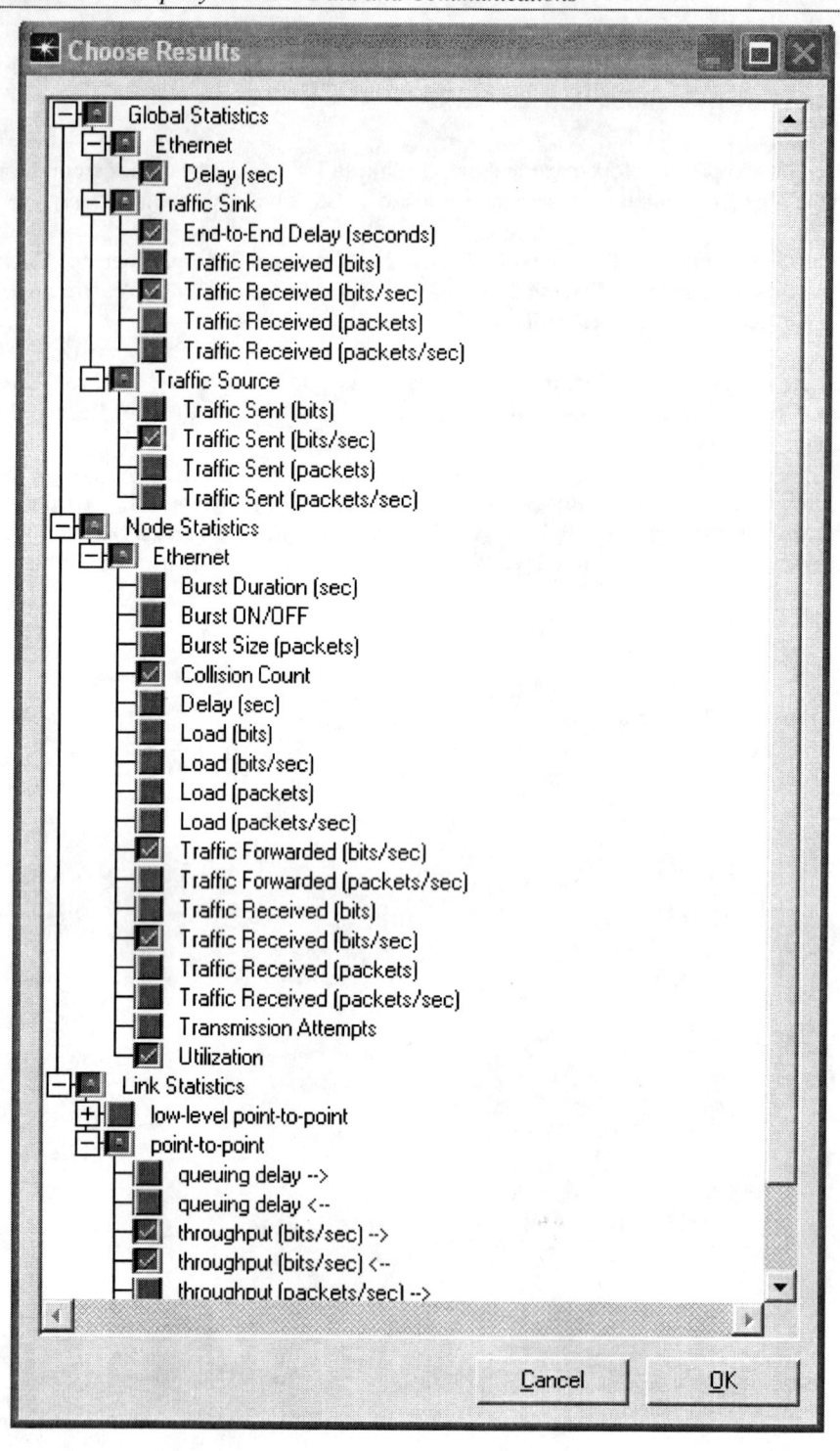

[Cancel] [OK]

5. From the Main Menu, choose **File/Save** to save your project.

Lab 9   100 Mbps Ethernet versus 100 Mbps FDDI

Running the Simulation

1. Click on the **Configure and Run** Button. Set the **Duration** value to **10** and the time unit to **Minute(s)**. Click the **Run** button to run the simulation. Depending on your processor, the simulation should take several minutes to run.

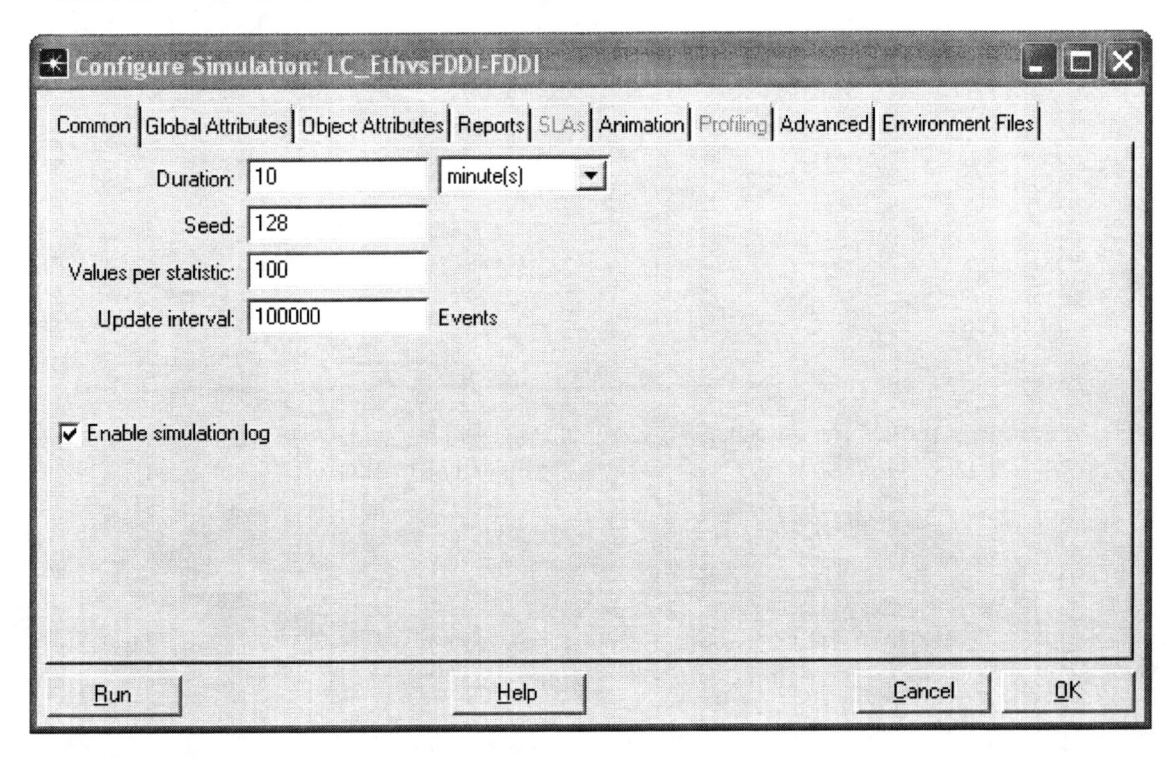

2. We will look at our results after simulating the FDDI LAN scenario. Choose **File/Save** from the Main Menu to save your project.

Creating a New Scenario

1. Now we want to see how an FDDI LAN matches in performance. From the Main Menu, choose **Scenarios/New Scenario**. Give your scenario a name such as **FDDI** and click **OK**. Another dialog box will appear. Choose **Create from Empty Scenario** and click **Next**.

2. Verify that **Create Empty Scenario** is selected in the **Initial Topology Dialog Box**. Click **Next**. You will now see the **Network Scale** dialog box.

3. Choose **Office** from the **Network Scale** list. Click **Next**. Choose **Meters** in the **size** drop-down box. Type **200** for the **X span** and **200** for the **Y span**. Click **Next**. You should see the **Select Technologies** dialog box.

4. Scroll down until you see **fddi** under the **Model Family** column. Under the **Include column**, click on the box next to **fddi**. This will change the "no" to "yes," indicating that fddi technologies will be included. Click **Next**.

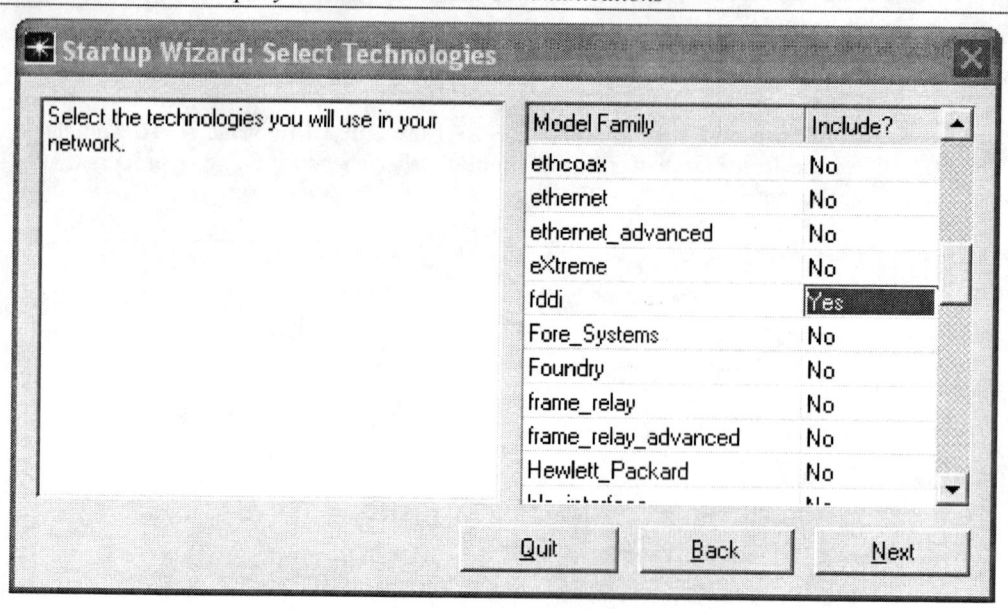

OPNET Lab Manual to accompany Business Data and Communications

[Startup Wizard: Select Technologies dialog showing Model Family list with fddi set to Yes, others No]

5. In the **Startup Wizard Review** dialog box, verify that the fddi technology was chosen and that the scale is Office 200m x 200m. Click **OK**. You should see the **Object Palette** Window.

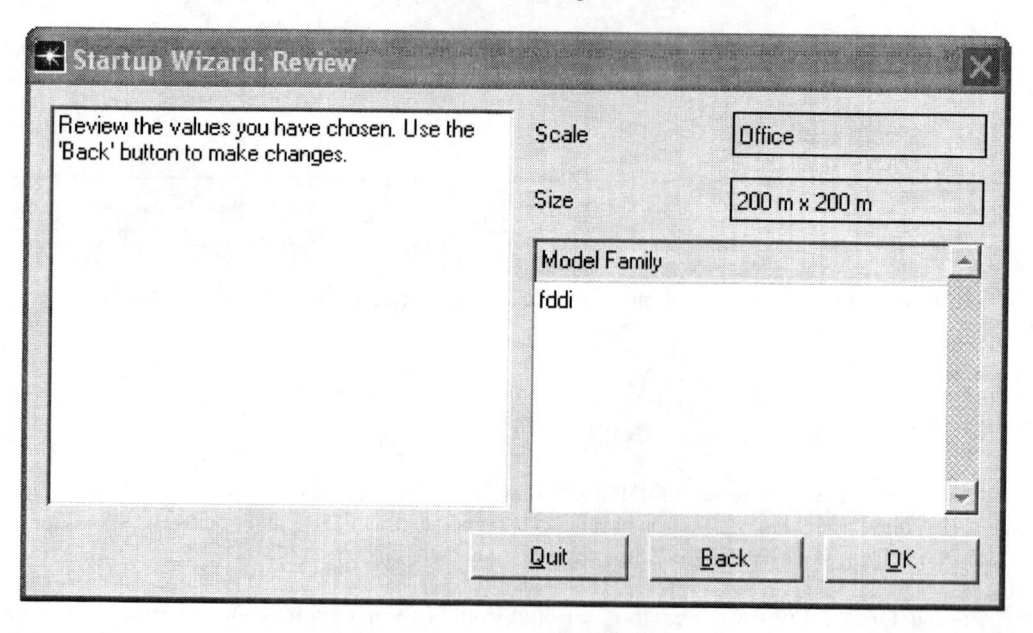

[Startup Wizard: Review dialog showing Scale: Office, Size: 200 m x 200 m, Model Family: fddi]

6. Again, we will use Rapid Configuration to create an FDDI LAN. We will create a ring network with eight Ethernet stations connected in point-to-point fashion. From the Main Menu, choose **Topology/Rapid Configuration**. A window will appear, prompting you to choose the topology or network shape. Choose **Ring** from the pull-down menu.

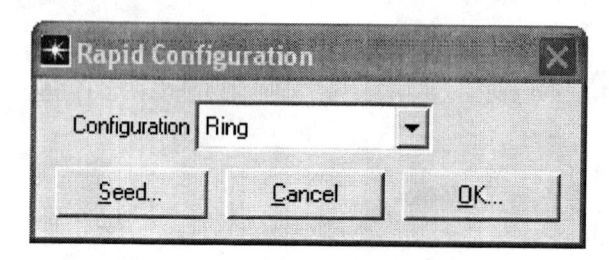

[Rapid Configuration dialog with Configuration: Ring]

7. A new window will appear which will allow you to choose the nodes in the network. For the **Node Model,** choose **fddi_station** from the pull-down menu. For the **Link Model,** choose **FDDI**. Change the

160

Lab 9   100 Mbps Ethernet versus 100 Mbps FDDI

number of nodes to **8**. For the **Type,** choose **Duplex.** This means that our links will be able to both send and receive at the same time. Set the **X** coordinate to **100**, the **Y** coordinate to **100,** and the **Radius** to **50**. This will put the LAN in the center of the workspace. Click **OK**.

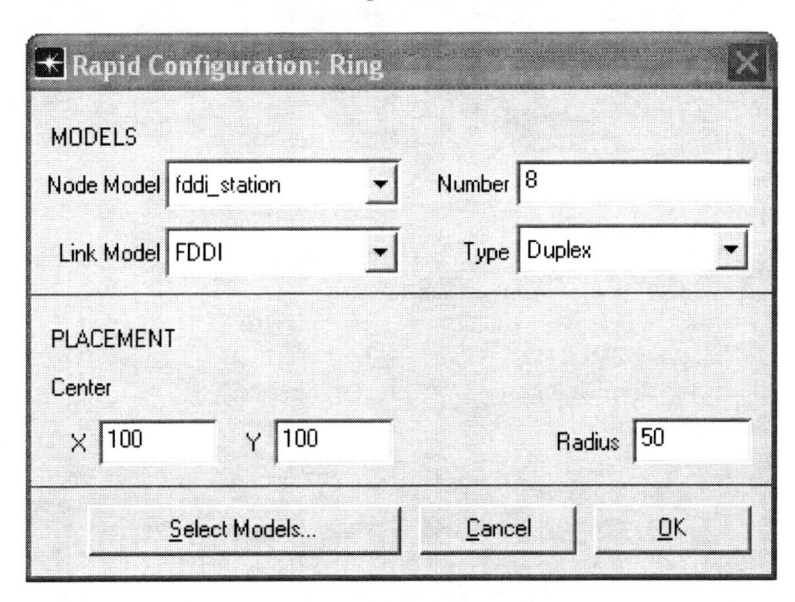

8. Right click on one of the **fddi_stations** and choose **Select Similar Nodes.** You will see a circle around the station indicating that it is selected. Right click on one of the selected **fddi_station** objects and choose **Edit Attributes.** A new window will appear.

9. Click on the box next to **Apply Changes to Selected Objects** at the bottom of the window. This will ensure that all 8 stations are configured in the same manner. Click on the **+** next to **Traffic Generation Parameters.** Click on the value next to **FDDI Asynchronous Traffic Mix** and choose **100%** from the pull-down menu.

10. Next, change the **Start Time (seconds)** to **constant(1),** the **ON State Time (seconds)** to **exponential(100),** the **OFF State Time (seconds)** to **exponential(0),** and the **Packet Interarrival Time** to **exponential(.02).** These are the same values that we used for our 100BaseT Ethernet scenario. Note that the default packet size for FDDI LANs is exponential(1024).

161

Attribute	Value
name	node_3
model	fddi_station
FDDI Asynchromous Traffic Mix	100%
⊟ Traffic Generation Parameters	(...)
Start Time (seconds)	constant (1)
ON State Time (seconds)	exponential (100)
OFF State Time (seconds)	exponential (0)
− **Packet Generation Arguments**	(...)
Interarrival Time (seconds)	exponential (.02)
Packet Size (bytes)	exponential (1024)
Segmentation Size (bytes)	No Segmentation
Stop Time (seconds)	Never

☑ Apply Changes to Selected Objects   ☐ Advanced

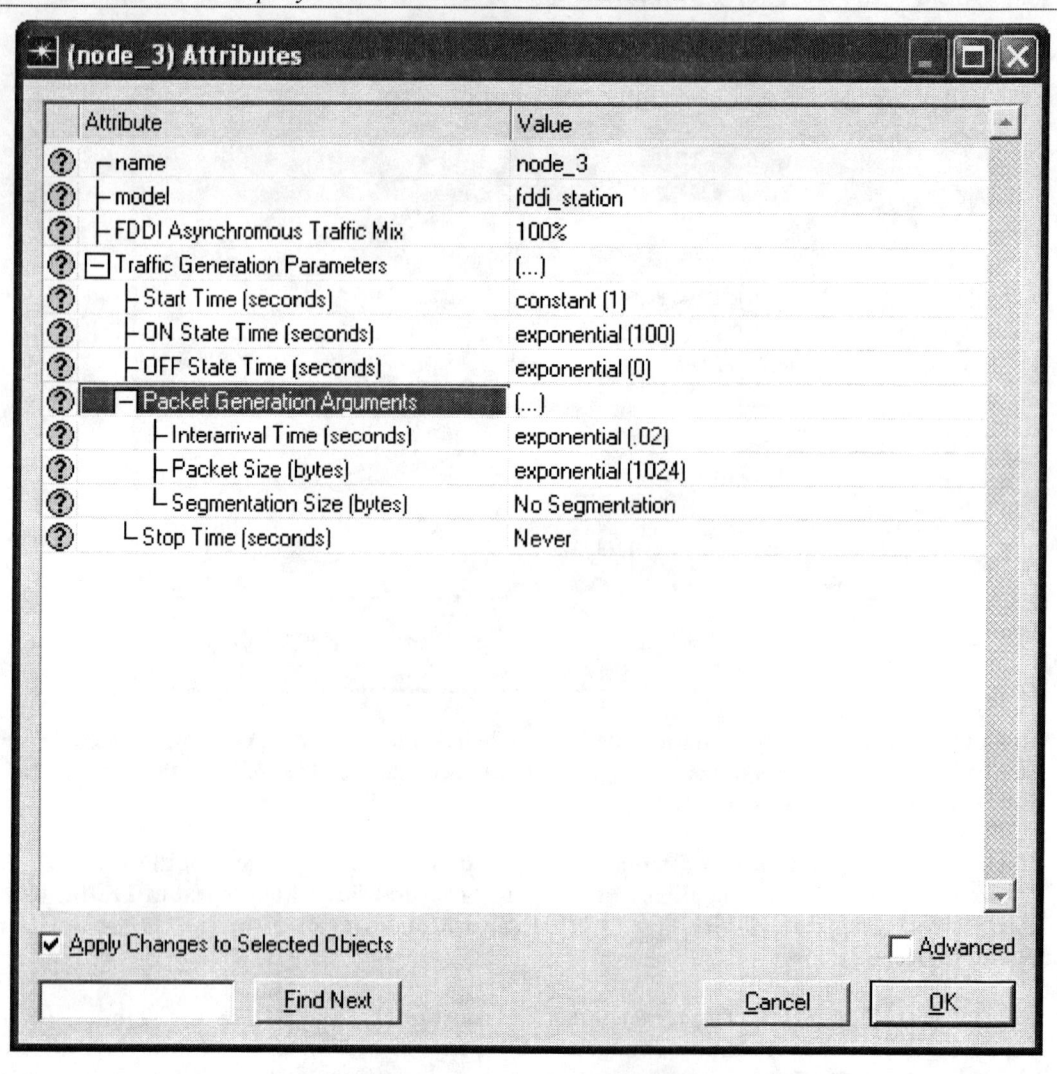

11. From the Main Menu, choose **File/Save** to save your project. Your project should resemble the one following:

Lab 9   100 Mbps Ethernet versus 100 Mbps FDDI

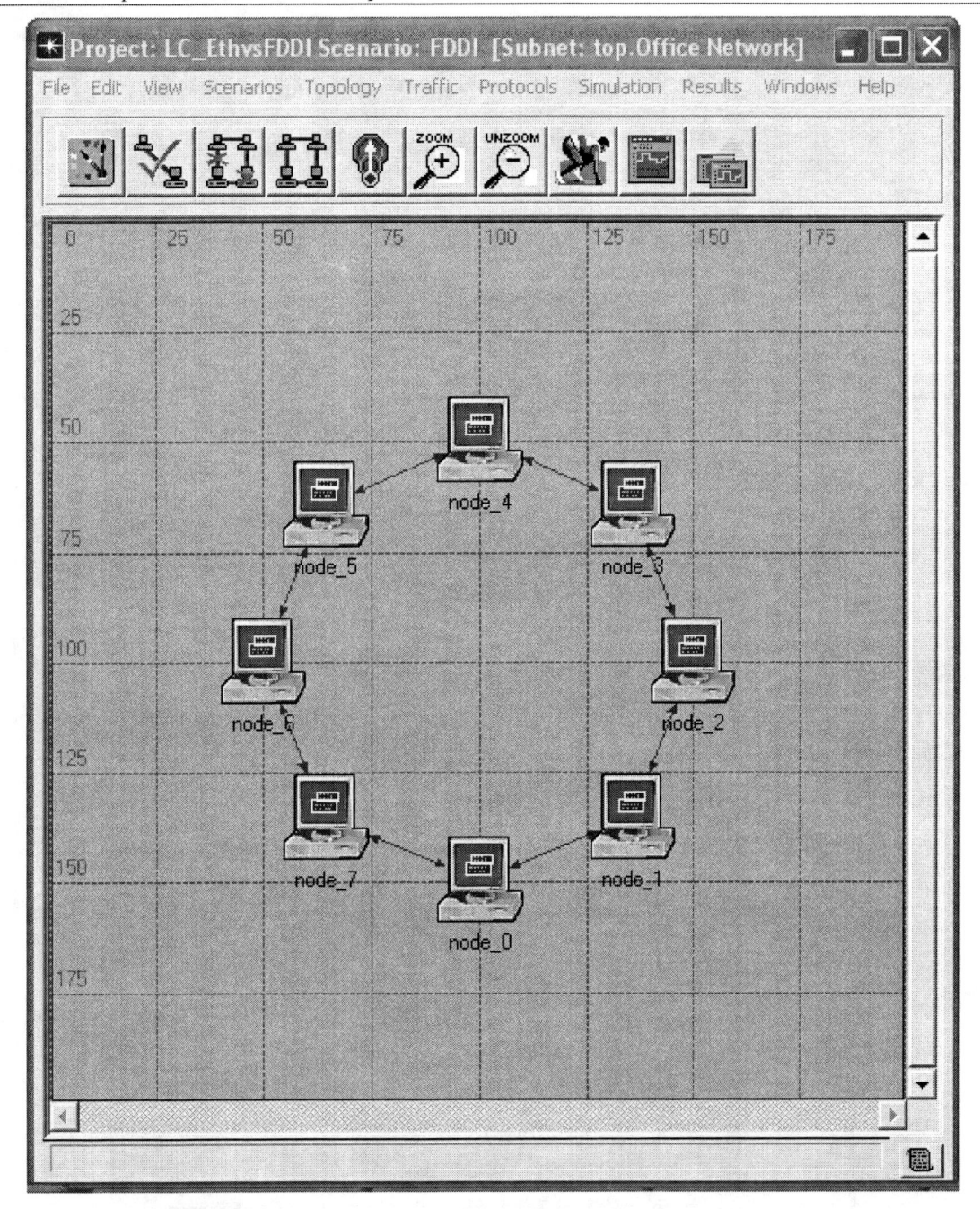

Configuring the FDDI simulation

1. Right click on the project workspace and select **Choose Individual Statistics** from the pull-down menu. Click on the **+** next to **Global** to expand the parameters. Click on **+** next to **FDDI**. Click on **Delay (sec)** and **Media Access Delay (sec)**.

2. Also, under **Global**, click on the **+** next to **Traffic Sink** to expand the parameters. Click to select **End-to-End Delay (seconds)** and **Traffic Received (bits/sec)**. Click on the **+** next to **Traffic Source** to expand the parameters. Click on **Traffic Sent (bits/sec)**.

3. Click on the **+** next to **Node Statistics** to expand the parameters. Click on the **+** next to **FDDI**. Click on **Delay (sec), Load (bits/sec),** and **Traffic Received (bits/sec).**

4. Click on the + next to **Link Statistics** to expand the parameters. Click on the + next to **point-to-point**. Click on **throughput → (bits/sec), throughput ← (bits/sec), utilization →, and utilization ←**. Verify that your parameters match those presented next and **click OK** to save these parameters.

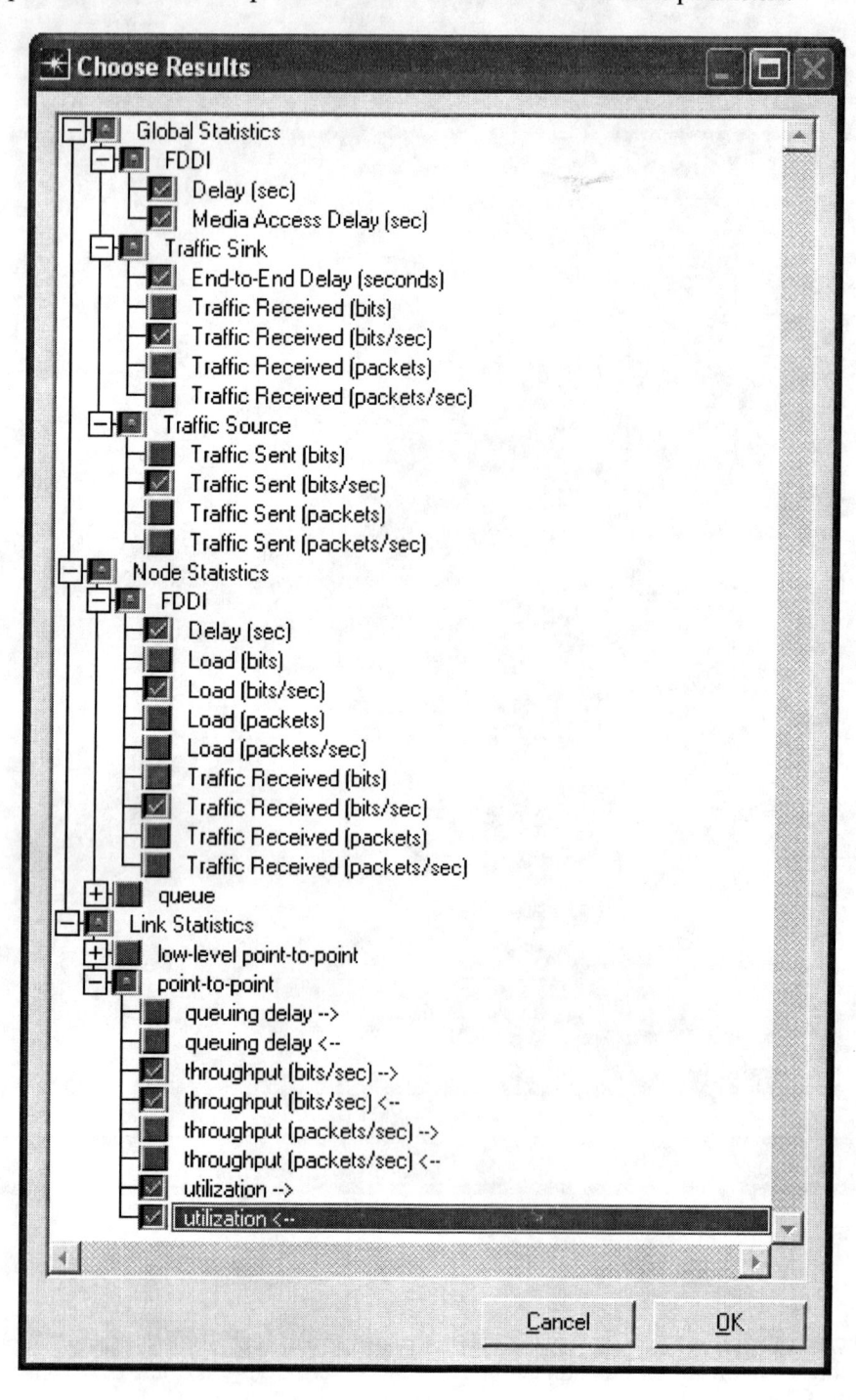

Choose Results

- Global Statistics
  - FDDI
    - ☑ Delay (sec)
    - ☑ Media Access Delay (sec)
  - Traffic Sink
    - ☑ End-to-End Delay (seconds)
    - ☐ Traffic Received (bits)
    - ☑ Traffic Received (bits/sec)
    - ☐ Traffic Received (packets)
    - ☐ Traffic Received (packets/sec)
  - Traffic Source
    - ☐ Traffic Sent (bits)
    - ☑ Traffic Sent (bits/sec)
    - ☐ Traffic Sent (packets)
    - ☐ Traffic Sent (packets/sec)
- Node Statistics
  - FDDI
    - ☑ Delay (sec)
    - ☐ Load (bits)
    - ☑ Load (bits/sec)
    - ☐ Load (packets)
    - ☐ Load (packets/sec)
    - ☐ Traffic Received (bits)
    - ☑ Traffic Received (bits/sec)
    - ☐ Traffic Received (packets)
    - ☐ Traffic Received (packets/sec)
  - [+] queue
- Link Statistics
  - [+] low-level point-to-point
  - point-to-point
    - ☐ queuing delay -->
    - ☐ queuing delay <--
    - ☑ throughput (bits/sec) -->
    - ☑ throughput (bits/sec) <--
    - ☐ throughput (packets/sec) -->
    - ☐ throughput (packets/sec) <--
    - ☑ utilization -->
    - ☑ utilization <--

[Cancel] [OK]

5. From the Main Menu, choose **File/Save** to save your project.

6. Click on the **Configure and Run** Button. Set the **Duration** value to **10** and the time unit to **Minute(s)**. Click the **Run** button to run the simulation. Depending on your processor, the simulation should take several minutes to run.

Comparing Simulation Results

1. From the Main Menu, choose **Results/Compare Statistics**.

2. To view end-to-end delay for both the Ethernet and FDDI LAN, click on the **+** next to **Global**. Next, click on **Traffic Sink** and **End-to-End Delay (seconds)**. Change the value **As Is** to **time_average**. Click on the **Show** button. You should see the following graph:

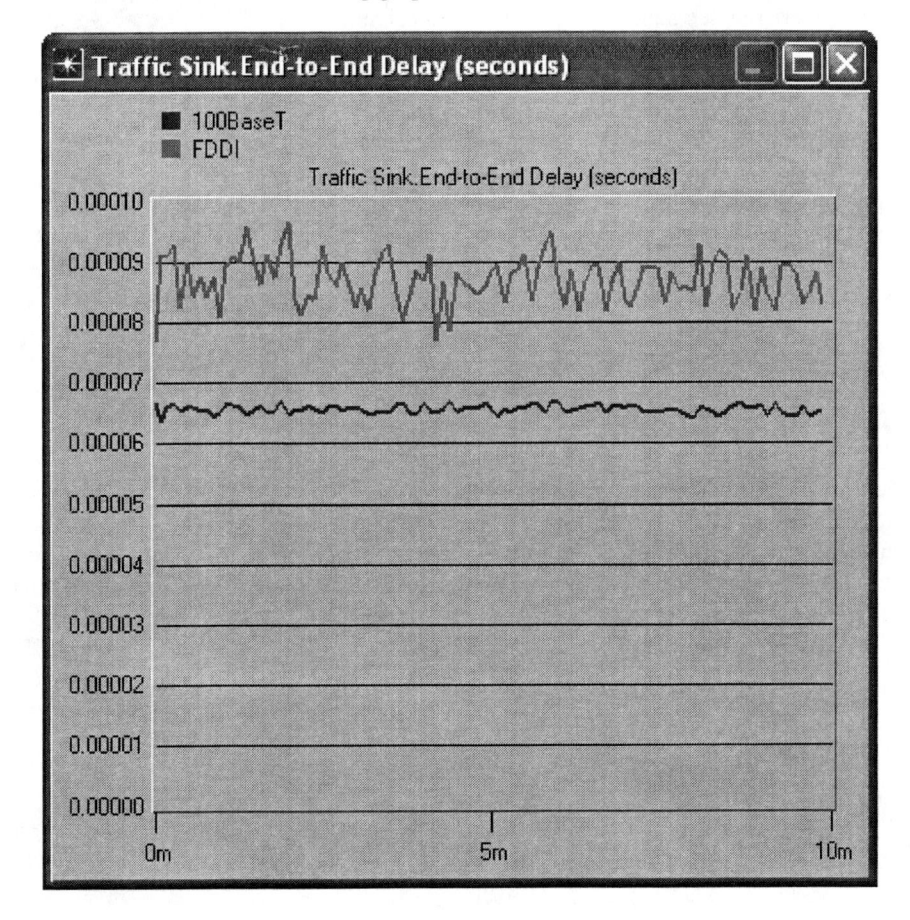

Click on the **X** to close the window. The **Close Analysis** window will appear asking if you want to hide the panel or permanently delete it. Click on the **Delete** button to delete the window.

You may discard the results of your simulations, rerun them, and manage your scenarios by doing the following:

Choose **Scenarios/Manage Scenarios** from the Main Menu. A window will appear with the names of your scenarios, the length of time that the simulation ran, and the status of the data.

By clicking on a scenario name, you may choose to **Discard Results** or **Collect Results**. By clicking on the values under the **Results** column, you may choose **recollect** or **collect** from the pull-down menu. Click **OK** to close the Results window.

From the Main Menu, choose **File/Save** to save your project.

Use the results you obtained in your simulations to answer the following questions:

Questions

1. What was the average number of bits/sec received in the Ethernet LAN scenario? Explain why there might be a difference between the number of bits sent and the number received.
2. What was the average number of bits/sec received in the FDDI LAN scenario? Explain why there might be a difference between the number sent and the number received. Explain how it is possible that the FDDI LAN received more bits/sec than the Ethernet LAN received.
3. What was the media access delay value for the FDDI LAN? Would this value likely increase or decrease if more stations were added to the ring? You may want to create a duplicate scenario, add more stations, and repeat the simulation to see if your thoughts are correct.
4. How does link throughput received compare in both scenarios?
5. How does link load compare in both scenarios?
6. Were their any collisions in the 100BaseT Ethernet scenario? If so, how many were there?
7. In our simulations packet generation had an exponential distribution with sizes between 46 and 1500 for the Ethernet LAN, and an exponential distribution centered at 1024 for the FDDI LAN. Change the packet size for both simulations to a constant size of 1024. Do fixed packet sizes impact results significantly? In particular, look at throughput and delay statistics.

Summary Questions

1. Which network had the highest end-to-end delay?
2. Which network had the highest throughput?
3. Based on your observations, which type of LAN performs better?
4. How does cost of network components and maintenance compare between Ethernet and FDDI? (You may need to do some additional research to answer this question.)

LAB 10: VIRTUAL PRIVATE NETWORKS

Objective
1. To create a Virtual Private Network (VPN) WAN.
2. To evaluate application response time in the presence and absence of a firewall.

Motivation
RecOnline sells recreational equipment over the Internet. Corporate headquarters are located in Pittsburgh, PA. RecOnline has a regional sales office in Southern California, and a business partner that manages and provides inventory located in Minnesota. The Minnesota site uses the File Transfer Protocol (FTP) to download customer merchandise orders from RecOnline headquarters. The Southern California site uses the corporate database located at headquarters to check stocked inventory and post credit card orders for their customers. For customer privacy, RecOnline has recently installed a Virtual Private Network (VPN) between their Southern California site and corporate headquarters. All sites communicate via email. Due to recent employee transfers, more email is being exchanged between the Minnesota site and headquarters. Krystal Brown of RecOnline would like to gather statistics for current application response times. She suspects that response time has been degrading due to the numerous email messages transferred between Minnesota and corporate headquarters. She suspects that these email messages are non-business related, and she is considering installing a firewall to block email from the Minnesota site.

Description
Virtual Private Networks (VPNs) can be used to provide a secure transfer of information over the public Internet. To create a VPN, a tunnel must be set up between a source and a destination site. Creating a tunnel involves configuring a router at each end site to encapsulate and de-encapsulate IP packets. Packets that arrive at the source end of the tunnel addressed to a network at the destination end will be encrypted and wrapped with a new IP header. The new header will differ from the original in that its source and destination addresses will reflect those of the routers at the source and destination ends of the tunnel. Intermediate routers forwarding the IP packet between the source and the destination end of the tunnel will be unaware of the encapsulated IP packet and its contents. When the packet arrives at the end of the tunnel, the router will "unwrap" the packet by removing its external IP header, and will deliver the internal packet to its original destination address. Although VPNs increase security, the addition of encryption and decryption, as well as wrapping and unwrapping packets at each end of the tunnel may lead to increased packet latency. Firewalls can also be used as a security mechanism. A firewall is a router that has been configured to "filter" or deny delivery of packets from particular applications or IP addresses.

In this lab, we will create an IP Wide Area Network (WAN) that carries information between RecOnline's corporate headquarters to its Southern California and Minnesota sites. A VPN will also be configured between the Southern California site and headquarters. Application response times will be evaluated in the presence and absence of a firewall.

Follow the directions listed next to create an OPNET model, run simulations, and view results. Questions regarding the lab appear at the end.

1. Start the **OPNET IT Guru Academic Edition Application**. Create a **New Project** by selecting **File/New/Project** from the Main Menu. Click **OK**.

2. Give the project a name such as **your_initials_VPN**. Give the scenario a name such as **noFirewall**. Click **OK**. You should see the **Startup Wizard Initial Topology** dialog box.

3. Verify that **Create Empty Scenario** is selected in the **Initial Topology Dialog Box**. Click **Next**. You will now see the **Network Scale** dialog box.

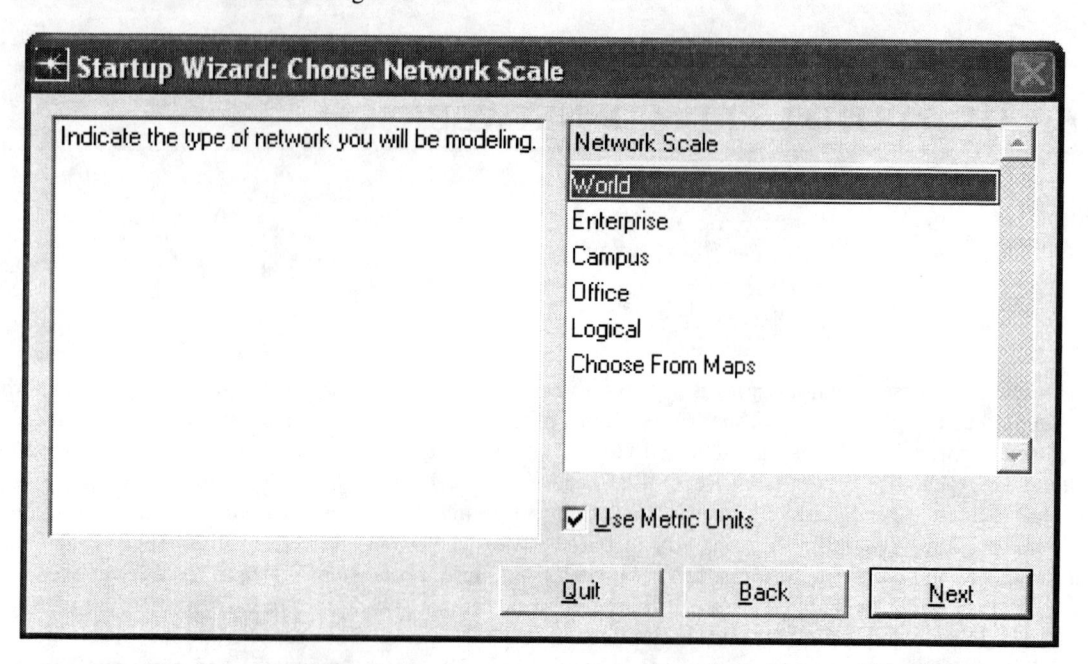

4. Choose **World** from the **Network Scale** list. Click **Next**. Choose **usa** as your map and click **Next**. You should see the **Select Technologies** dialog box.

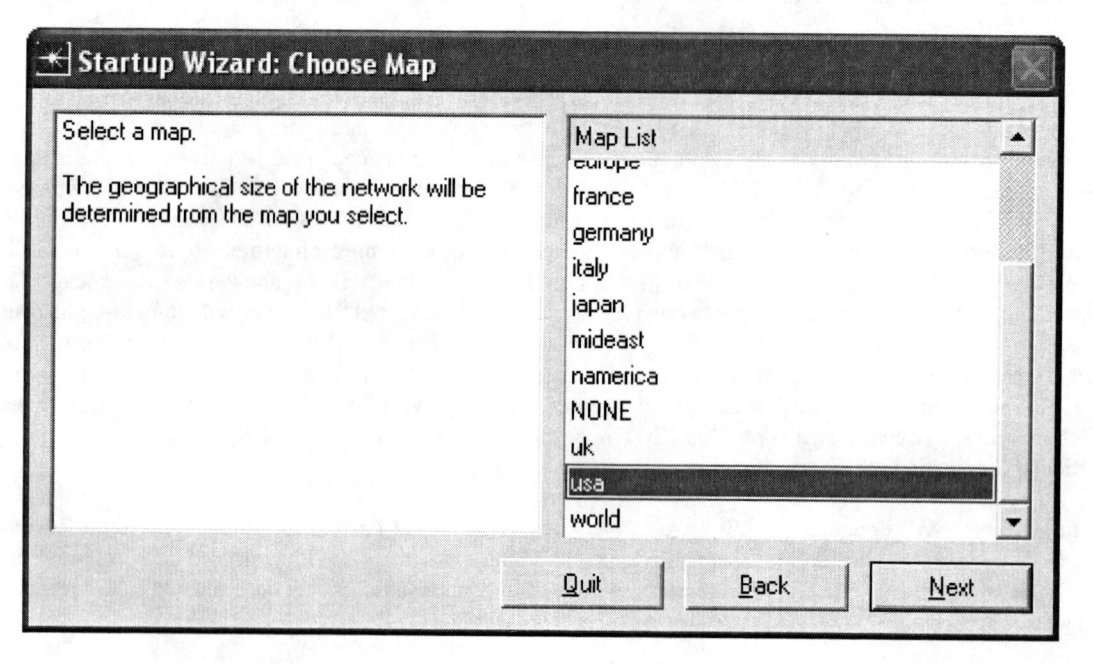

5. Scroll down until you see **internet_toolbox** under the **Model Family** column. Under the **Include column**, click on the box next to **internet_toolbox**. This will change the "no" to "yes" indicating that Internet technologies will be included. Click **Next**.

Lab 10   Virtual Private Networks

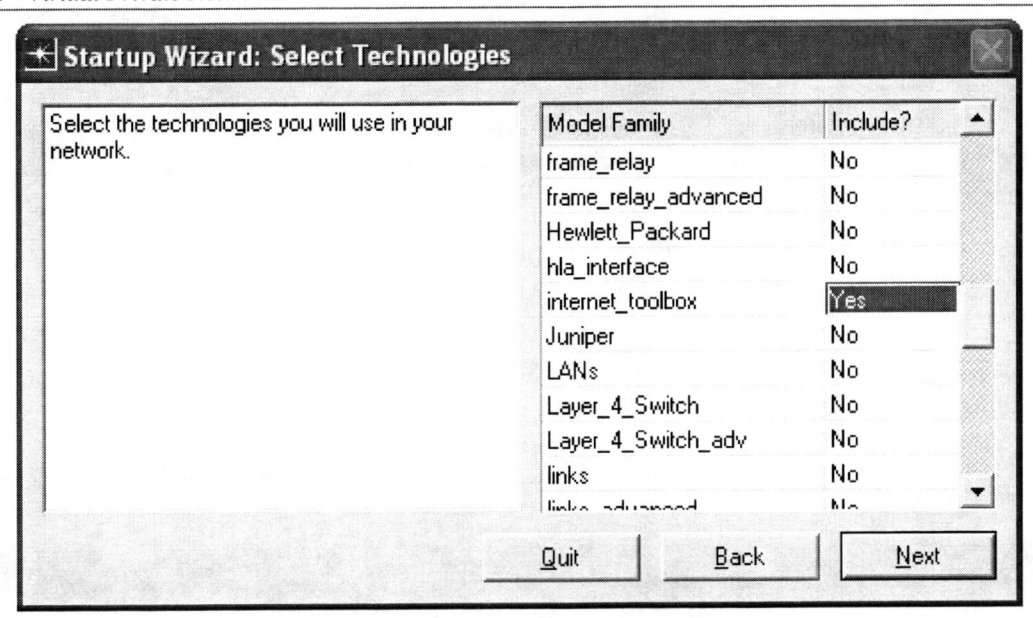

6. In the **Startup Wizard Review** dialog box, verify that the **internet_toolbox**, technologies are chosen and that the scale is World and the map is usa. Click **OK**. You should see the **Object Palette** window.

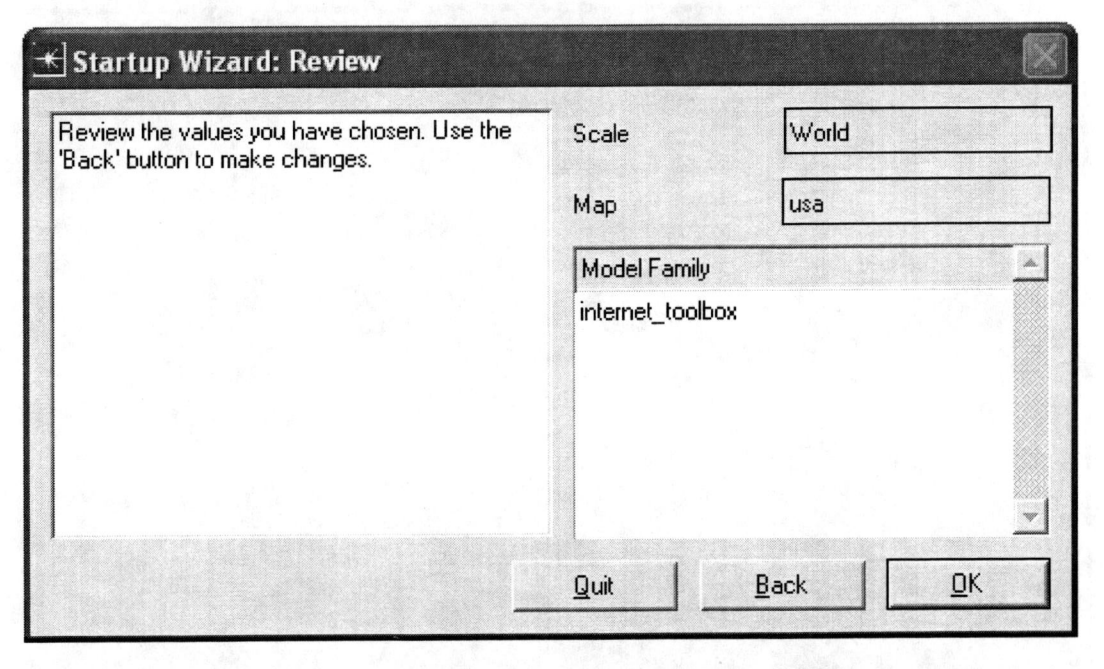

7. From the **Object Palette**, select a **100BaseT_Lan** object and click to paste it in the area of Southern California. Paste another copy in the area of Minnesota. Right click to release the object. Right click on the Southern California 100BaseT_LAN and choose **Set Name** from the pull-down menu. Name the LAN **SouthernCaliforniaOffice**. Repeat this step and name the Minnesota LAN **MinnesotaOffice.**

8. From the **Object Palette**, click on the **subnet** object to select it. Click to paste it in the area of Pennsylvania. Right click to release the subnet object. Right click on the subnet object, and choose **Set Name** from the pull-down menu. Name the subnet **Headquarters**.

9. We will begin by creating the Pennsylvania Headquarter's subnet, which will consist of a 100BaseT LAN, a switch and three servers. The switch will be connected to a router that will connect to another gateway router that will connect to an IP cloud. Double click on the **Headquarters** subnet object to descend into the subnet. From the **Object Palette**, click on the **ethernet16_switch** and click to paste it in the area of

Pittsburgh. Right click to release the object. Right click on the switch, and choose **Set Name** from the pull-down menu. Name the switch **HQSwitch**.

10. From the **Object Palette**, select an **ethernet_server** object. Paste **three** copies of the server object around the switch. Right click to release the object. One at a time, right click on the server objects, and choose **Set Name** from the pull-down menu. Name the servers **DBServer, EmailServer**, and **FTPServer.**

11. Next, from the **Object Palette**, click on the **100BaseT_Lan** subnet object. Paste a copy of it near the switch. Right click to release the object. Right click on the 100BaseT_LAN and choose **Set Name** from the pull-down menu. Name the LAN **HQLan**.

12. Again, from the **Object Palette**, click on the **100BaseT Link** object. Connect the **100BaseT LAN, DBServer, EmailServer**, and **FTPServer** to the switch. Right click to release the object.

13. Choose **File/Save** from the Main Menu to save your project. Your subnet should resemble the following one.

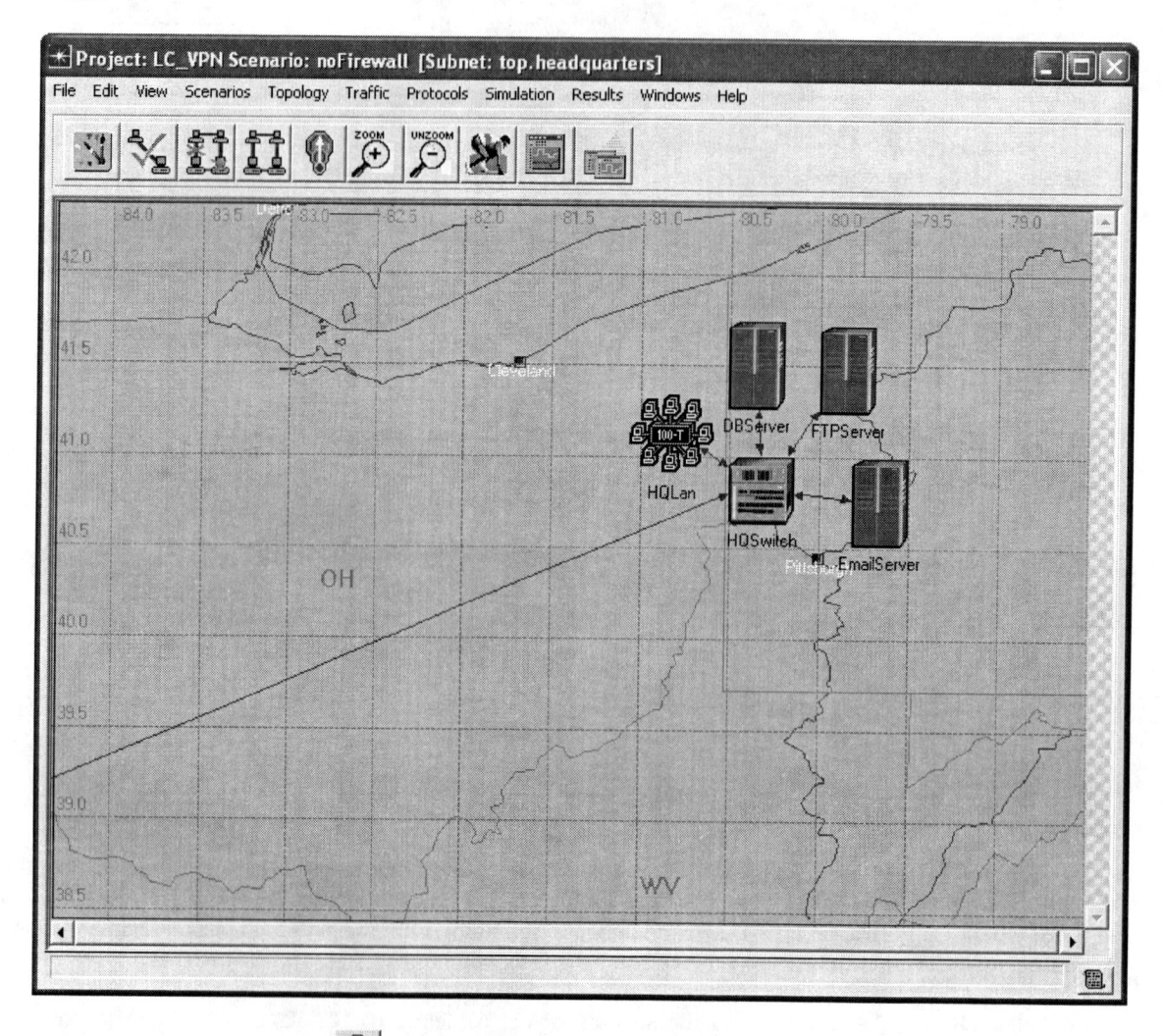

14. Click on the **Arrow Button** to go to the next higher level in the project hierarchy.

15. Now we will create the IP WAN that will connect the Southern California office and the Minnesota office to Pennsylvania headquarters. From the **Object Palette**, click on the **ethernet4_slip8_gtwy** object. Paste a copy of it near the Southern California LAN. Right click on the **ethernet4_slip8_gtwy,** and choose **Set Name** from the pull-down menu. Name the router **Router1**.

Lab 10 Virtual Private Networks

16. From the **Object Palette**, click on the **ip32_cloud** object. Paste a copy of it to the right of **Router1**. Right click on the **ip32_cloud**, and choose **Set Name** from the pull-down menu. Name the cloud **IPCloud**.
17. Again, from the **Object Palette**, click on the **ethernet4_slip8_gtwy** object. Paste one copy of it underneath the Minnesota LAN, and another copy of it to the right of the first. Right click on the first **ethernet4_slip8_gtwy**, and choose **Set Name** from the pull-down menu. Name the router **Router2**. Repeat the step to name the second router **Router3**. Your project should resemble the figure below.

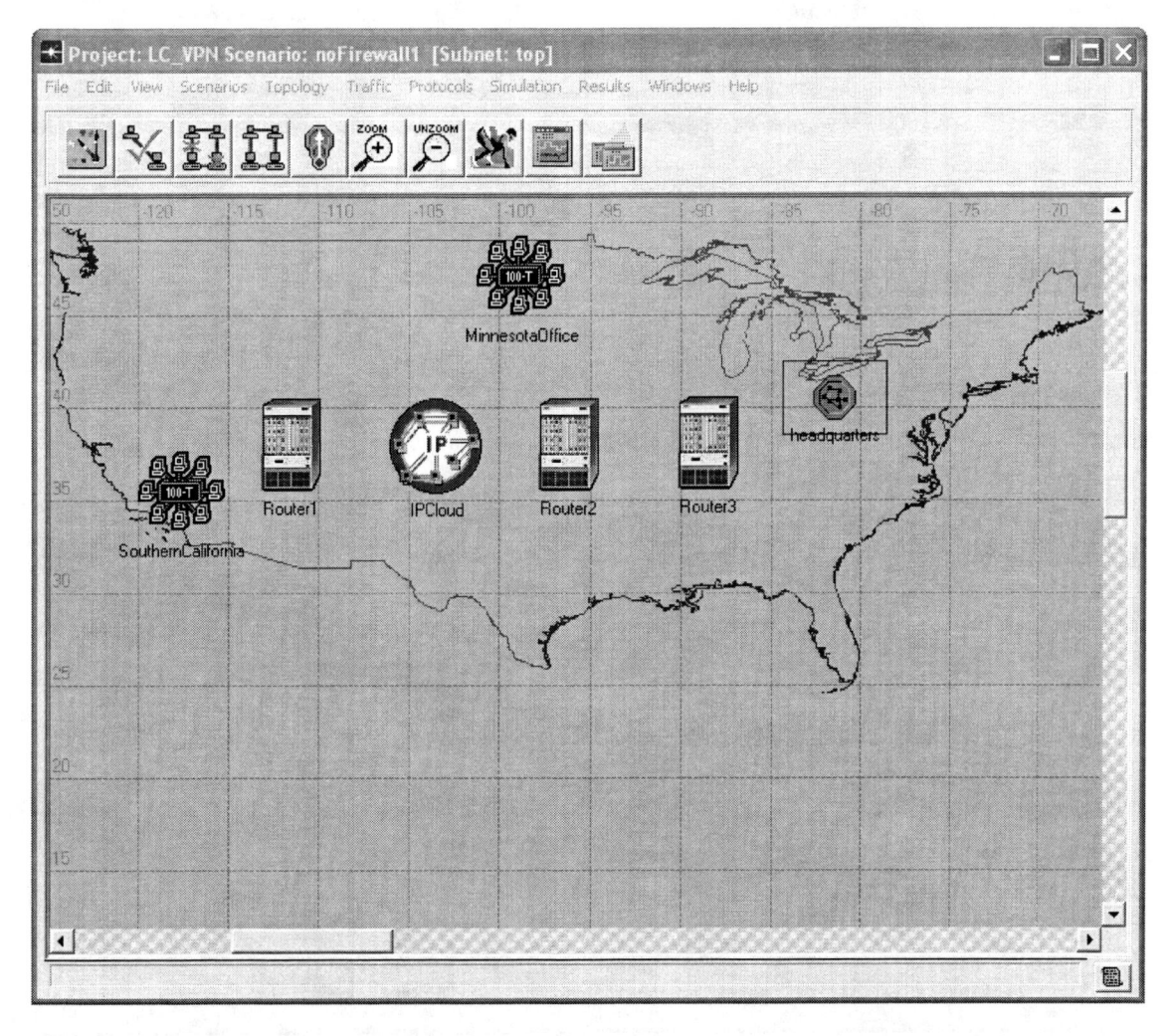

18. Now we will connect each LAN and the headquarters switch to the nearest router with a 100BaseT link. From the **Object Palette**, click on the **100BaseT Link** object. Connect the **SouthernCalifornia** LAN to **Router1**. Connect the **MinnesotaOffice** to **Router2**. Connect **Router3** to the **Headquarters.HQswitch**. Right click to release the link object.

19. We will connect the routers and the IP cloud with Point-to-Point DS1 links. From the **Object Palette**, click on the **PPP_DS1 link** object. Connect **Router1** to the **IPCoud**, connect the **IPCloud** to **Router2**, and connect **Router2** to **Router3**.

20. Finally, we will add an **Application Config, Profile Config,** and a **VPN Config** object to the workspace. The Application Config object will allow us to choose the applications we want to include in our simulation. We will use database, file transfer, and email applications. The Profile Config object allows us to use default traffic profiles for different types of users or to create new profiles. We will create two new profiles: one for the Southern California LAN and the other for the Minnesota LAN. The VPN Config object will allow us to create a VPN between the SouthernCalifornia LAN and the Pennsylvania headquarters.

From the **Object Palette**, click on the **Profile Config** object, and click to place it on the project grid. Right click to release the object. Right click on the **Profile Config** object, and choose **Set Name**. Name the object **ProfileConfig**. Next, click on the **Application Config** object, and click to paste it to the grid. Right click to release the object. Right click on the **Application Config** object, and choose **Set Name**. Name the object **ApplicationConfig**. Next, from the **Object Palette** click on the **IP VPN Configuration** object, and click to paste it to the project grid. Right click to release the object. Right click on the **IP VPN Config** object, and choose **Set Name**. Name the object **VPNConfig**. Verify that your project looks like the following figure:

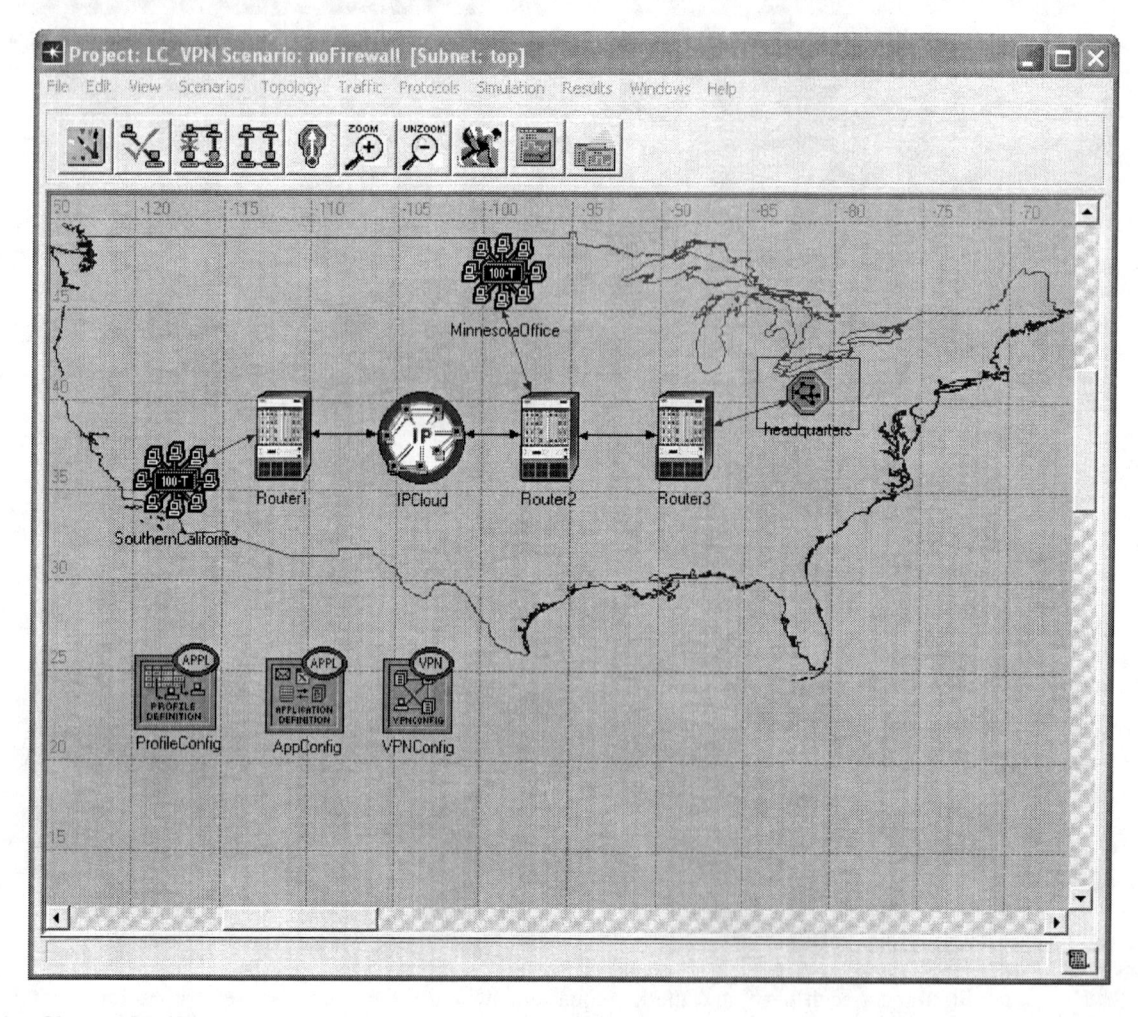

21. Choose **File/Save** from the Main Menu to save your project.

Configuring the Network

1. We will start by configuring the application object to provide the application we are interested in using in our simulation. We will use the predefined **Database (Heavy)**, **File Transfer (Heavy)**, and **Email (Heavy)** applications. Right click on the **AppConfig** object and choose **Edit Attributes**. Click on the value next to **Application Definitions**. Choose **Default** from the pull-down menu.

2. Click on the + next to **Application Definitions** to expand the parameters. Click on the + next to **Database Access (Heavy)**. Click on the + next to **Description,** then click on the value next to **Database** and choose **Edit** from the pull-down menu. A new window will appear.

Lab 10   Virtual Private Networks

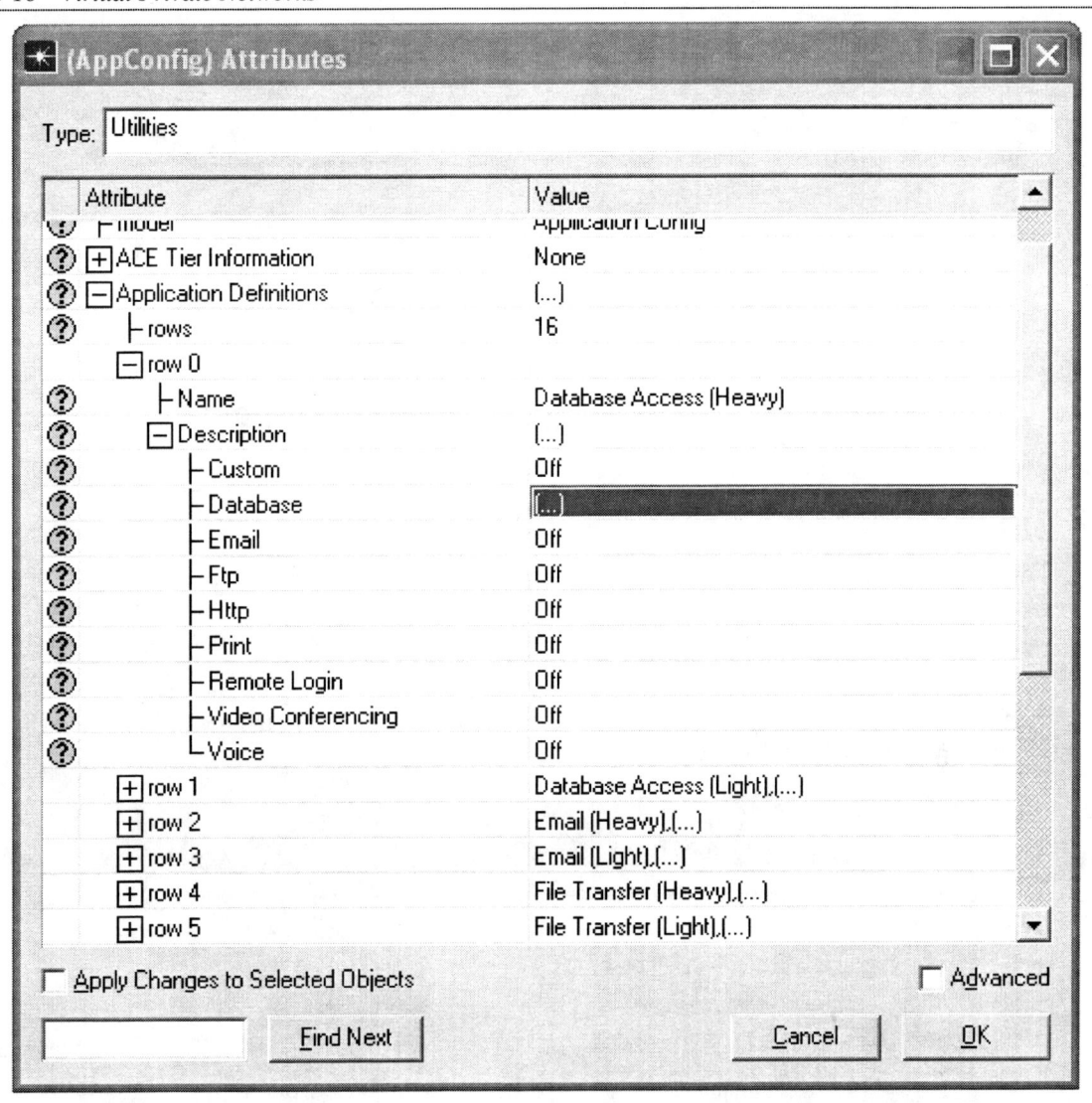

Attribute	Value
⊢ model	Application Config
⊞ ACE Tier Information	None
⊟ Application Definitions	(...)
⊢ rows	16
⊟ row 0	
⊢ Name	Database Access (Heavy)
⊟ Description	(...)
⊢ Custom	Off
⊢ Database	(...)
⊢ Email	Off
⊢ Ftp	Off
⊢ Http	Off
⊢ Print	Off
⊢ Remote Login	Off
⊢ Video Conferencing	Off
⊢ Voice	Off
⊞ row 1	Database Access (Light),(...)
⊞ row 2	Email (Heavy),(...)
⊞ row 3	Email (Light),(...)
⊞ row 4	File Transfer (Heavy),(...)
⊞ row 5	File Transfer (Light),(...)

3. We will modify the database description so that all transactions are queries. Click on the value next to **Transaction Mix (Queries/Total Transactions)** and choose **100%** from the pull-down menu. We will decrease the time between transactions. Click on the value next to **Transaction Interarrival Time**. A new window will appear. Change the value to **exponential(5)**. Click **OK** to close the window.

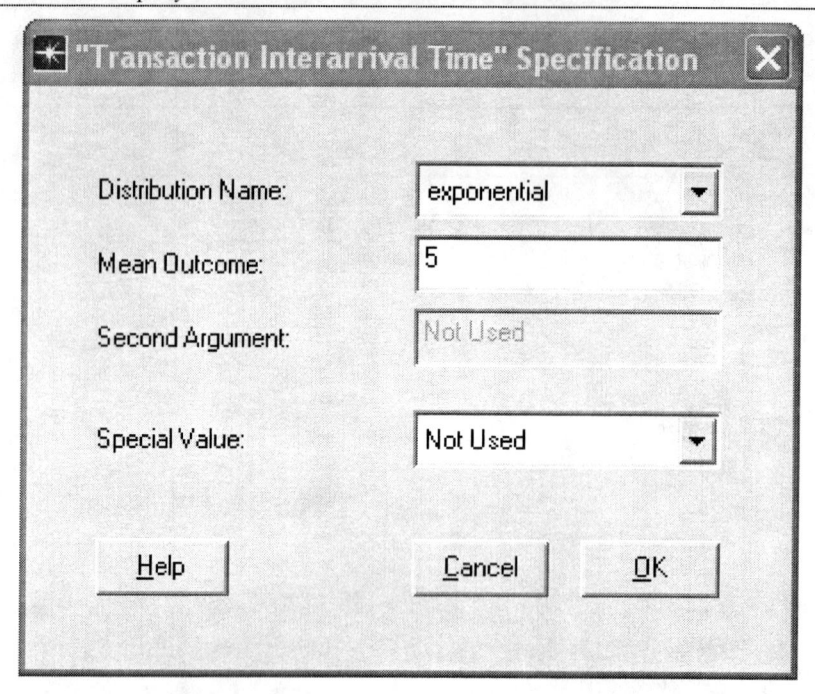

4. The values in the Database Table should now resemble those indicated in the next window image. Click **OK** to close the Database Table window.

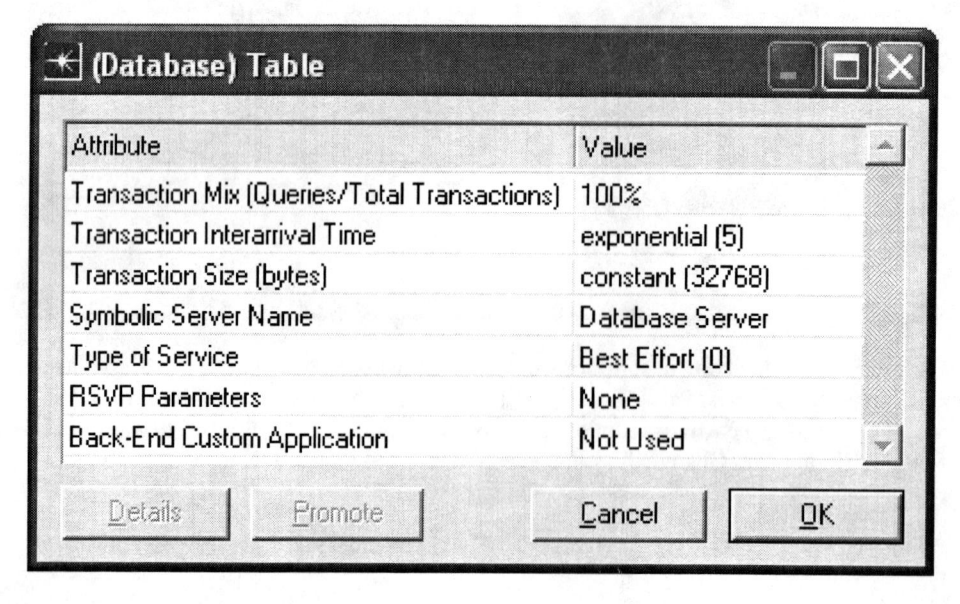

5. Click on the + next to **Email (Heavy)**. Click on the + next to **Description**. Then click on the value next to **Email** and choose **Edit** from the pull-down menu.

6. First, we will modify the send arrival time. Click on the value next to **Send Arrival Time (Seconds)**. A new window will appear. Use the pull-down menu to change the **Special Value** field to **Not Used**. Set the **Distribution Name** to **exponential** and the **Mean Outcome** value to **1000**. This means that the amount of time between each email message will follow an exponential distribution with a mean of 1,000 seconds. Click **OK**.

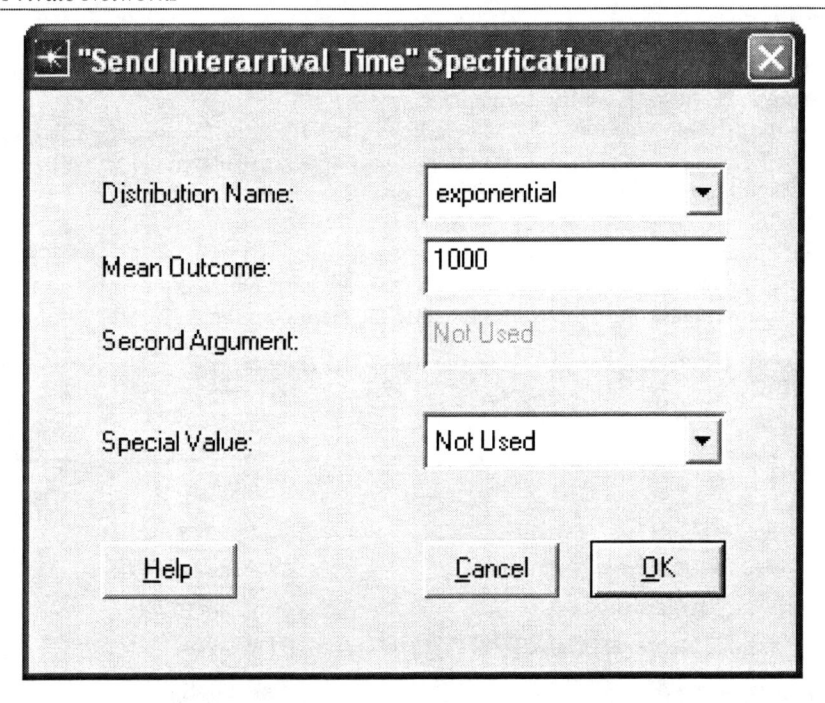

Lab 10   Virtual Private Networks

[Dialog: "Send Interarrival Time" Specification — Distribution Name: exponential; Mean Outcome: 1000; Second Argument: Not Used; Special Value: Not Used]

7. We do not care about the Receive Interarrival Time because we will not be receiving messages from the server; our client will only upload messages. To enforce this, click on the value next to **Receive Group Size**. A new window will appear. Use the pull-down menu to change the **Special Value** field to **Not Used**. Set the **Distribution Name** to **constant** and the **Mean Outcome** value to **0** and click **OK**.

8. Next, we will change the size of the Email message to 1 MB. We want a fairly large email messages that will create many TCP segments. Click on the value next to **E-mail Size (bytes)**. A new window will appear. Use the pull-down menu to change the **Special Value** field to **Not Used**. Set the **Distribution Name** to **constant** and the **Mean Outcome** value to **1000000** and click **OK**. You should now have the values listed in the following table.

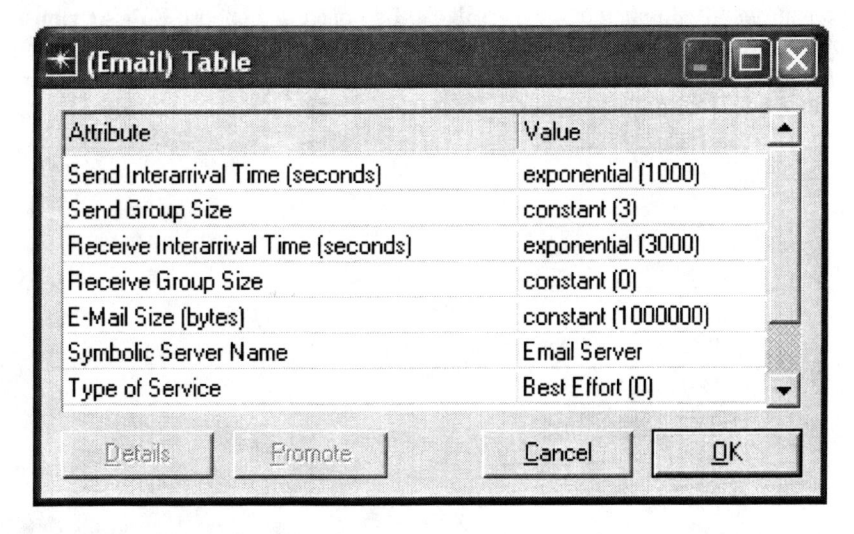

Attribute	Value
Send Interarrival Time (seconds)	exponential (1000)
Send Group Size	constant (3)
Receive Interarrival Time (seconds)	exponential (3000)
Receive Group Size	constant (0)
E-Mail Size (bytes)	constant (1000000)
Symbolic Server Name	Email Server
Type of Service	Best Effort (0)

9. Click **OK** to close the Email Table window.

10. Click on the + next to **File Transfer (Heavy)**. Click on the + next to **Description**. Then click on the value next to **ftp** and choose **Edit** from the pull-down menu.

11. First, we will modify the command mix. It is currently sent to 50% meaning 50%, of the commands will be GET-file commands. Click on the value next to **Command Mix (Get/Total)** and change it to 100%. Now all of our commands will be GET-file commands.
12. We will now modify the value which controls how often file transfers occur. Click on the value next to **Inter-Request Time (seconds).** A new window will appear. Change the value to **exponential(10)** and click **OK**. This will create file transfer requests every 10 seconds.

13. Next, we will change the size of the file that is transferred. We will set it to be 500 bytes. Click on the value next to **File Size (bytes).** A new window will appear. Change the value to **constant(50000)** and click **OK**. You should now have the values listed in the following image.

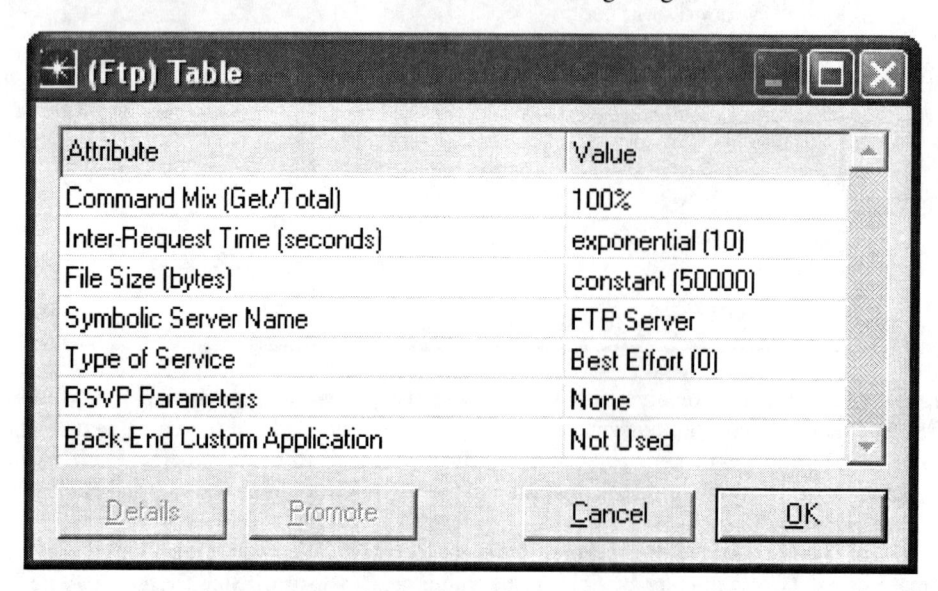

Attribute	Value
Command Mix (Get/Total)	100%
Inter-Request Time (seconds)	exponential (10)
File Size (bytes)	constant (50000)
Symbolic Server Name	FTP Server
Type of Service	Best Effort (0)
RSVP Parameters	None
Back-End Custom Application	Not Used

14. Click **OK** to close the Ftp Table window. Click **OK** to close the Application Attributes window.

15. Now we will create a traffic profiles for the Southern California and Minnesota sites. The first profile, used by the Southern California site, will be called Regional. It will include the **Database (Heavy)** and **Email (Heavy)** applications. Right click on the **ProfileConfig** object and choose **Edit Attributes.** Click on the + next to **Profile Configuration.** Click on the value next to **rows** and choose **Edit**. Change the number of rows to **2**. Click on the value next to the first added row and choose **Expand Row**. Next to **Profile Name** type **Regional**. Choose **Simultaneous** under **Operation Mode**.

Lab 10 Virtual Private Networks

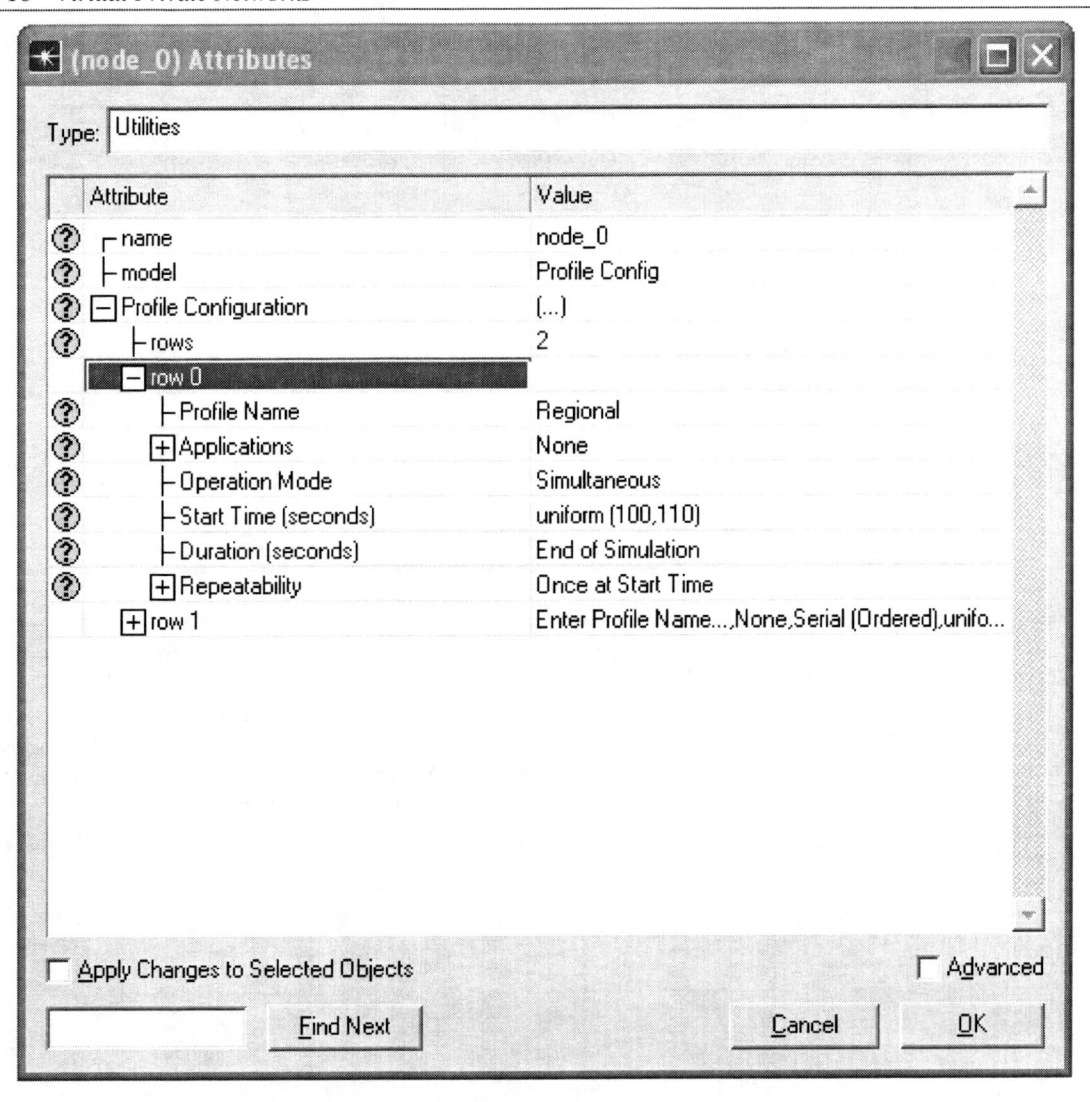

16. Click on the value next to **Applications** and choose **Edit** from the pull-down menu. A new window will appear. Next to **Rows** at the bottom of the window, choose **2** from the pull-down menu. Two new rows will appear. Click on the value under the Name column in the first row and choose **Database Access (Heavy)** from the pull-down menu. In the second row, click on the value under **Name** and choose **Email (Heavy)** from the pull-down menu. Click **OK.**

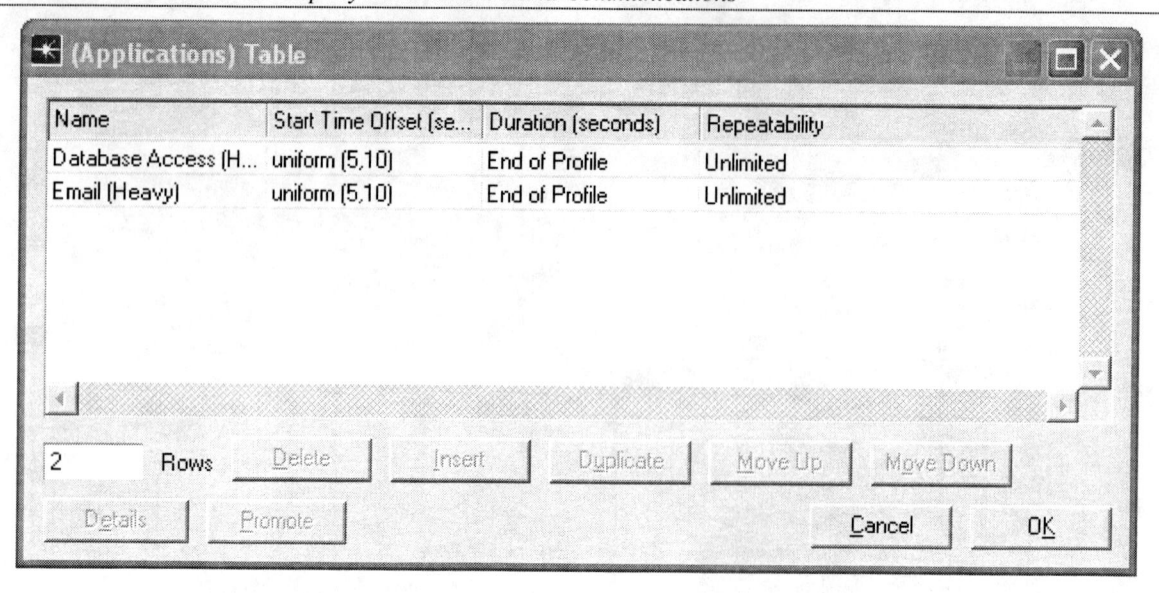

(Applications) Table

Name	Start Time Offset (se...	Duration (seconds)	Repeatability
Database Access (H...	uniform (5,10)	End of Profile	Unlimited
Email (Heavy)	uniform (5,10)	End of Profile	Unlimited

2 Rows

17. Now we will add a second profile for the Minnesota partner site that uses FTP and email applications. Click on the value next to the second profile row and choose **Expand Row**. Next to **Profile Name,** type **Partner**. Choose **Simultaneous** under **Operation Mode**.

18. Click on the value next to **Applications** and choose **Edit** from the pull-down menu. A new window will appear. Next to **Rows** at the bottom of the window, choose **2** from the pull-down menu. Two new rows will appear. Click on the value under the **Name** column in the first row and choose **File Transfer (Heavy)** from the pull-down menu. In the second row, click on the value under **Name** and choose **Email (Heavy)** from the pull-down menu, and click **OK.**

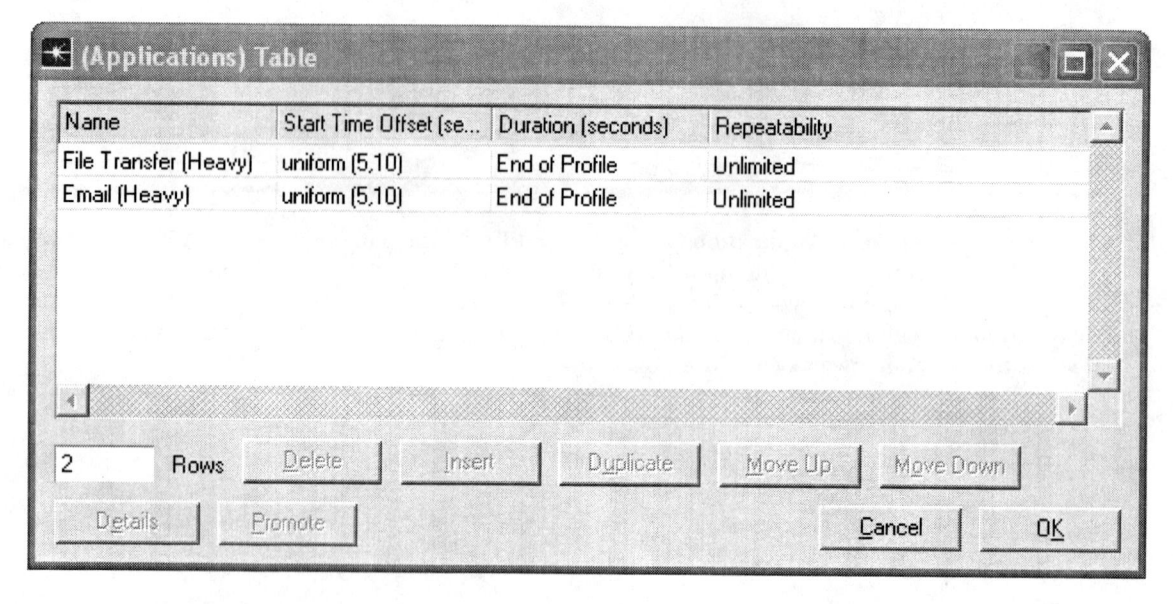

(Applications) Table

Name	Start Time Offset (se...	Duration (seconds)	Repeatability
File Transfer (Heavy)	uniform (5,10)	End of Profile	Unlimited
Email (Heavy)	uniform (5,10)	End of Profile	Unlimited

2 Rows

19. The Profile Configurations should look like those the ones presented next. Click **OK** to close the Profile Configuration window.

Lab 10 Virtual Private Networks

(profileConfig) Attributes

Type: Utilities

Attribute	Value
⌐ name	profileConfig
├ model	Profile Config
⊟ Profile Configuration	(...)
├ rows	2
⊟ row 0	
├ Profile Name	Regional
⊟ Applications	(...)
├ rows	2
⊞ row 0	Database Access (Heavy),uniform (5,10),End of ...
⊞ row 1	Email (Heavy),uniform (5,10),End of Profile,Unlimi...
├ Operation Mode	Simultaneous
├ Start Time (seconds)	uniform (100,110)
├ Duration (seconds)	End of Simulation
⊞ Repeatability	Once at Start Time
⊟ row 1	
├ Profile Name	Partner
⊟ Applications	(...)
├ rows	2
⊞ row 0	Email (Heavy),uniform (5,10),End of Profile,Unlimi...
⊞ row 1	File Transfer (Heavy),uniform (5,10),End of Profil...
├ Operation Mode	Simultaneous
├ Start Time (seconds)	uniform (100,110)
├ Duration (seconds)	End of Simulation
⊞ Repeatability	Once at Start Time

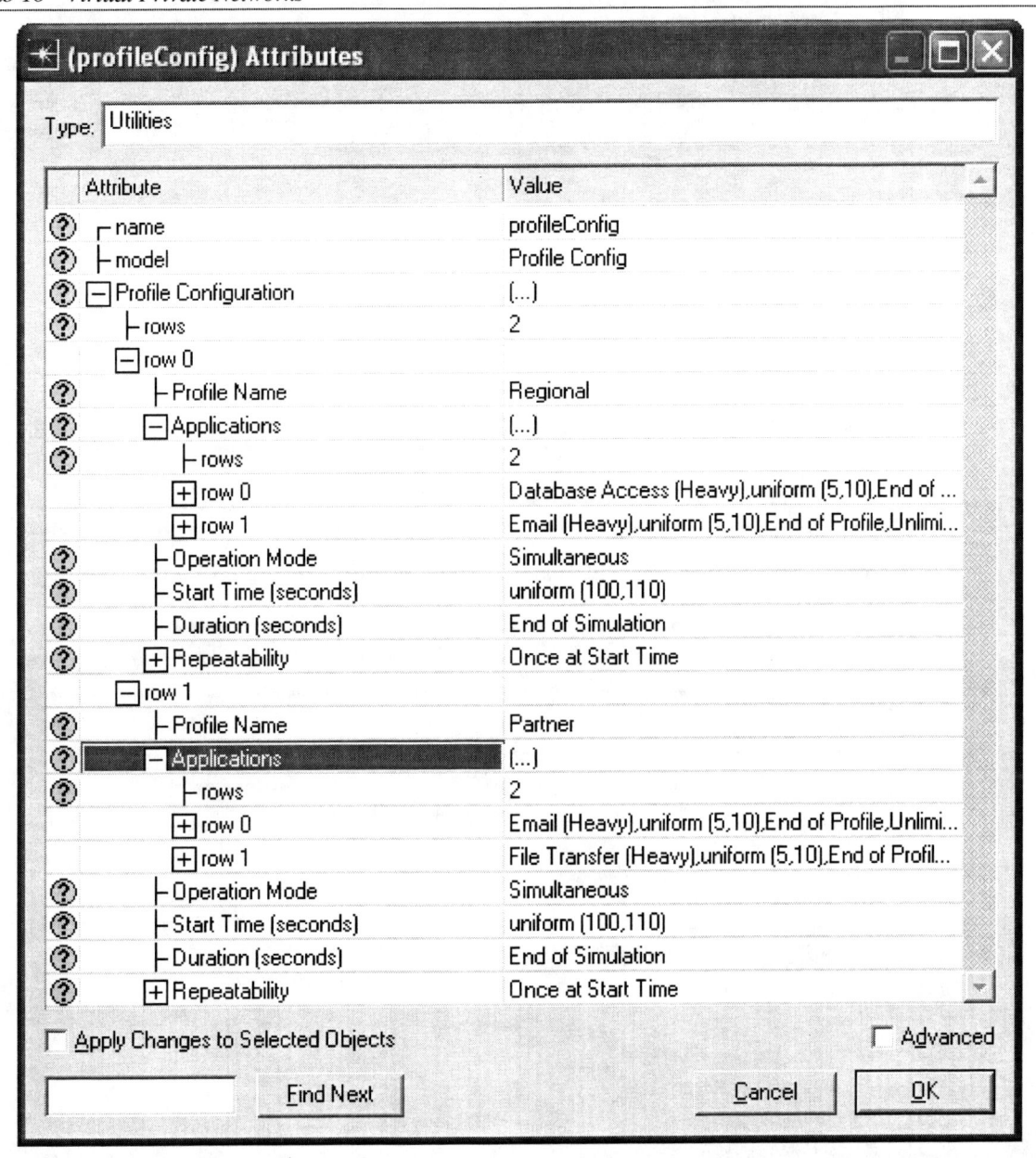

☐ Apply Changes to Selected Objects   ☐ Advanced

[Find Next]   [Cancel]   [OK]

20. Now we will give the Southern California site the Regional profile. Right click on the **SouthernCalifornia LAN** object and choose **Edit Attributes**. Next, click on the + next to **Application: Supported Profiles** to expand the parameters. Click on the value next to **Application: Supported Profiles** and choose **Edit**. A new window will appear.

21. Click on the value listed next to **Rows** and choose **1** from the pull-down menu. You will see a row added to the table. Now click on the value under **Profile Name**. Choose **Regional** from the pull-down menu. Click **OK** to close the window.

(Application: Supported Profiles) Table

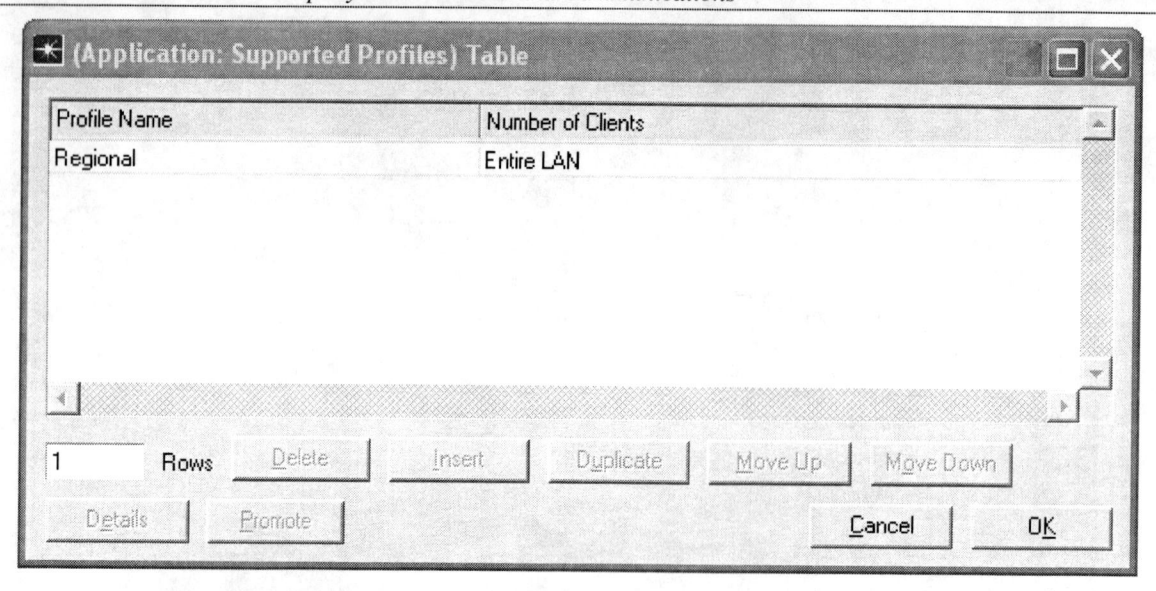

Profile Name	Number of Clients
Regional	Entire LAN

22. The client profile definitions should look like those shown next. Click **OK** to close the **Attribute** window.

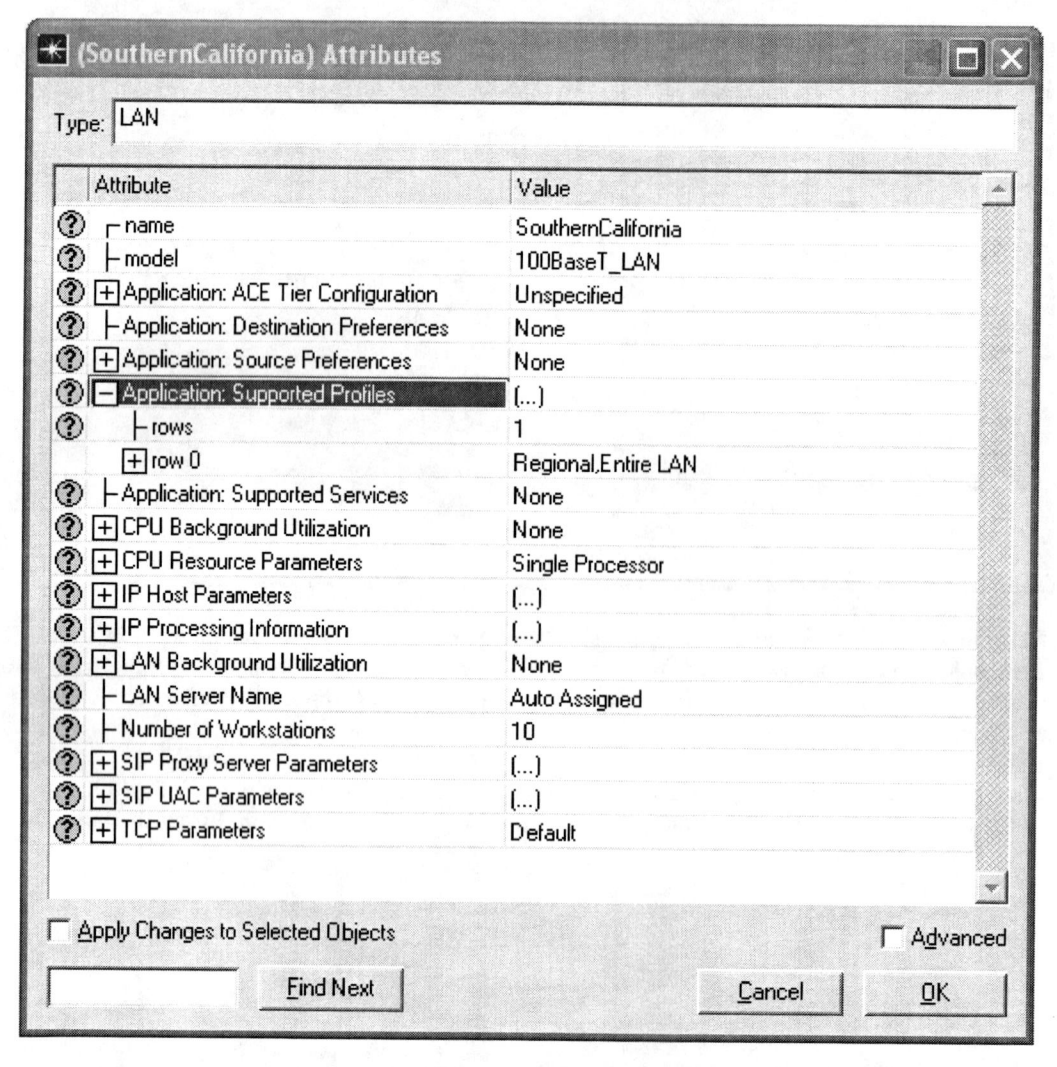

(SouthernCalifornia) Attributes

Type: LAN

Attribute	Value
name	SouthernCalifornia
model	100BaseT_LAN
Application: ACE Tier Configuration	Unspecified
Application: Destination Preferences	None
Application: Source Preferences	None
Application: Supported Profiles	(...)
rows	1
row 0	Regional,Entire LAN
Application: Supported Services	None
CPU Background Utilization	None
CPU Resource Parameters	Single Processor
IP Host Parameters	(...)
IP Processing Information	(...)
LAN Background Utilization	None
LAN Server Name	Auto Assigned
Number of Workstations	10
SIP Proxy Server Parameters	(...)
SIP UAC Parameters	(...)
TCP Parameters	Default

23. We will now give the Minnesota site the Partner profile. Right click on the **Minnesota LAN** object and choose **Edit Attributes.** Next, click on the **+** next to **Application: Supported Profiles** to expand the

Lab 10   Virtual Private Networks

parameters. Click on the value next to **Application: Supported Profiles** and choose **Edit**. A new window will appear.

24. Click on the value listed next to **Rows** and choose **1** from the pull-down menu. You will see a row added to the table. Now click on the value under **Profile Name.** Choose **Partner** from the pull-down menu. Click **OK** to close the window.

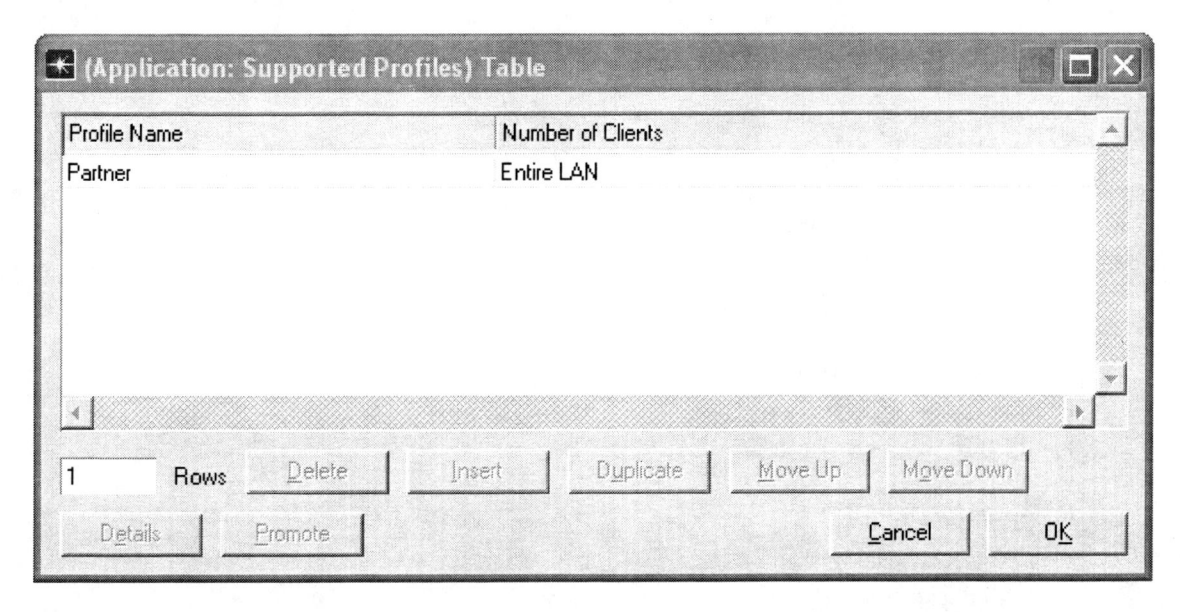

25. The client profile definitions should appear exactly as the selection sin the **Attributes** window image shown next. Click **OK** to close the **Attribute** window.

(MinnesotaOffice) Attributes

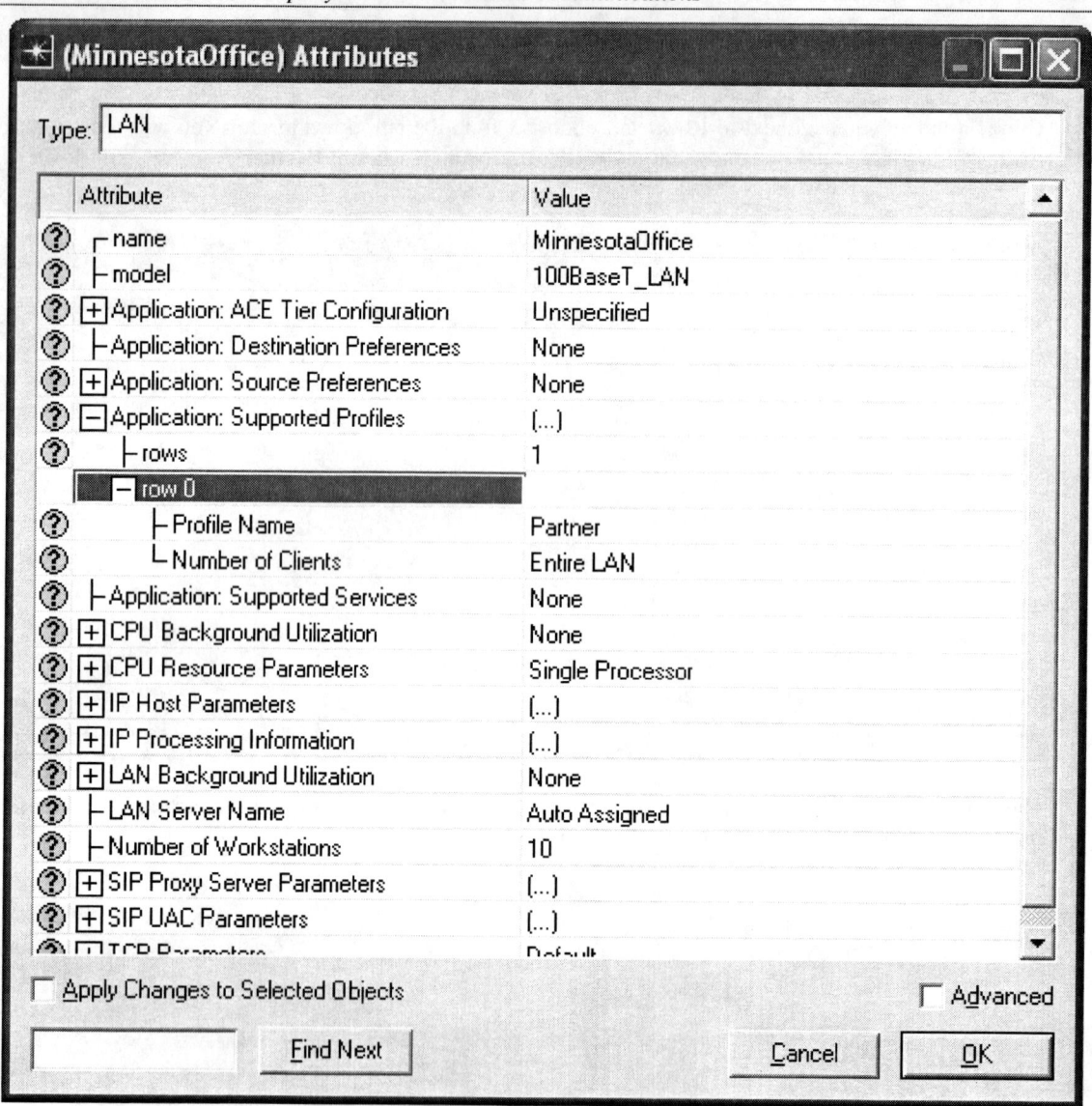

Type: LAN

Attribute	Value
name	MinnesotaOffice
model	100BaseT_LAN
⊞ Application: ACE Tier Configuration	Unspecified
Application: Destination Preferences	None
⊞ Application: Source Preferences	None
⊟ Application: Supported Profiles	(...)
rows	1
⊟ row 0	
Profile Name	Partner
Number of Clients	Entire LAN
Application: Supported Services	None
⊞ CPU Background Utilization	None
⊞ CPU Resource Parameters	Single Processor
⊞ IP Host Parameters	(...)
⊞ IP Processing Information	(...)
⊞ LAN Background Utilization	None
LAN Server Name	Auto Assigned
Number of Workstations	10
⊞ SIP Proxy Server Parameters	(...)
⊞ SIP UAC Parameters	(...)
⊞ TCP Parameters	Default

☐ Apply Changes to Selected Objects   ☐ Advanced

[Find Next]   [Cancel]   [OK]

26. Next, we will configure the corporate headquarters LAN to provide database, file transfer, and email services. Double click on the headquarters subnet to descend into the subnet. Click on the **DBServer** object and choose **Edit Attributes.** Next, click on the value next to **Application: Supported Services** and choose **Edit.** A new window will appear.

Lab 10 Virtual Private Networks

(DBServer) Attributes

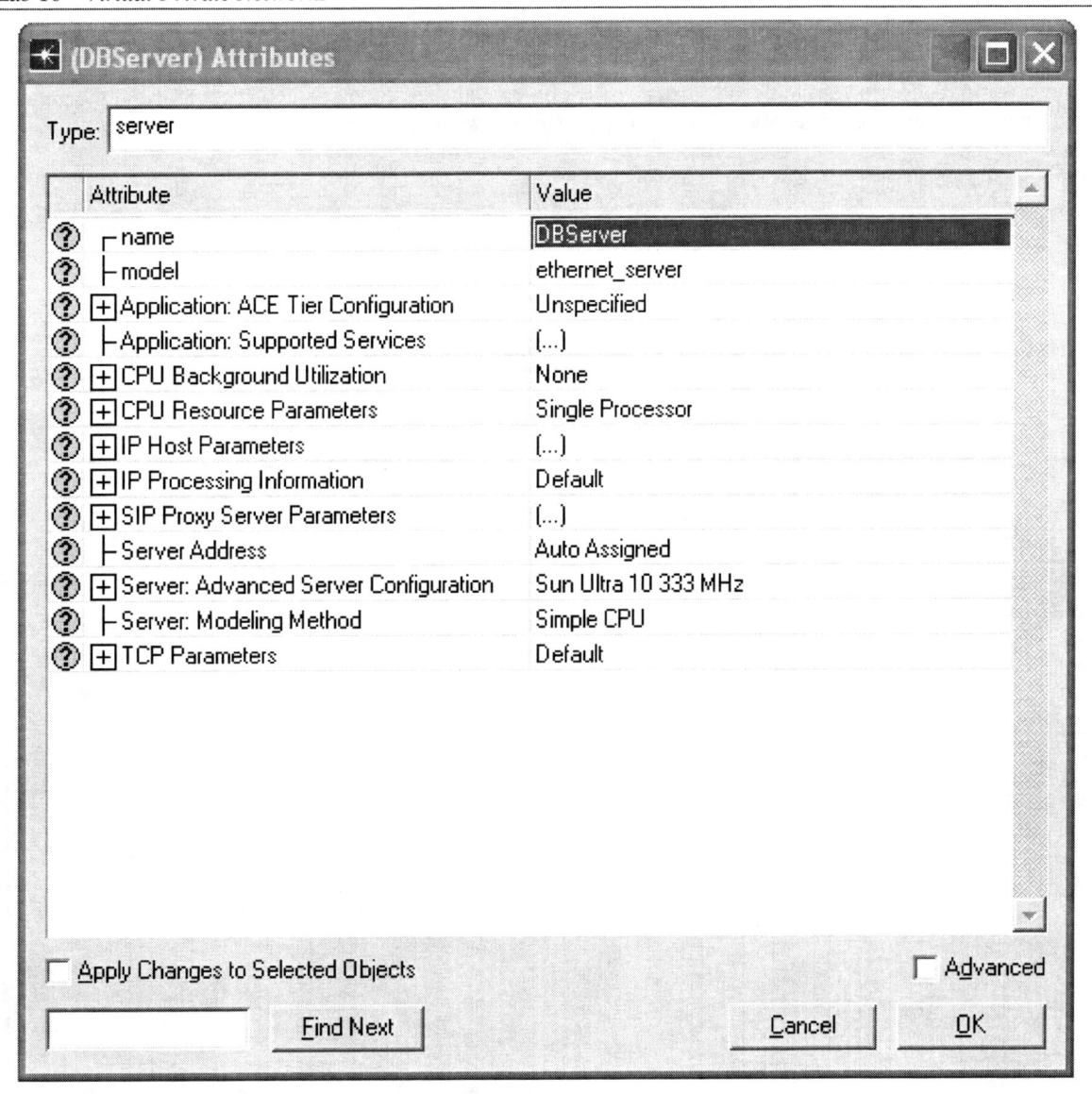

Type: server

Attribute	Value
ⓘ ─ name	DBServer
ⓘ ├ model	ethernet_server
ⓘ ⊞ Application: ACE Tier Configuration	Unspecified
ⓘ ├ Application: Supported Services	(...)
ⓘ ⊞ CPU Background Utilization	None
ⓘ ⊞ CPU Resource Parameters	Single Processor
ⓘ ⊞ IP Host Parameters	(...)
ⓘ ⊞ IP Processing Information	Default
ⓘ ⊞ SIP Proxy Server Parameters	(...)
ⓘ ├ Server Address	Auto Assigned
ⓘ ⊞ Server: Advanced Server Configuration	Sun Ultra 10 333 MHz
ⓘ ├ Server: Modeling Method	Simple CPU
ⓘ ⊞ TCP Parameters	Default

☐ Apply Changes to Selected Objects ☐ Advanced

[Find Next]   [Cancel]   [OK]

27. Click on the value listed next to **Rows,** and choose **1** from the pull-down menu. A new row will be added to the table. Now click on the value under **Application Name.** Choose **Database Access (Heavy)** from the pull-down menu. Click **OK** to close the Application: Supported Services Table. Click **OK** again to close the Attribute window.

28. Click on the **EmailServer** object and choose **Edit Attributes.** Next, click on the value next to **Application: Supported Services,** and choose **Edit.** Click on the value listed next to **Rows,** and choose **1** from the pull-down menu. A new row will be added to the table. Now click on the value under **Name.** Choose **Email (Heavy)** from the pull-down menu. Click **OK** to close the Application: Supported Services Table. Click **OK** again to close the Attribute window.

29. Click on the **FTPServer** object and choose **Edit Attributes.** Next, click on the value next to **Application: Supported Services,** and choose **Edit.** Click on the value listed next to **Rows,** and choose **1** from the pull-down menu. A new row will be added to the table. Now click on the value under **Name.** Choose **FTP (Heavy)** from the pull-down menu. Click **OK** to close the Application: Supported Services Table. Click **OK** again to close the Attribute window.

30. Click on the **Arrow Button** to go up to the next higher level of the project hierarchy.

31. Finally, we will create a VPN between the Southern California **Router1** and the headquarters **Router3**. Right click on the **VPNConfig** object and choose **Edit Attributes**. Click on the + next to **VPN Configuration** to expand the parameters. Click on the value next to **rows** and choose **1** from the pull-down menu. A new row will appear. Click on the + next to **row 0** to expand the parameters.

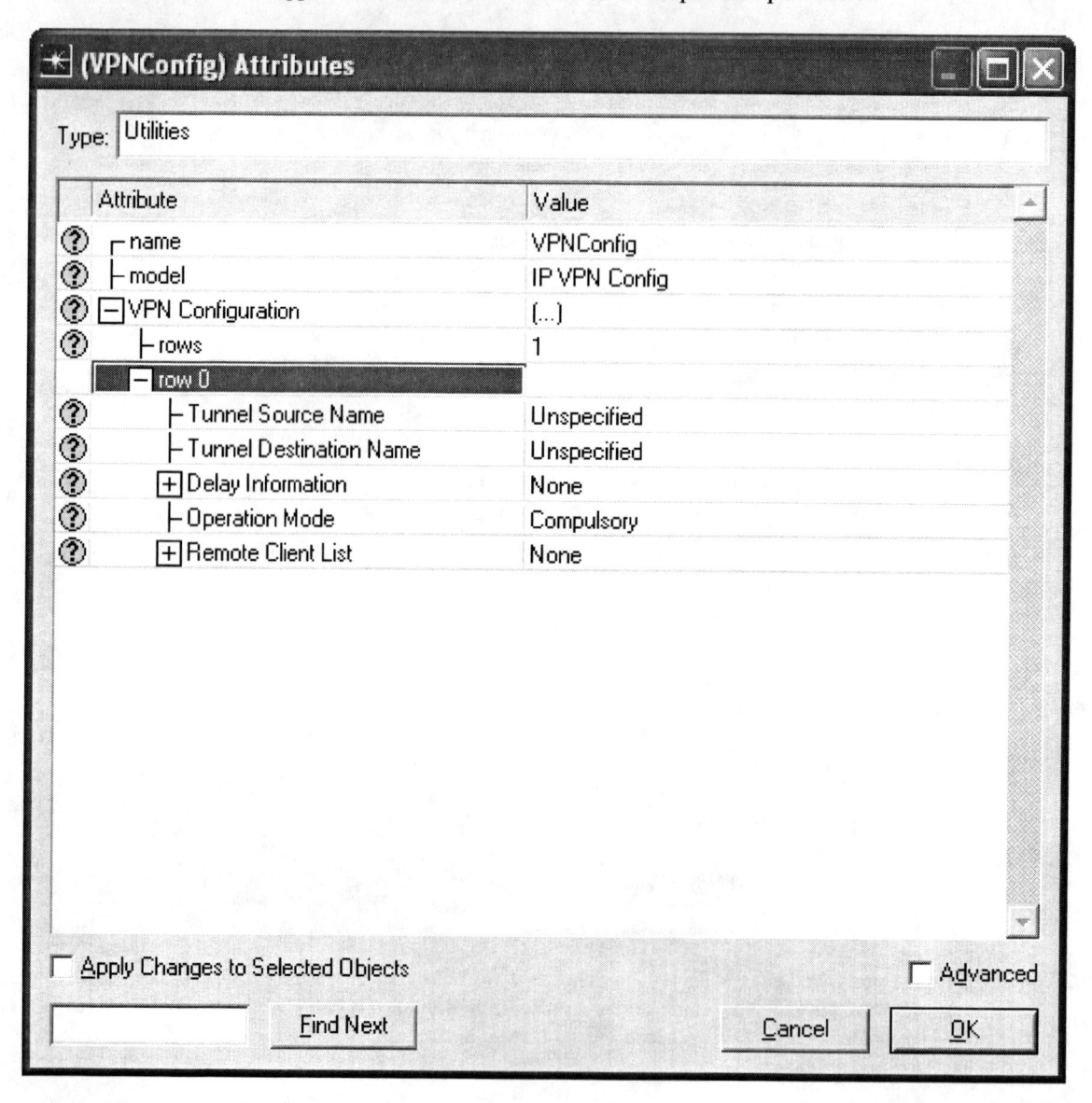

32. Click on the value next to **Tunnel Source Name** and choose Edit from the pull-down menu. Type in the name **Router1**. Click on the value next to **Tunnel Destination Name** and choose **Edit** from the pull-down menu. Type in the name **Router3**.

33. We will leave configuring the delay value as an exercise. To finish configuring the tunnel, we need to specify the remote client. This will be the Southern California site. Click on the + next to **Remote Client List** to expand the parameters. Click on the value next to **rows** and choose **1** from the pull-down menu. A new row will appear. Click on the + next to **row 0** to expand the parameters. Click on the value next to **Client Node Name** and choose **Edit** from the pull-down menu. Type the name **SouthernCalifornia**. Click **OK** to close the **VPNConfig** window.

Lab 10 Virtual Private Networks

(VPNConfig) Attributes

Type: Utilities

Attribute	Value
name	VPNConfig
model	IP VPN Config
VPN Configuration	(...)
rows	1
row 0	
Tunnel Source Name	Router1
Tunnel Destination Name	Router3
Delay Information	None
Operation Mode	Compulsory
Remote Client List	(...)
rows	1
row 0	
Client Node Name	SouthernCalifornia

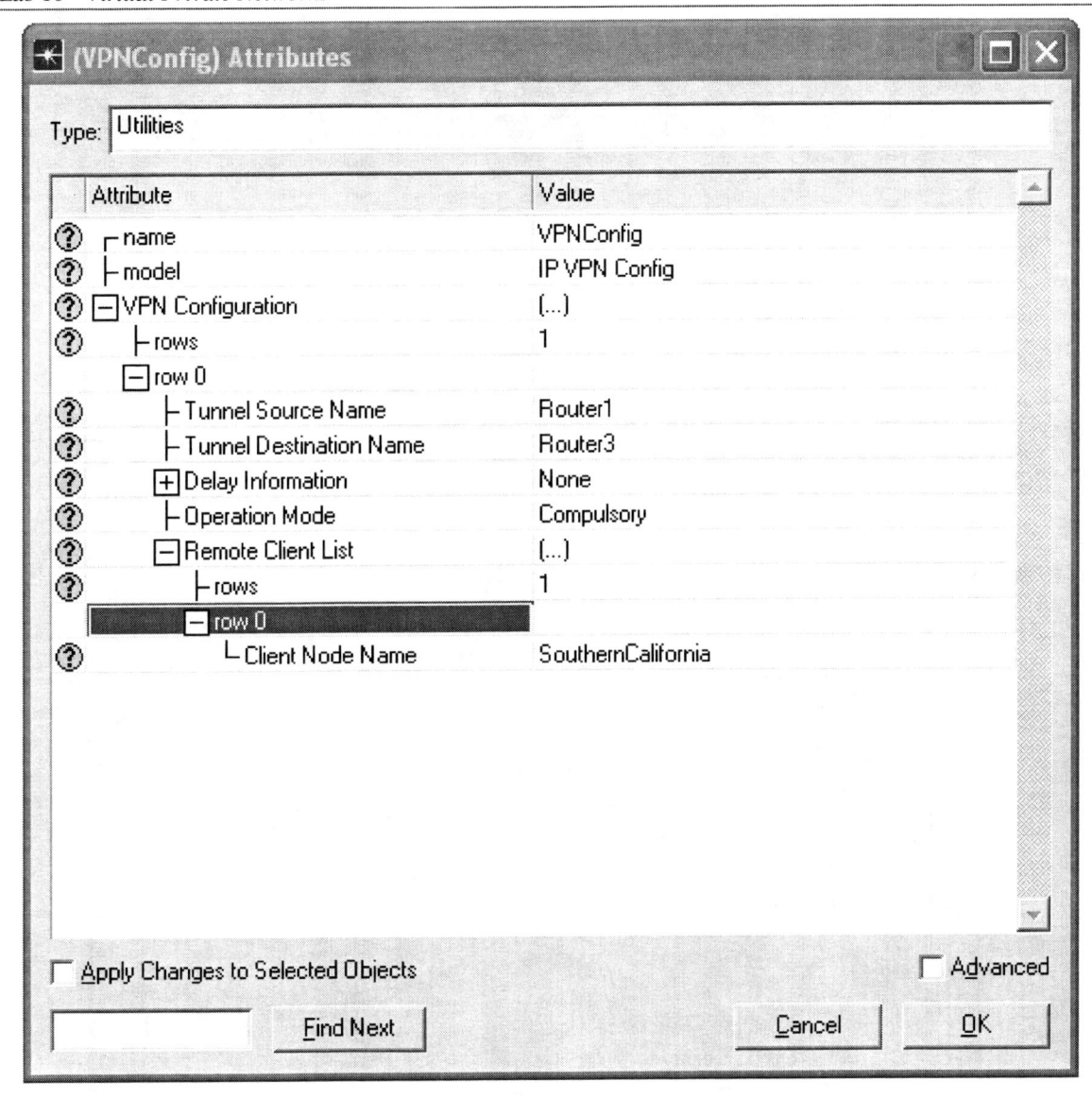

34. Choose **File/Save** from the Main Menu to save your project.

Configuring the Simulation and Choosing Statistics

Now we will choose response time statistics for the applications we are using in our project.

1. Right click somewhere on the project workspace and select **Choose Individual Statistics** from the pull-down menu. Click on the + next to **Global** to expand the parameters. First, we will select statistics for the database application. Click on the + next to **DB Query** to expand the options. Click on the box next to **Response Time (Ssec), Traffic Received (bytes/sec),** and **Traffic Sent (bytes/sec).**

2. Repeat the steps to choose the same statistics for the File Transfer (FTP) application. Click on the + next to **FTP** to expand the options. Click on the box next to **Download Response Time (sec), Traffic Received (bytes/sec), Upload Response Time (sec),** and **Traffic Sent (bytes/sec).**

3. Repeat the steps to choose the same statistics for the Email application. Click on the + next to **email** to expand the options. Click on the box next to **Download Response Time (sec), Traffic Received (bytes/sec), Upload Response Time (sec),** and **Traffic Sent (bytes/sec).**

OPNET Lab Manual to accompany Business Data and Communications

Choose Results

- Global Statistics
  - ACE
  - BGP
  - Bridge
  - Cache
  - Custom Application
  - DB Entry
    - ☑ Response Time (sec)
    - ☑ Traffic Received (bytes/sec)
    - ☐ Traffic Received (packets/sec)
    - ☑ Traffic Sent (bytes/sec)
    - ☐ Traffic Sent (packets/sec)
  - DB Query
    - ☑ Response Time (sec)
    - ☑ Traffic Received (bytes/sec)
    - ☐ Traffic Received (packets/sec)
    - ☑ Traffic Sent (bytes/sec)
    - ☐ Traffic Sent (packets/sec)
  - EIGRP
  - Email
    - ☑ Download Response Time (sec)
    - ☑ Traffic Received (bytes/sec)
    - ☐ Traffic Received (packets/sec)
    - ☑ Traffic Sent (bytes/sec)
    - ☐ Traffic Sent (packets/sec)
    - ☑ Upload Response Time (sec)
  - Ethernet
  - Ftp
    - ☑ Download Response Time (sec)
    - ☑ Traffic Received (bytes/sec)
    - ☐ Traffic Received (packets/sec)
    - ☑ Traffic Sent (bytes/sec)
    - ☐ Traffic Sent (packets/sec)
    - ☑ Upload Response Time (sec)

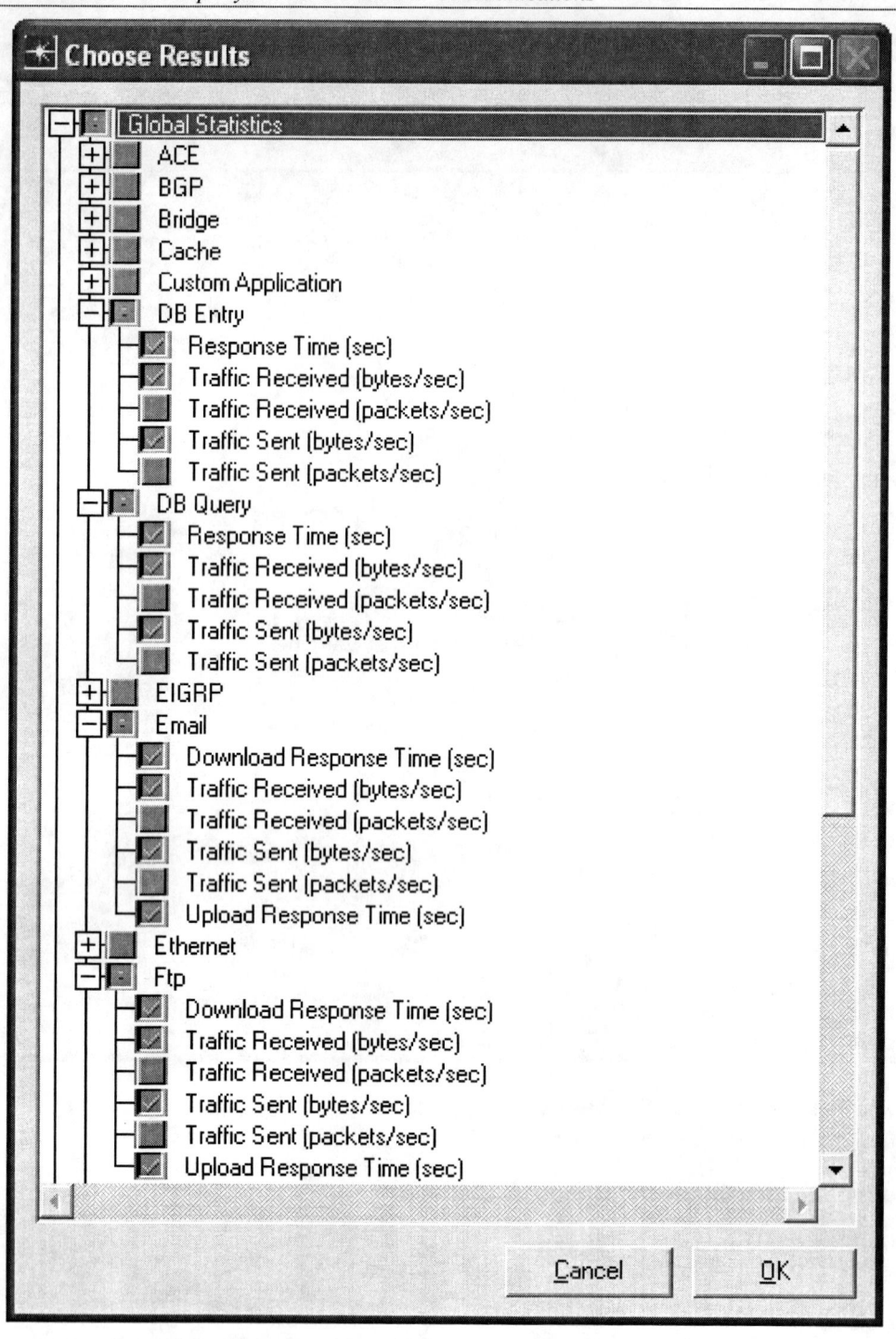

[Cancel] [OK]

4. Click on the **+** next to **Node Statistics** to expand the parameters. Select all statistics for the VPN Tunnel Node by clicking on the box next to **IP VPN Tunnel**. Click **OK** to close the **Choose Results** window.

Running the Simulation

1. Click on the **Configure and Run** Button. Change the **Duration** of the simulation to **5**. Choose **Minute(s)** from the pull-down menu. Click the **Run** button to run the simulation. Depending on your processor, the simulation should take several minutes to complete.

Lab 10 Virtual Private Networks

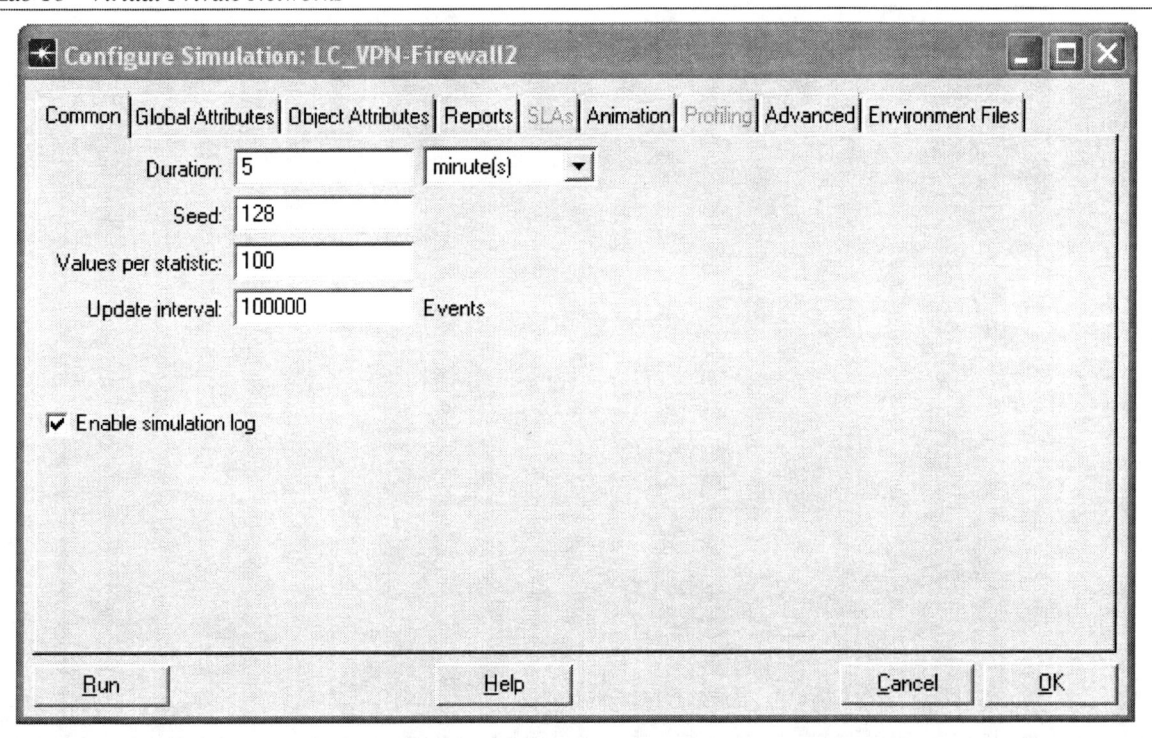

2. To view results, choose **Results/View Results** from the Main Menu. To view results for **FTP** response time, click on the **+** next to **FTP,** and click next to **Download Response Time (seconds).** Choose to view the graph as **time_average.** Click on the **Show** button to display the graph in a new window. Your graph should resemble the following figure:

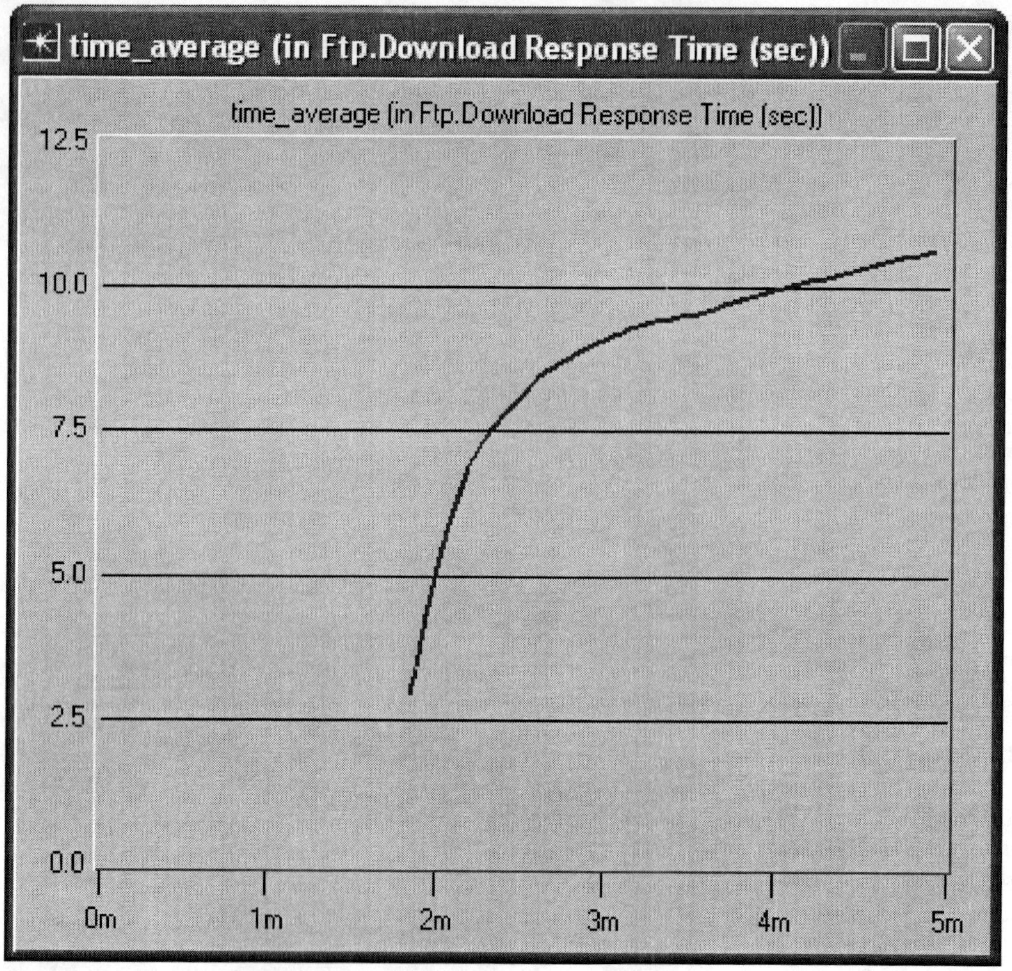
time_average (in Ftp.Download Response Time (sec))
time_average (in Ftp.Download Response Time (sec))
12.5
10.0
7.5
5.0
2.5
0.0
0m
1m
2m
3m
4m
5m

Click on the **X** to close the window. The **Close Analysis** window will appear asking you if you want to hide the panel or permanently delete it. Click on the **Delete** button to delete the window.

Creating the Firewall Scenario

Now, we will create a new scenario in which the router that the Minnesota site is attached to is replaced with a firewall. We will configure the firewall to block all incoming email traffic. Email traffic from the Southern California site, however, will still be delivered because it is being sent through the VPN tunnel to Router 3. We expect that by removing some of the email traffic, response time for database and file transfer application will improve.

1. From the Main Menu, choose **Scenarios/Duplicate Scenario**. Give the new scenario a name such as **firewall.**

2. Right click on **Router2** to select it, and choose **Edit Attributes** from the pull-down menu. Click on the value next to **model** and choose **ethernet2_slip8_firewall** from the pull-down menu. Change the **name** value to **Firewall**.

3. Click on the + next to **Proxy Server Information** to expand the parameters. Click on the + next to the row labeled **Email** to expand the parameters. Click on the value next to **Proxy Server Deployed,** and choose **No** from the pull-down menu. Note that you may also choose a latency value for the firewall. It will typically take some extra time for a firewall to filter packets. For now, we will leave this value at 0 meaning that the filtering will not add extra delay. Click **OK** to close the window.

Lab 10 Virtual Private Networks

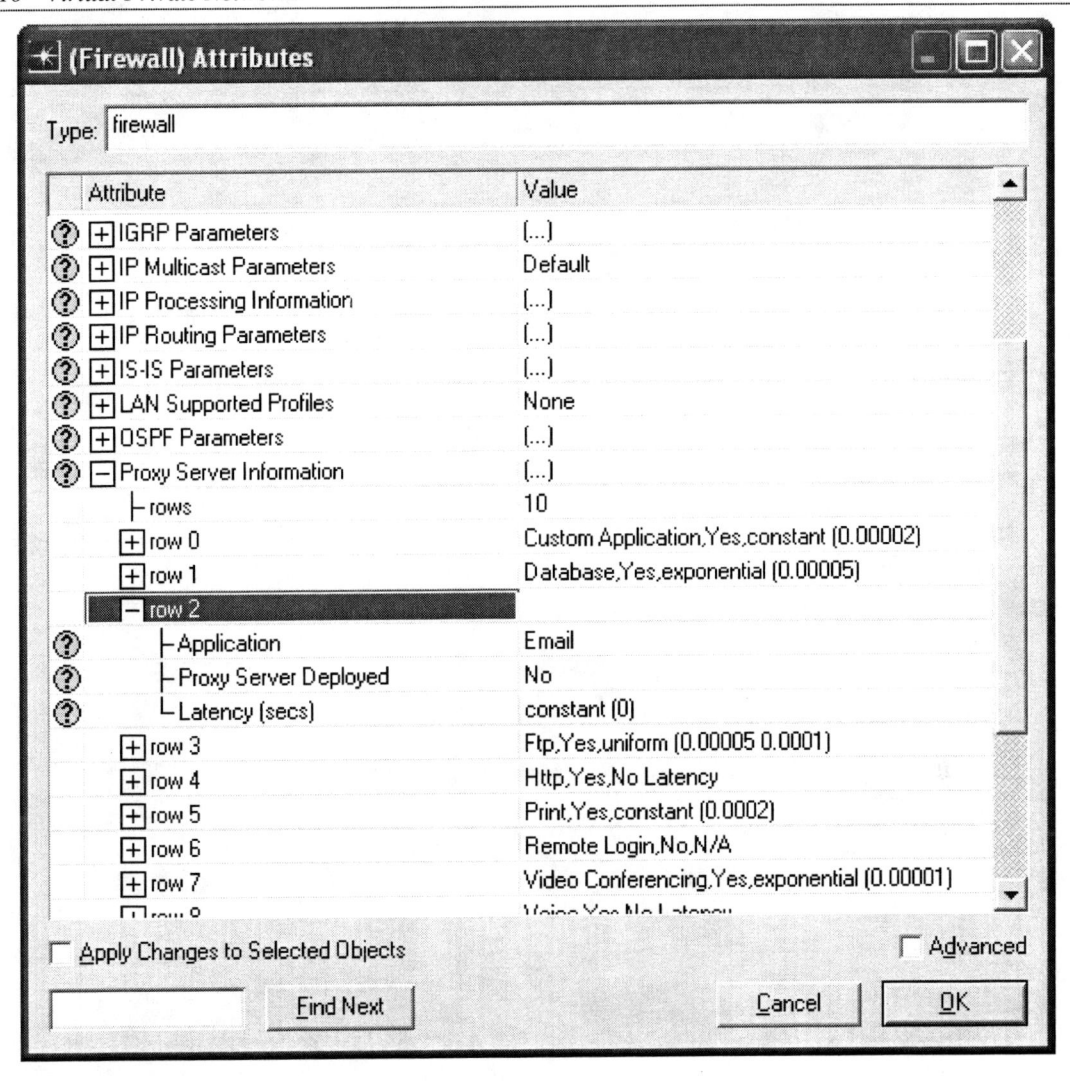

![Firewall Attributes dialog box showing Type: firewall, with attributes including IGRP Parameters, IP Multicast Parameters (Default), IP Processing Information, IP Routing Parameters, IS-IS Parameters, LAN Supported Profiles (None), OSPF Parameters, Proxy Server Information with rows 10, row 0: Custom Application,Yes,constant (0.00002), row 1: Database,Yes,exponential (0.00005), row 2 expanded showing Application: Email, Proxy Server Deployed: No, Latency (secs): constant (0), row 3: Ftp,Yes,uniform (0.00005 0.0001), row 4: Http,Yes,No Latency, row 5: Print,Yes,constant (0.0002), row 6: Remote Login,No,N/A, row 7: Video Conferencing,Yes,exponential (0.00001)]

4. Choose **Simulation/Run Discrete Event Simulation** to gather results for the new scenario.

5. Choose **File/Save** from the Main Menu to save your project.

Comparing Results

Now we can view the results and see if response time for the FTP application improved with the addition of the firewall. To view results, choose **Results/Compare Results** from the Main Menu. Click on the + next to **FTP** and click next to **Download Response Time (seconds)**. Select **time_average** from the pull-down menu. Click on the **Show** button to display the graph in a new window. Your graph should resemble the following figure:

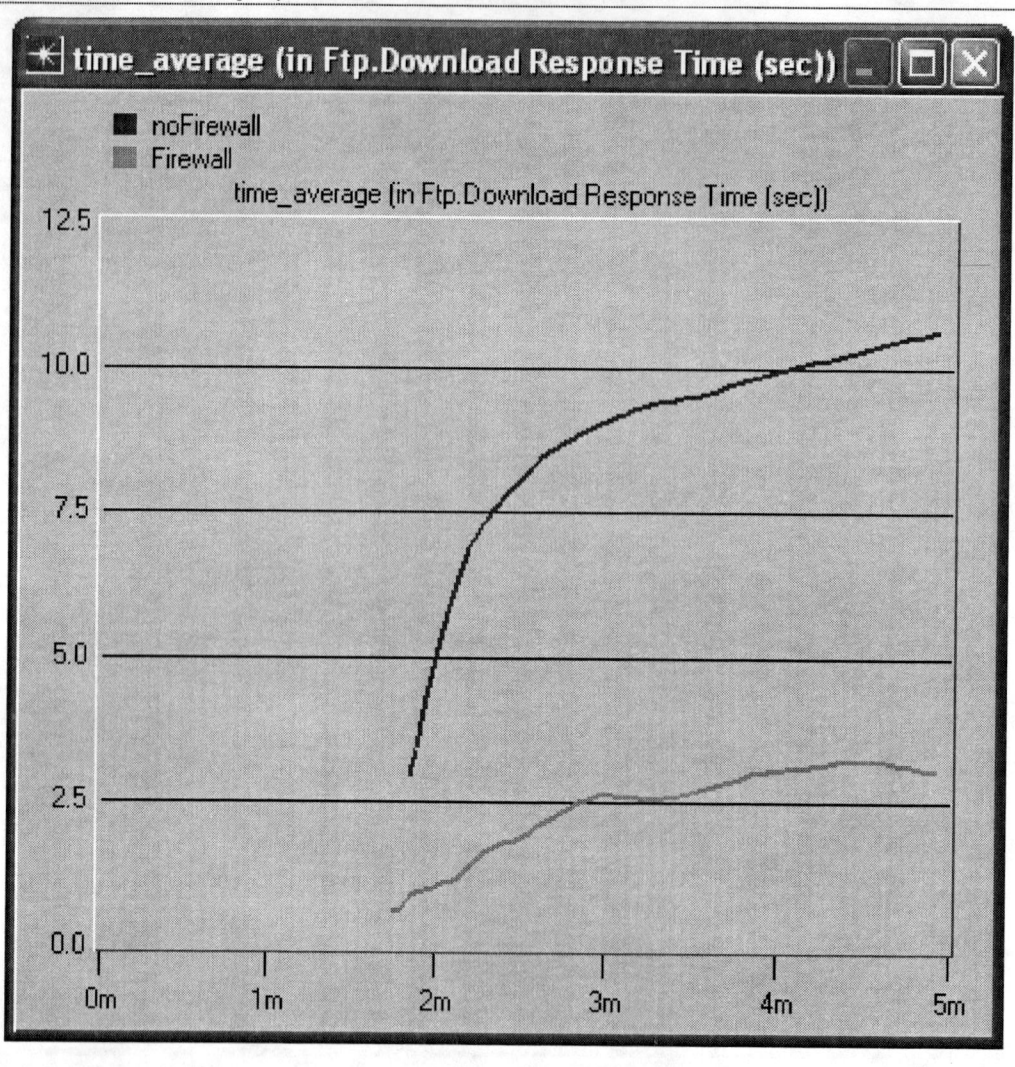

Questions

1. What is the average response time for the FTP application in the Firewall and NoFirewall scenarios?
2. What is the average response time for Database Queries in the Firewall and NoFirewall scenarios?
3. Based on your results, would you recommend that RecOnline use a firewall? Explain your answer.
4. Look at Traffic Sent (bits/second) for the email application. What is the difference in bits/sent for the Firewall and NoFirewall scenarios?
5. We would expect that the email server Load (sessions/sec) would decrease in the Firewall scenario. Verify that this is true. What is the difference in load?
6. Look at the Load (requests/sec for the FTP server). Is there a significant difference in load between the two scenarios? Explain your results.
7. Explain why there is an increase in VPN tunnel delay for Router3 in the Firewall scenario.
8. Look at the statistics for Database Query (bits/sec) at the Southern California site. Does the firewall significantly affect the number of bits sent for this application?
9. Duplicate the NoFirewall scenario and change the VPN Configuration to include a .2 second encryption and .2 decryption delay. Rerun your simulation and compare response time statistics with those from the original NoFirewall scenario. Describe your results.
10. Duplicate the Firewall scenario and change the Firewall configuration to include a .5-second latency for email filtering. Rerun your simulation and compare results with those from the original Firewall scenario. Does increasing the firewall latency increase response time for all three applications? Describe your results.